TREATISE ON ANALYTICAL CHEMISTRY

PART I
THEORY AND PRACTICE
SECOND EDITION

TREATISE ON ANALYTICAL CHEMISTRY

PART I

THEORY AND PRACTICE

SECOND EDITION

VOLUME 11

Edited by JAMES D. WINEFORDNER

Department of Chemistry, University of Florida

Associate Editor: MAURICE M. BURSEY

Department of Chemistry, University of North Carolina at Chapel Hill

Editor Emeritus: I. M. KOLTHOFF

Department of Chemistry, University of Minnesota

WILEY

AN INTERSCIENCE® PUBLICATION

JOHN WILEY & SONS
New York — Chichester — Brisbane — Toronto — Singapore

An Interscience ® Publication

Copyright © 1989 by John Wiley & Sons, Inc.

Library of Congress Cataloging in Publication Data:

(Revised for vol. 11, Part I)

Kolthoff, I. M. (Izaak Maurits), 1894–
  Treatise on analytical chemistry.

    Pt. 1, v. 8–   : edited by Philip J. Elving;
associate editor, Edward J. Meehan.
    Pt. 1, v. 11–   : edited by James D. Winefordner;
associate editor, Maurice M. Bursey.
    "An Interscience publication."
    Contents: pt. 1. Theory and practice
    1. Chemistry, Analytic. I. Elving, Philip
Juliber, 1913–   . II. Meehan, Edward J.
III. Winefordner, James D. (James Dudley), 1931–
IV. Title.

QD75.2.K64   1978   543   78–1707
ISBN 0–471–50938–8 (pt. 1. v. 11)

Printed in the United States of America

10 9 8 7 6 5 4 3 2 1

TREATISE ON ANALYTICAL CHEMISTRY

PART I
THEORY AND PRACTICE

VOLUME 11

AUTHORS OF VOLUME 11

D. L. DONOHUE
G. A. EADON
CURTISS D. HANSON
W. W. HARRISON

ROY A. KELLER
ERIC L. KERLEY
A. C. MILLER
DAVID H. RUSSELL

Authors of Volume 11

D. L. Donohue

*Analytical Chemistry Division, Oak Ridge
National Laboratory, Oak Ridge,
Tennessee, Chapter 3*

G. A. Eadon

*Wadsworth Center for Laboratories and
Research, State of New York Department
of Health, Albany, New York, Chapter 1*

Curtiss D. Hanson

*Department of Chemistry, Texas A & M
University, College Station, Texas,
Chapter 2*

W. W. Harrison

*Department of Chemistry, University of
Florida, Gainesville, Florida
Chapter 3*

Roy A. Keller

Department of Chemistry, State University of New York, College at Fredonia, Fredonia, New York, Chapter 4

Eric L. Kerley

Department of Chemistry, Texas A & M University, College Station, Texas, Chapter 2

A. C. Miller

Zettlemoyer Center for Surface Studies, Lehigh University, Bethlehem, Pennsylvania Chapter 5

David H. Russell

Department of Chemistry, Texas A & M University, College Station, Texas, Chapter 2

Preface to the Second Edition of the Treatise

In the mid-1950s, the plan ripened to edit a "Treatise on Analytical Chemistry" with the objective of presenting a comprehensive treatment of the theoretical fundamentals of analytical chemistry and their implementation (Part I) as well as of the practice of inorganic and organic analysis (Part II); an introduction to the utilization of analytical chemistry in industry (Part III) was also considered. Before starting this ambitious undertaking, the editors discussed it with many colleagues who were experts in the theory and/or practice of analytical chemistry. The uniform reaction was most skeptical; it was not thought possible to do justice to the many facets of analytical chemistry. Over several years, the editors spent days and weeks in discussion in order to define not only the aims and objectives of the Treatise but, more specifically, the order of presentation of the many topics in the form of a table of contents and the tentative scope of each chapter. In 1959, Volume 1 of Part I was published. The reviews of this volume and of the many other volumes of Part I as well as of those of Parts II and III have been uniformly favorable, and the first edition has become recognized as a contribution of classical value.

Even though analytical chemistry still has the same objectives as in the 1950s or even a century ago, the practice of analytical chemistry has been greatly expanded. Classically, qualitative and quantitative analysis have been practiced mainly as "solution chemistry." Since the 1950s, "solution analysis" has involved to an ever increasing extent physicochemical and physical methods of analysis, and automated analysis is finding more and more application, for example, its extensive utilization in clinical analysis and production control. The accomplishments resulting from automation are recognized even by laymen, who marvel at the knowledge gained by automated instruments in the analysis of the surfaces of the moon and of Mars. The computer is playing an ever increasing role in analysis and particularly in analytical research. This revolutionary development of analytical methodology is catalyzed by the demands made on analytical chemists, not only industrially and academically but also by society. Analytical chemistry has always played an important role in the development of inorganic, organic, and physical chemistry and biochemistry, as well as in that of other areas of the natural sciences such as mineralogy and geochemistry. In recent years, analytical chemistry—often of a rather sophisticated nature—has become increasingly important in the medical and biological sciences, as well as in the solving of such social problems as environmental pollution, the tracing of toxins, and the dating of art and archaeological objects, to mention only a few. In the area of atmospheric science, ozone reactivity and persistence in the stratosphere is presently a topic of great priority; extensive analysis is required both for monitoring atmospheric constituents and for investigating model systems.

One example of the increasing demands being made on analytical chemists is the growing need for speciation in characterizing chemical species. For example, in reporting that lake water contains dissolved mercury, it is necessary to report in which oxidation state it is present, whether as an inorganic salt or complex, or in an organic form and in which form.

As a result of the more or less revolutionary developments in analytical chemistry, portions of the first edition of the Treatise are becoming—and, to some extent, have become—out-of-date, and a revised, more up-to-date edition must take its place. In recognition of the extensive development and because of the increased specialization of analytical chemists, the editors have fortunately secured for the new edition the cooperation of experts as coeditors for various specific fields.

In essence, it is the objective of the second edition of the Treatise, as it was of the first edition (whose preface follows this one), to do justice to the theory and practice of contemporary analytical chemistry. It is a revision of Part I, which mirrors the development of analytical chemistry. Like the first edition, the second edition is not an extensive textbook; it attempts to present a thorough introduction to the methods of analytical chemistry and to provide the background for detailed evaluation of each topic.

I. M. KOLTHOFF
J. D. WINEFORDNER

Minneapolis, Minnesota
Gainesville, Florida

Preface to the First Edition of the Treatise

The aims and objectives of this Treatise are to present a concise, critical, comprehensive, and systematic, but not exhaustive, treatment of all aspects of classical and modern analytical chemistry. The Treatise is designed to be a valuable source of information to all analytical chemists, to stimulate fundamental research in pure and applied analytical chemistry, and to illustrate the close relationship between academic and industrial analytical chemistry.

The general level sought in the Treatise is such that, while it may be profitably read by the chemist with the background equivalent to a bachelor's degree, it will at the same time be a guide to the advanced and experienced chemist—be he in industry or university—in the solution of his problems in analytical chemistry, whether of a routine or of a research character.

The progress and development of analytical chemistry during most of the first half of this century has generally been satisfactorily covered in modern textbooks and monographs. However, during the last fifteen or twenty years, there has been a tremendous expansion of analytical chemistry. Many new nuclear, subatomic, atomic, and molecular properties have been discovered, several of which have already found analytical application. In the development of techniques for measuring these and also the more classical properties, the revolutionary progress in the field of instrumentation has played a tremendous role.

It has been difficult, if not impossible, for anyone to digest this expansion of analytical chemistry. One of the objectives of the present Treatise is not only to describe these new properties, their measurement, and their analytical applicability, but also to classify them within the framework of the older classifications of analytical chemistry.

Theory and practice of analytical chemistry are closely interwoven. In solving an analytical chemical problem, a thorough understanding of the theory of analytical chemistry and of the fundamentals of its techniques, combined with a knowledge of and practical experience with chemical and physical methods, is essential. The Treatise as a whole is intended to be a unified, critical, and stimulating treatment of the theory of analytical chemistry, of our knowledge of analytically useful properties, of the theoretical and practical fundamentals of the techniques for their measurement, and of the ways in which they are applied to solving specific analytical problems. To achieve this purpose, the Treatise is divided into three parts: I, analytical chemistry and its methods; II, analytical chemistry of the elements; and III, the analytical chemistry of industrial materials.

Each chapter in Part I of the Treatise illustrates how analytical chemistry draws on the fundamentals of chemistry as well as on those of other sciences: it stresses for its particular topic the fundamental theoretical basis insofar as it affects the analytical approach, the methodology and practical fundamentals used both for the development of analytical methods and for their implementation for analytical service, and the critical factors in their application to both organic and inorganic materials. In general, the practical discussion is confined to fundamentals and to the analytical interpretation of the results obtained. Obviously then, the Treatise does not intend to take the place of the great number of existing and exhaustive monographs on specific subjects, but its intent is to serve as an introduction and guide to the efficient utilization of these specialized monographs. The emphasis is on the analytical significance of properties and of their measurement. In order to accomplish the above aims, the editors have invited authors who are not only recognized experts for the particular topics, but who are also personally acquainted with and vitally interested in the analytical applications. Only in this way can the Treatise attain the analytical flavor which is one of its principal objectives.

Part II is intended to be very specific and to review critically the analytical chemistry of the elements. Each chapter, written by experts in the field, contains in addition to a critical and concise treatment of its subject, critically selected procedures for the determination of the element in its various forms. The same critical treatment is contemplated for Part III. Enough information is presented to enable the analyst both to analyze and to evaluate a product.

The response in connection with the preparation of the Treatise from all colleagues has been most enthusiastic and gratifying to the editors. It is obvious that it would have been impossible to accomplish the aims and objectives cited in the Preface without the wholehearted cooperation of the large number of distinguished authors whose work appears in this and future volumes of the Treatise. To them and to our many friends who have encouraged us we express our sincere appreciation and gratitude. In particular, considering that the Treatise aims to cover all of the aspects of analytical chemistry, the editors have found it desirable to solicit the advice of some colleagues in the preparation of certain sections of the various parts of the Treatise. They would like at this time to acknowledge their indebtedness to Professor Ernest B. Sandell of the University of Minnesota for his interest and active cooperation in the organizing and detailed planning of the Treatise.

I. M. KOLTHOFF
P. J. ELVING

Minneapolis, Minnesota
Ann Arbor, Michigan

PART I. THEORY AND PRACTICE

CONTENTS—VOLUME 11

2. Recent Developments in Experimental Fourier Transform–Ion Cyclotron Resonance

3. Spark Source Mass Spectrometry

TREATISE ON ANALYTICAL CHEMISTRY

PART I
THEORY AND PRACTICE
SECOND EDITION

Chapter 1

MASS SPECTROMETRY OF ORGANIC AND BIOLOGICAL COMPOUNDS

By G. A. EADON

Wadsworth Center for Laboratories and Research, State of New York Department of Health, Albany, New York

Contents

1

I. INTRODUCTION

Before modern spectroscopic techniques became available, chemists were nevertheless often able to deduce the structures of fairly complex molecules. For example, a pure compound might be isolated from a complex mixture of natural products by repeated recrystallization. After initial attempts to estimate the compound's molecular weight using cryoscopic or osmometric techniques, the chemist might then attempt to degrade the molecule into smaller, more readily characterized fragments. If these fragments were not themselves directly identifiable, a second generation of still smaller fragments might be produced. Alternatively, the parent molecule might be degraded with a different set of reagents to produce a different group of fragmentation products. Eventually, the fragments, the building blocks of the molecule, were identified. Then, the chemist would apply knowledge of the mechanisms and courses of degradation reactions to deduce the structure of the parent molecule.

It is obvious that the process just described is extremely laborious and time consuming. A mass spectrometer can be thought of as a device for accomplishing many of these steps automatically and in a few minutes. The unknown molecule is treated with a "reagent" that converts it into an ion; typically, a fraction of these ions will remain intact until collection and detection. Others will fragment, breaking apart into smaller neutral and charged particles. The mass spectrometer "works up" the "reaction mixture," separating charged species from neutral species, and characterizes the products according to their mass-to-charge ratios. Thus, in a typical case, a chemist can in a matter of a few minutes learn the exact molecular weight of a molecule (by consideration of the m/z ratios of intact ions that were collected) and also learn the mass-to-charge ratios of the degradation products that are produced. Since the "rules" governing compound fragmentation are now reasonably well understood, the chemist can, in favorable cases, attempt to deduce the structure of the parent molecule from this information. Alternatively, the relative abundances of different m/z ions can be used as a "fingerprint." Either manual or computer-assisted searches can be used to match the fragmentation pattern of the unknown to that of a known "reference" spectrum. Even instrument operation and data collection can be relegated to the computer. The mass spectrometer is readily coupled to a gas chromatograph or a high-pressure liquid chromatograph, permitting the direct analysis of complex mixtures. Most important, the mass spectrometer is extraordinarily sensitive; in favorable cases, subpicogram detection limits ($< 10^{-12}$ g) can be obtained. It is not surprising that this very versatile and powerful analytical technique is finding increased application to problems in analytical, organic, inorganic, and biological chemistry.

A. HISTORY

The early development of mass spectrometry has been reviewed in some detail (28, 68a). The aim of this section is to more briefly trace the evolution of mass spectrometers and their applications toward their current state of development.

Like most spectroscopic techniques, mass spectrometry was developed and first used by physicists. In the last half of the nineteenth century, physicists began study of the properties of electrical discharges in gas. Initially, their attention was directed toward "cathode rays," now understood to consist of electrons. In 1886, Goldstein (73) observed *"kanal strahlen,"* or canal rays, emitted within the discharge tube and traveling in an opposite direction from the cathode rays. The canal rays, of course, consist of positively charged ions generated by collisions between electrons (cathode rays) and residual gases present in the discharge tube. Numerous experiments were performed to study the deflection of these beams by electric and magnetic fields. Soon, these experiments led to the construction of rudimentary "mass spectrometers," devices capable of generating ions and determining their relative m/z. For example, Fig. 1 depicts an early apparatus constructed by Thompson (180). Cathode rays generated in the discharge tube traveled between the anode A and the cathode B, occasionally colliding with and ionizing residual gases. The resulting positive ions were collimated by the narrow tube C and thus directed between coincident magnetic and electric fields at D. The ions were deflected onto the fluorescent screen E into bands whose position was related to the mass-to-charge ratio of the positive ions. For example, when hydrogen was present in the discharge tube, Thompson observed two bands of relative mass-to-charge ratio $2:1$ (Fig. 2). Thompson correctly deduced that the bands corresponded to ionized molecular and atomic hydrogen. More sophisticated equipment was soon constructed. By 1912, Thompson was able to demonstrate the existence of two isotopes of neon (179). Aston's "mass spectrograph" (7), constructed in 1919, focused a collimated beam of ions of a particular mass onto the same point on a photographic plate regardless of their velocities

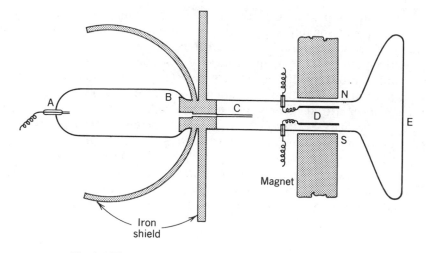

Fig. 1.   Thompson's apparatus for study of positive rays (30).

Fig. 2.   Two parabolic bands observed when hydrogen gas was introduced into discharge tube of Thompson's apparatus (179).

(Fig. 3); Thompson's apparatus had produced parabolas. Thus, the apparatus was markedly more sensitive and capable of higher resolution. Other instrumental advances followed (8, 9, 43). However, Dempster's design (Fig. 4) is readily recognizable as the precursor of the modern-day low-resolution magnetic mass spectrometer (56). Positive ions are produced in an electron bombardment source with low translational energies and accelerated through a large potential drop. The resulting beam of ions is nearly monoenergetic. After passing through a slit S, the beam enters a uniform magnetic field. The pathway followed by a particular ion under these circumstances depends on the ion's mass-to-charge ratio, the accelerating voltage, and the magnetic field strength. (These factors are discussed in more detail in a subsequent section.) For technical reasons, Dempster scanned the accelerating voltage to bring various ions into focus at the collector electrode. Contemporary magnetic sector instruments usually scan the magnetic field and hold the accelerating voltage constant. Dempster and his instrument made many important studies on isotope abundances (57).

Mass spectrometry made many early and important contributions to fundamental problems in physics. However, the technique's first practical application

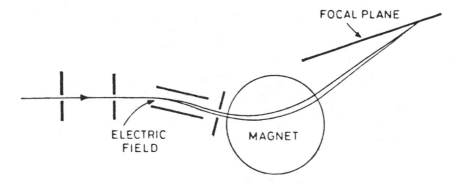

Fig. 3.   Aston's "mass spectrograph" (30).

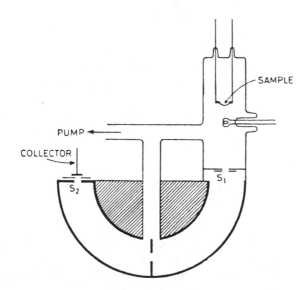

Fig. 4.   Dempster's mass spectrometer (30).

was to a problem in chemistry. The catalytic crackers of the petroleum industry produce a hydrocarbon product whose composition must be accurately monitored. Before the 1940s, this was accomplished by careful fractional distillation and analysis of the resulting fractions based on their infrared spectra and their refractive indices. It soon became apparent that direct mass spectrometric examination of the cracker output was more efficient (35, 87, 185, 186). For example, in 1943 it was estimated that analyses of a nine-component mixture of five- and six-carbon hydrocarbons by fractionation/refractive index would require 240 h. In contrast, the mass spectrometer could accomplish the task in 1 h of instrument time, 4 man-hours total. Once an important practical application had been demonstrated for the mass spectrometer, the production and rapid improvement of commercial instrumentation began.

Aliphatic hydrocarbons were highly studied by early mass spectroscopists. They were of particular interest to the petroleum industry. Further, they are relatively volatile and thermally stable, necessary characteristics in view of the crude inlet systems in use during the early days of mass spectrometry. However, in retrospect, the attention devoted to hydrocarbons may have discouraged the application of mass spectrometry to functionalized molecules. Naively, it might be expected that hydrocarbons would exhibit rather simple mass spectral behavior. In fact, it is often extremely difficult to rationalize the fragmentation reactions of hydrocarbons. A classical example is the formation of an ethyl ion from 2-methylpropane $[CH_3CH(CH_3)_2]$ (37). In an effort to explain the apparently deep-seated rearrangements commonly observed in hydrocarbon spectra, concepts such as "sudden death" and "hydrogen soup" gained some currency. The former postulated that collision of a molecule with a 70-eV electron imparted sufficient energy to rupture most of the covalent bonds in a molecule, leading to a collapse of molecular structure. The latter implies an intact carbon skeleton surrounded by a swarm of hydrogen atoms (38). By the early fifties, however, sufficient functionalized molecules had been examined to demonstrate that molecular structure before ionization had important influence on the course of fragmentation reactions. With the growing realization that these fragmentations were often readily rationalized using well-established principles of organic chemistry, mass spectrometry entered the modern era.

B. LITERATURE OF MASS SPECTROMETRY

Numerous books discuss in detail fundamental aspects of mass spectrometry and its applications; a few such references are listed in the General Bibliography. A number of compilations of mass spectra are available to assist the researcher in identifying unknown compounds; these compilations are discussed in Section IV.A.1. Scientists interested in remaining abreast of recent developments in mass spectroscopy have a number of resources available. *Mass Spectrometry Bulletin*, *Gas Chromatography–Mass Spectrometry Abstracts*, and *Chemical Abstracts Selects—Mass Spectrometry* (referenced in the General Bibliography) provide abstracts of recent articles but are not exhaustive in their coverage. The most current "Fundamental Review—Mass Spectrometry" (published biannually in the journal *Analytical Chemistry* (General Bibliography) contain very brief but critical reviews of current work organized under separate headings (e.g., "Metastable Ions," "Clinical Chemistry," and "Prostaglandins and Related Compounds"). *Specialist Periodical Reports in Mass Spectrometry*, published biannually by the Chemical Society of London, provides more lengthy discussions organized by chapters (e.g., "Trends in Instrumentation" and "Natural Products"), but its coverage of the literature is less than timely. *Mass Spectrometry Reviews*, published quarterly by Wiley (New York), covers rapidly developing areas of mass spectrometry with reviews of individual topics of about 50 pages.

Important journals devoted exclusively to mass spectrometry include *International Journal of Mass Spectrometry and Ion Processes* (Elsevier Publishing Company, Amsterdam), which emphasizes physical and instrumental aspects. The orientation of *Biomedical and Environmental Mass Spectrometry* (Heyden & Sons, London) is apparent from its title. *Organic Mass Spectrometry* (Heyden & Sons, London) is more broadly based than its name suggests and typically includes papers on theory, ion structures, organometallic chemistry, instrumental advances, and so on.

II. PRINCIPLES

Mass spectrometers must usually perform four major functions during the generation of a mass spectrum. The samples must be vaporized and introduced into the ionization chamber in a controlled fashion. A fraction of these sample molecules must be converted to positive or negative ions. The resulting ions and their fragmentation products must be mass analyzed. Finally, the mass-analyzed ions must be detected and their relative abundances determined. Much of the versatility of the mass spectrometer can be attributed to the fact that there are a number of distinct and complimentary techniques for accomplishing each of these tasks.

A. SAMPLE INTRODUCTION

1. Batch Inlet Systems

Gases and very volatile liquids can be studied using an unheated batch inlet system (Fig. 5). The sample is introduced into an evacuated glass or stainless steel reservoir and allowed to enter the ion source at a controlled rate through a molecular leak. The leak might consist of a perforated gold or steel foil or ceramic frit.

At room temperature the vapor pressures of most liquids and solids are insufficient to produce a useful mass spectrum after passage through such a leak. Therefore, batch inlet systems capable of operation at temperatures up to 350°C are commonly used. Since many organic and inorganic molecules undergo catalytic decomposition on contact with hot metal surfaces, it is customary to construct these systems entirely of glass.

Samples can be introduced into the batch inlet system by injection through a gas-tight septum using a microliter syringe. A potential problem with the procedure is that bleed from the septum may interfere with recording the sample's mass spectrum. Another technique uses a sintered-glass disk impermeable to mercury or gallium but permeable to liquid samples. One side of the disk is covered with a layer of the liquid metal, and the inlet side is evacuated. The sample is then applied to the surface of the disk through a micropipette, and it

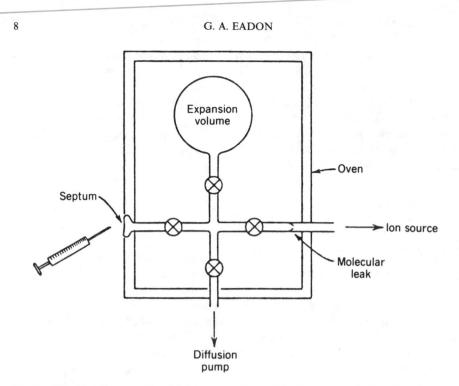

Fig. 5.　Simplified diagram of batch inlet system with provision for sample injection through septum (124a).

rapidly diffuses into the reservoir. A third technique that minimizes contamination uses two hollowed-out Teflon slugs. The sample is placed between the slugs, which are then forced through a tightly fitted inlet port and into the sample reservoir. The lower plug falls into the reservoir, exposing the bottom of the upper plug to the reservoir. The upper slug remains in the inlet port to preserve the vacuum in the reservoir.

Batch inlets find their most important applications in cases where a sample is to be studied for more than 2–3 min. For example, consider a fairly typical inlet system with a volume V of 1 liter and a leak whose conductance C is 10^{-4} liters/s for a particular compound and temperature. Since

$$P/P^0 = e^{-Ct/V} \tag{1}$$

it can readily be calculated that ca. 8 min will be required before the pressure within the sample reservoir (and thus the pressure in the ionization chamber) decreases by 5%. Thus, applications requiring repeated or slow scans become possible. Typical examples might include mechanistic studies, metastable ion investigations, and measurement of ion abundance ratios. Caution should be exercised in interpreting the latter experiments since conductance varies inversely with molecular weight if a molecular leak is used. Some other applications of batch inlet systems include determination of reference spectra of pure compounds and admission of reference compounds for tuning and calibration purposes at low resolution and for mass markers at high resolution.

Batch inlets have two important disadvantages. First, they waste large amounts of sample. Although the mass required to produce a useful mass spectrum via batch inlet will vary with reservoir size and leak conductance, typical values are 10^{-4}–10^{-6} g. The origin of this lack of sensitivity is fairly obvious. Suppose that a typical mass spectral scan requires 10 s for completion. During that period of time, the inlet system just described will consume at most 0.1% of the total sample. The same source pressure could be produced and thus the same sensitivity obtained if a sample weighing 0.1% as much as the batch sample could be vaporized completely during the 10-s scan.

A second aspect of batch inlet systems is that they are unsuitable for many compounds. Compounds must have an appreciable vapor pressure (e.g., 0.01 torr) at an accessible temperature if they are to produce sufficient pressure within the source to permit recording a mass spectrum. Further, if a heated inlet system is required, the compound must be thermally stable.

2. Direct-Probe Inlet

A direct insertion probe (Fig. 6) permits controlled introduction of the entire sample into the ionization chamber. One design utilizes a ceramic tip as the sample holder. The tip can be coated lightly with the pure solid or liquid or it can be dipped into a solution of the compound in a volatile solvent. A second design uses glass capillaries as sample containers. These are mounted at the tip of the probe.

It is undesirable to expose the ionization chamber to pressures near atmospheric. The probe is therefore first inserted into a vacuum lock, where a

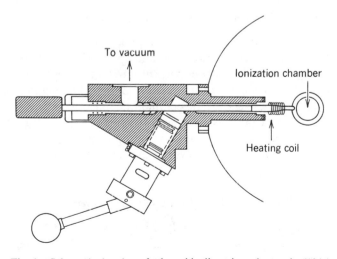

Fig. 6.   Schematic drawing of a heatable direct insertion probe (124a).

pressure of 10^{-3} torr is achieved. Then, the vacuum lock is opened to the ionization chamber and the probe inserted further until the ceramic tip or capillary is butted up against a small hole in the ionization chamber. A skilled operator can usually assure that essentially the entire sample will be admitted into the ionization chamber.

Some simple probes have no provision for auxiliary heating or cooling. Most compounds will vaporize at a satisfactory rate only over a narrow temperature range. Thus, a reasonable strategy with such a probe is to insert the probe with the ionization region/source cold; if the volatilization rate is insufficient under these conditions to produce a usable mass spectrum, the instrument's source and ionization chamber heaters can be used. A more versatile probe design incorporates resistive heating to permit independent temperature control (up to 400°C) and liquid nitrogen cooling to permit use of the probe on relatively volatile samples that might otherwise be lost in the vacuum lock or might volatilize too rapidly in the ionization chamber.

Much smaller sample sizes are necessary when the direct insertion probe is used. A skilled operator can control the rate of evaporation so that a high, nearly constant sample pressure is maintained for periods of time comparable to the required scan time. Since essentially the entire sample can be admitted into the ionization chamber, full-scan mass spectra can be obtained with 10^{-7}–10^{-9} g or less of material.

The second advantage of the direct insertion probe is that less volatile samples and/or much lower temperatures can be used. The batch system's leak interposed between the reservoir and the ion source results in a large pressure drop. A high pressure must be maintained in the reservoir to produce an adequate source pressure. In contrast, the pressure drop between the probe tip and the ionization chamber is negligible. The critical source pressure necessary for producing a mass spectrum is thus more readily attained.

3.   Gas Chromatographic Inlet Systems

If the effluent from a gas chromatographic column is admitted to the ionization chamber of a mass spectrometer and appropriate provisions are made to limit or manage the flow of carrier gas, very useful mass spectral data can be obtained. The narrow peak widths commonly observed in high-resolution gas chromatography result in very low detection limits. The separation produced by the gas chromatograph permits analyses of complex mixtures without prior separation. The importance of this inlet system to contemporary mass spectrometry is such that whole texts are devoted to it (113b). It will therefore not be discussed further here.

4.   Liquid Chromatographic Inlet Systems

Coupling of a high-pressure liquid chromatograph (HPLC) to a mass spectrometer represents a formidable technical problem. Mass spectrometry as most commonly practiced requires high vacuum in the source region

$(10^{-4}-10^{-5}$ torr). Liquid flow rates used in HPLC are most commonly 1–2 ml/min. If vaporized, this flow would correspond to 100–1000 ml/min of gas at standard temperature and pressure (STP), orders of magnitude more than the gas flows typically used in gas chromatography. Thus, either the system can be designed to transmit only a small fraction of the HPLC effluent to the mass spectrometer (with an inevitable loss of sensitivity) or a procedure must be devised to remove the solvent from the HPLC effluent prior to its admission to the ion source. The latter approach is complicated by the diverse solvents and solutes typically used in HPLC. It is, however, notable that HPLC finds many of its most important applications in the analysis of nonvolatile or thermally labile molecules, those unsuitable for GC. Since the most commonly used ionization techniques require vaporization of the sample molecules, many substances of interest will not be successfully analyzed by HPLC/MS. Any sample enrichment scheme used must take into account the thermal lability of many of the target molecules.

Recent reviews describe a number of approaches that have been used to couple the HPLC and the mass spectrometer (6, 46a, 123). This chapter will briefly discuss the techniques that have found widest use.

The first commercially marketed liquid chromatography–mass spectrometry (LC/MS) accessory uses a moving-belt interface (Fig. 7) (124). The effluent from the LC is applied as small droplets to a moving belt typically 0.3 cm wide and either of stainless steel or polyimide. The effluent droplets are carried on the surface of the moving belt, first past an infrared reflector, then through two vacuum locks, and finally to an entrance port of the ion source where the sample is flash vaporized. An additional heater is used to evaporate any residual solute from the belt before more effluent is applied. After optimization of heater temperatures, it is reported that sample transfer efficiencies of 30–40% can be obtained (123). Poorer results will be obtained if the analyte boils below 180°C since considerable loss will occur in the vacuum locks. The nature of the solvent used imposes a more serious limitation. Aqueous solutions tend to evaporate irregularly (i.e., superheat and then "bump" in the ionization region). Drastically reduced flow rates (e.g., 0.1 ml/min) must often be used with such solvent systems. Since this is usually an insufficient flow for HPLC as most commonly performed, a splitter must be used with a consequent decrease in sensitivity. This

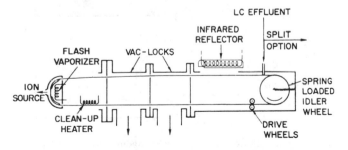

Fig. 7.   Schematic of moving-belt LC/MS interface (123).

is especially troublesome because aqueous solutions are commonly used with the very popular reverse-phase columns. The advent of micro-HPLC using 1-mm-i.d. columns and solvent flow rates several orders of magnitude below the 1 ml/min typical of conventional HPLC may ameliorate this problem, however (81a).

An advantage of the moving-belt interface is that because the sample is introduced into the source region adsorbed on the belt in a nearly solventless state, a range of complimentary ionization techniques can be used (e.g., as discussed later, electron impact, chemical ionization, or fast-atom bombardment). On the other hand, the necessity to vaporize the solvent and, with the more commonly used ionization techniques, to thermally desorb the sample molecules limits the range of volatilities and thermal stabilities suitable for this technique.

A complimentary commercially available and widely used HPLC/MS interface, the direct liquid introduction (DLI) system, is of simpler construction. The effluent stream is split, if necessary, and only the volume that the pumping system of a chemical ionization (CI) mass spectrometer can accommodate is admitted. Typically, for conventional HPLC, less than 1% of the sample is used in this method, resulting in detection limits in the microgram range. However, a much larger percentage of the eluent from a micro-HPLC system can be admitted, resulting in detection limits one or two orders of magnitude lower (81a). The sample flows past a replaceable diaphragm with a 5–15-μm orifice that is mounted in the tip of a direct insertion probe (Fig. 8). The probe is easily inserted and removed through a standard vacuum lock. This is a desirable feature since the orifice frequently becomes plugged.

A major advantage of the DLI system is that the analyte enters the ion source in solution and thus is ionized without directly contacting a heated surface. This permits ionization of thermally sensitive materials. A disadvantage of the high solvent concentrations in the source is the prevalence of CI-like conditions. Thus, little fragmentation of the parent ion is usually noted, and little structural information other than molecular weight is obtained.

A third technique for inputting HPLC effluents into the mass spectrometer has recently gained increasing popularity as adaptations to a variety of mass

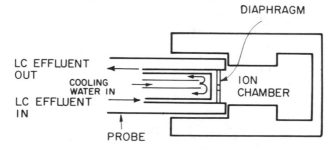

Fig. 8.   Schematic diagram of direct insertion probe modified for direct liquid introduction of HPLC effluent.

spectrometers have been commercialized. The thermospray technique pumps the aqueous eluent containing an electrolyte directly through a heated capillary (the "vaporizer") into a heated ion source (180e) (Fig. 9). (Ammonium acetate is the most popular electrolyte; its comparative volatility minimizes deposit formation in the source, thus decreasing instrument maintenance.) The resulting aerosol beam consists of neutral droplets as well as positively and negatively charged droplets that contain unequal amounts of the electrolyte's ions. As the droplets traverse the heated source, evaporation of solvent continues and the droplets shrink. Ultimately, the electric field at the droplet surface becomes sufficient to eject an ion from the droplet, a process akin to field ionization discussed in the next section. The ionized sample molecule can be expelled directly. Alternatively, a CI-like reaction can occur between an electrolyte ion and a neutral sample molecule. This mode of operation, referred to as 'filament-off" thermospray, produces spectra with dominant molecular ions and thus little structural information. It is especially useful for analysis of high-water-content eluents that are typical of reverse-phase HPLC and are difficult to handle with older HPLC/MS interfaces.

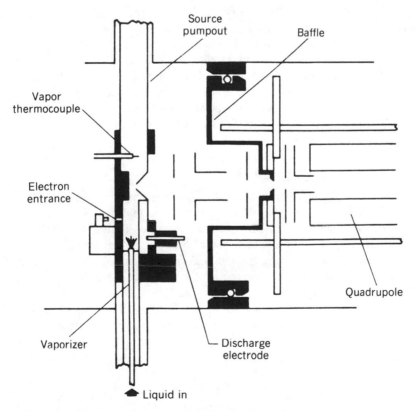

Fig. 9.   Simplified diagram of commercial thermospray HPLC/MS interface mounted in quadrupole mass spectrometer. Reprinted with permission from Vestec Corporation.

When nonaqueous HPLC conditions are required, the thermospray system can be operated as a "hot DLI" source (69a). In this so-called filament-on mode, an auxiliary conventionally generated electron beam produces CI-like spectra in the absence of a supporting electrolyte. The filament-on mode is not suitable for routine use with aqueous eluents because the resulting highly oxidizing conditions will shorten filament life. However, a third mode of operation, known as the discharge ionization mode, is available for such systems. Here, additional ionization is produced by passing an electric arc through the heated vapors.

The thermospray technique can thus be seen to be applicable to a wide range of HPLC eluents. Its routine operation and maintenance do not require unusual skill and its wide commercial availability have enhanced its acceptability. However, the efficiency of the ionization processes for a particular analyte appear to be sensitive to a broad range of experimental parameters (e.g., temperatures of the vaporizer tip, the ion source block, and the vapor, eluent flow rate and composition, etc.). This differential sensitivity to different compounds can complicate qualitative work, and short-term variations in sensitivity can complicate quantitative work.

The continuing efforts to commercialize new devices demonstrates that a universally and routinely applicable HPLC/MS interface has not been discovered. Recent developments include the monodisperse aerosol generation interface (MAGIC), whose claimed advantages include the ability to produce true electron impact spectra and quantitative linear response down to the nanogram range (188a), as well as the related "Thermobeam" interface (32b) and the Plasmaspray (which directs the aerosol from a thermospray unit through a plasma discharge to break up the species in the aerosol and provide more structural information) (32b). Detailed discussion of these techniques is beyond the scope of this chapter.

B. IONIZATION

An ideal ionization technique would convert a high percentage of sample molecules to ions. It would be suitable for thermally labile and involatile compounds. A significantly intense molecular ion would invariably be produced, but sufficient fragmentation would occur to assist in characterizing molecular structure. The spread of kinetic energies among ions leaving the source would be minimal. Sample introduction and removal would be rapid; cross-contamination peaks would be negligible.

Unfortunately, no single ionization technique excels at all of these tasks. However, the commonly used ionization techniques are rather complimentary to one another; judiciously chosen combinations of techniques can come close to meeting these ideals.

1. Electron Impact

If an electron is traveling with sufficient energy and if it approaches a sample molecule M: sufficiently closely, an electron will be repelled out of the molecule

M to generate the cation radical M^{+} [Eq. (2)]:

$$M: + e^{-} \longrightarrow M^{+} + 2e^{-} \tag{2}$$

This process, known as electron-impact-induced ionization or simply electron ionization, forms the basis of the most commonly used ionization technique.

Figure 10 depicts a simplified schematic drawing of an electron impact (EI) source. A tungsten or rhenium filament is heated to incandescence ($> 2000°C$) by passage of an electric current (the "filament current"). Since the housing of the ionization chamber is maintained at a potential 10–100 V higher than the filament, the electrons acquire kinetic energy until they enter the nearly field-free region inside the ionization chamber. Since the potential difference between the ionization chamber and the filament (the "ionizing voltage") is readily varied, the operator can control the kinetic energy of the electrons. The emission current corresponds to the total number of electrons per second emitted from the filament. Typically, perhaps three-fourths of the electrons emitted from the filament will collide with the walls of the ionization chamber. The remainder will pass completely through the ionization chamber and be collected at the trap (corresponding to the "trap current"). Under typical operating conditions about one or two electrons per 10^{6} emitted make an ionizing collision, and a similar fraction of the sample molecules passing through the ionization chamber is ionized. Nevertheless, EI is one of the ionization techniques of choice when high sensitivity is required. Complete EI mass spectra are routinely attainable on

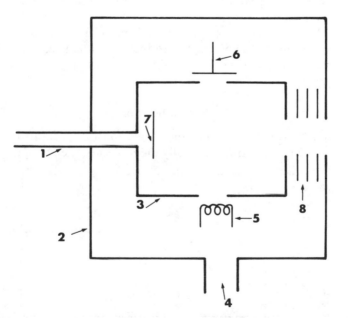

Fig. 10.   Simplified diagram of electron impact source: (1) sample inlet tube; (2) source housing; (3) ionization chamber housing; (4) line to high-vacuum source; (5) filament; (6) trap; (7) repeller; (8) focusing elements.

1–10 ng of sample; selected ion monitoring (a technique in which only a few intense ions are continuously monitored) permits detection limits three or more orders of magnitude lower.

The ionization potential of a molecule can be defined as the minimum amount of energy that must be imparted to induce expulsion of an electron. Figure 11 depicts an idealized plot of the probability of ion production by EI (i.e., the collision cross section) versus the kinetic energy of the impacting electrons. Obviously, if the colliding electrons have kinetic energies below the ionization potential of the molecule under investigation, no ions will be produced. As the kinetic energy of the electrons equals and then exceeds the ionization potential, ions will be produced in increasing abundance. In fact, careful study of ion abundance versus electron beam energy in this "threshold" region can produce good estimates of a molecule's ionization potential (118, 184). Ions produced by impact of relatively low energy electrons typically contain less internal excitation than ions produced by impact of higher energy electrons. Molecular ions will be relatively more intense and fragmentation less extensive under these circumstances; often, the predominant fragmentation reactions will differ at high and low ionizing voltage. These effects can be used to assist in the interpretation of unknown spectra as discussed in a later section. However, low-voltage studies inevitably result in poorer sensitivity, as consideration of Figure 11 demonstrates. Further, the fragmentation patterns observed in low-voltage spectra are somewhat dependent on the exact ionizing voltage used and source residence time. Thus, low-ionizing-voltage spectra are most useful as sources of confirmatory data.

Between 30 and 100 eV, the probability of an ionizing collision is relatively large and the amount of fragmentation produced essentially constant. For both reasons, EI mass spectra are conventionally recorded at high ionizing voltages, typically 70 eV.

2. Chemical Ionization

Most chemists would agree that the single most important datum to be obtained from a mass spectrum is usually the molecular weight of the molecule

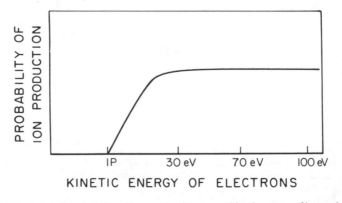

Fig. 11.   Idealized plot of probability of ion production versus kinetic energy of impacting electrons.

under investigation. Not uncommonly, spectra obtained using EI-induced ionization do not provide this information.

Electron impact tends to produce ions with considerable excess internal energy; extensive fragmentation may occur, resulting in a weak or nonexistent molecular ion peak. In 1966, an alternative ionization technique was described that offered a solution to this problem (68, 147, 148). It has been sufficiently successful that the technique of chemical ionization (CI) is now the second most widely used ionization technique.

In CI, a high pressure (0.5–2 torr) of reagent gas and a low pressure (10^{-3} torr) of the analyte are simultaneously present within a modified electronic ionization chamber. The reagent gas (e.g., CH_4) is ionized and fragments in a conventional manner [Eq. (3)]:

$$CH_4 + e \longrightarrow CH_4^{+\cdot} \longrightarrow CH_3^+, CH_2^{+\cdot}, CH^+ \cdots + 2e^- \qquad (3)$$

Since the pressure of reagent gas within the ionization chamber is high, numerous collisions will occur between these ions and neutral methane. These ion–molecule collisions will result in reactions leading to the formation of secondary reaction products [Eq. (4)]:

$$CH_4^{+\cdot}, CH_3^+, CH_2^{+\cdot}, CH^+ + CH_4 \longrightarrow CH_5^+, C_2H_5^+, C_3H_7^+ \qquad (4)$$

The relatively stable and long-lived secondary products eventually undergo ion–molecule reactions with the analyte to produce a variety of ionized products. Since the reactant ions are essentially thermalized as a result of frequent collisions and since many reactions are not highly exothermic, the resulting analyte ions often will not possess sufficient internal excitation to undergo further fragmentation. As a result, they will survive to generate peaks characteristic of the intact analyte molecule and thus directly related to its molecular weight. Perhaps the most useful species from the point of view is the $(M+1)^+$ or MH^+ peak [also called the quasi-molecular ion in the earlier literature; this usage is frowned upon by the IUPAC Sub-Commission on Mass Spectrometry (196a)], which involves transfer of a proton from the powerful Lewis acid CH_5^+.

$$M + CH_5^+ \longrightarrow MH^+ + CH_4 \qquad (5)$$

Naturally, the facility of the process and thus the intensity of the product ion peak will depend on the basicity of the analyte molecule. In favorable cases, essentially only a single product ion is observed, resulting in extremely high sensitivity. In other cases, however, additional products of ion–molecule reactions may be produced, for example (67),

$$C_2H_5^+ + M \xrightarrow{\text{x}} MC_2H_5^+ \qquad (6)$$

$$C_3H_7^+ + M \xrightarrow{\text{x}} MC_3H_7^+ \qquad (7)$$

or fragmentation may occur.

One of the most useful features of CI is that a variety of reagent gases can be used, often to good effect. The proton affinity of a species X [defined as the negative of the enthalpy change for protonation of X; Eq. (8)] and the hydride affinity of HX^+ [defined as the negative of the enthalpy change when HX^+ accepts a hydride ion; Eq. (9)] are illustrated as

$$X + H^+ \longrightarrow XH - \Delta H \tag{8}$$

$$HX^+ + H^- \longrightarrow H_2X - \Delta H \tag{9}$$

Proton affinities of some reagent gases and simple subtrate molecules are given in Table 1; they can be estimated for most sample molecules by consideration of model compounds (91). Ion–molecule reactions such as the protonation of X by other species do not occur unless they are exothermic; further, the extent of fragmentation produced will depend on the degree of exothermicity involved in formation of the parent ion. Thus, by judicious choice of reagent gas, the relative abundances of the parent ion and its fragmentation products can be varied or one component in a mixture can be selectively ionized. For example, the commonly used reagent gas isobutane $[(CH_3)_3CH]$ generates abundant tert-butylcarbonium ions $[(CH_3)_3C^+]$. Transfer of a proton from the tert-butylcarbonium ion is ca. 66 kcal less exothermic than proton transfer from CH_5^+ (92). Thus, isobutane CI spectra typically exhibit less extensive fragmentation than methane CI spectra (Fig. 12). The reagent gas ammonia produces ions of even higher proton affinity. Most classes of compounds are not protonated by such ions. Thus, ammonia reagent gas can be used for preferential ionization of nitrogen-containing compounds in a mixture (189, 190).

TABLE 1

Proton Affinities of Some Common Reagent Gases and Simple Substrate Molecules[a]

Molecule	Proton Affinity (kcal/mol)
CH_4	132
HOH	166.5
C_6H_6	181.3
CH_3OH	181.9
CH_3CHO	186.6
C_2H_5OH	188.3
CH_3COOH	190.2
n-BuCHO	192.6
$(CH_3)_2C=CH_2$	195.9
CH_3COCH_3	196.7
NH_3	204.0
$(CH_3)_3N$	225.1

[a] Data taken from Lias, S. G., Liebman, J. F., and Levin, R. D., *J. Phys. Chem. Ref. Data*, **13**, 695 (1984).

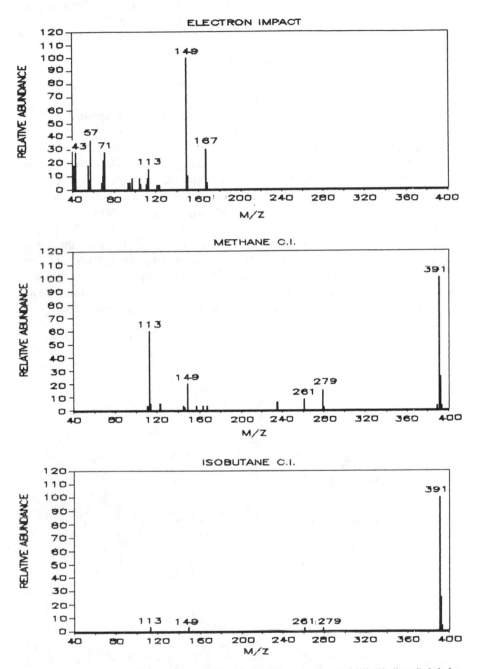

Fig. 12.   Comparison of EI, methane CI, and isobutane CI mass spectra of di(2-ethylhexyl)phthalate (MW 390) recorded at 150°C. Reprinted with permission from *Anal. Chem.*, **43**, 1785 (1971). Copyright (1971) American Chemical Society.

Other useful reagent gases are water and deuterium oxide (91). The latter has proven useful for the determination of the number of active (readily exchangeable) hydrogens in organic molecules. Another group of reagent gases induce ionization as a result of charge exchange (63):

$$He^{\cdot+} + M: \longrightarrow He: + M^{\cdot+} \qquad (10)$$

The molecular ions produced are, of course, similar to those generated by direct EI. However, the utility of this technique is due to the ability to alter the amount of excess energy in the analyte molecular ion by altering the reagent gas. Thus, for example, helium and neon have very high ionization potentials and recombination energies (Table 2). Charge exchange with most organics will be exothermic by ca. 10 eV, resulting in the formation of excited molecular ions and very extensive fragmentation. Argon and nitrogen exhibit intermediate ionization potentials and produce mass spectra very similar to those obtained by direct EI. Finally, the ionization potential of nitrous oxide is such that charge exchange will be nearly thermoneutral for typical organic compounds. The resulting molecular ions have little excess internal energy and tend to produce intense molecular ion peaks. Some workers (90, 98) have advocated tailor-made mixtures designed to give optimally intense molecular ions and structurally informative fragmentations.

TABLE 2

Ionization Potentials of Some Common Reagent Gases and Simple Substrate Molecules[a]

Molecule	Ionization Potential (eV)
C_6H_6	9.25
NO	9.25
$(CH_3)_2C=CH_2$	9.58
CH_3COCH_3	9.70
C_4H_9CHO	9.82
CH_3CHO	10.2
CH_3COOH	10.3
C_2H_5OH	10.5
CH_3OH	10.8
CH_4	12.7
N_2	15.6
Ar	15.8
Ne	21.6
He	24.6

[a] Data taken from Franklin, J. L., J. G. Dillard, H. M. Rosenstock, J. T. Herron, K. Draxl, and F. H. Field, "Ionization Potentials, Appearance Potentials and Heats of Formation of Gaseous Positive Ions," U.S. Department of Commerce, National Bureau of Standards, Washington, D.C., 1969.

The design of an ionization chamber for the CI technique is generally similar to that of a conventional EI source. The main modifications required are a more sophisticated inlet system (capable of admitting regulated amounts of both reagent gas and analyte) and provisions for containing a relatively high pressure within the ionization chamber without allowing background pressure in the instrument to increase sufficiently to produce a deterioration in the instrument's performance or dangerous arcing. Pertinent modifications include use of faster vacuum pumps and construction of a more nearly gas-tight ionization chamber. Since the latter requires shrinking the ion exit slit below optimum for EI analysis, combined EI–CI sources are typically 10 times less sensitive in the EI mode than a conventional EI source. Recently, the trend has been toward tandem or dual EI–CI sources, which can be rapidly interconverted from the exterior of the instrument. A variety of clever designs are commercially available (119). Another commercially available option on certain quadrupole instruments is pulsed positive- and negative-ion CI (95). In this technique, the potentials in a CI source are pulsed to alternately extract positive ions and negative ions. Thus, spectra of both positively charged and negatively charged ions are produced essentially simultaneously. Negative-ion CI often produces high abundance ions reflective of molecular weight, especially when the substrate has electronegative substituents; the technique is discussed in the following section.

3. Negative Ions from Electron Impact and Chemical Ionization

Three distinct mechanisms account for the formation of negative ions by direct interaction of electrons with sample molecules in a conventional EI source (18, 140):

Resonance capture:	$XY + e^- \longrightarrow XY^{\overset{.}{-}}$	(11a)
Dissociative resonance capture:	$XY + e^- \longrightarrow X^{\cdot} + Y^-$	(11b)
Ion pair production:	$XY + e^- \longrightarrow X^+ + Y^- + e^-$	(11c)

Resonance capture [Eq. (11a)] is of particular interest since it alone can result in the production of intact molecular ions. At low pressures, the process can occur only with electrons in a very narrow kinetic energy range dependent on the energy of the lowest unoccupied molecular orbital and usually near zero. Resonance capture is therefore not a favored process for most molecules under the conditions that prevail in a conventional EI source. If 70-eV electrons are used, the requisite thermalized electrons can arise only by interaction of higher energy electrons with the ion chamber walls or with molecules (e.g., through positive ion formation or through ion pair production [Eq. (11c)]). If low-energy electrons are used, however, the spread in kinetic energies characteristic of the electrons emitted from an electrically heated filament is sufficiently wide that only a small fraction of the electrons actually have the appropriate kinetic energy to participate in resonance capture.

The energy requirements for dissociative electron capture [0–15-eV electrons Eq. (11b)] and ion pair production [> 10 eV, Eq. (11c)] are less demanding since additional particles are produced to carry away excess kinetic energy. Nevertheless, negative-ion formation by direct electron impact has not proven to be a generally useful technique. Negative ions formed in a conventionally operated EI source are typically 10^{-3}–10^{-4} times less abundant than positive ions. Further, the behavior of the negative ions produced (i.e., the relative importance of the processes depicted in Eqs. (11a)–(11c) depends sharply on source temperature, pressure, and electron energy. The combination of lack of sensitivity and poor reproducibility severely limits the value of this technique.

Von Ardenne et al. (4, 5) developed an elaborate instrumental solution to these problems, the Duoplasmatron Ion Source. An externally produced plasma containing a large population of low-energy electrons is allowed to interact with the gaseous sample in this technique. In fact, the procedure showed considerable promise for producing usefully intense, reproducible negative-ion mass spectra. However, interest in the technique was diminished by Dougherty's demonstration that intense beams of negative ions could be simply obtained using a standard CI source at ca. 1 torr total pressure (57, 58, 198). The simplest applications of negative-ion chemical ionization mass spectrometry (NICIMS) use relatively inert species such as methane, isobutane, or nitrogen as the reagent gas. The first function of the reagent gas is to thermalize the high-energy electrons emitted from the filament. In their passage through the reagent gas, these electrons will be converted to secondary electrons by ionization processes; the secondary electrons will be slowed further by collision with other reagent molecules. A second function of the reagent gas may be to stabilize the molecular anions formed by resonance capture [Eq. (11a)]. Unless the internal excitation resulting from electron capture is removed by collision with a third body, the molecular anion will eventually fragment or the extra electron will be detached. As a result of both effects, a high pressure of moderating reagent gas drastically increases the production of molecular anions. In many cases, the sensitivity of NICIMS is several orders of magnitude greater than positive-ion CIMS (94). Of course, the sensitivity of NICIMS will depend strongly on the electron affinity of the sample molecule.

The technique just described can be referred to as "electron capture NICIMS." An alternative mode of analysis is "reactant ion NICIMS," represented by

Attachment: $R^- + XY \longrightarrow RXY^-$ (12)

Electron transfer: $R^- + XY \longrightarrow R^{\cdot} + XY^{\overline{\cdot}}$ (13)

Proton abstraction (X = H)
or nucleophilic displacement: $R^- + XY \longrightarrow RX + Y^-$ (14)

In this scheme, the reagent gas serves to generate a high concentration of the reactive species R^-, which then reacts with the sample molecule by attachment [Eq. (12)], electron transfer [Eq. (13)], proton abstraction or nucleophilic displacement [Eq. (14)], or other processes (101). A few reactant ions and their functions are Cl^- [attachment to generate intense $(M + Cl)^-$ ions from carboxylic acids, amides, phenols, polysaccharides, etc. (94, 198)]; O_2^- [proton abstraction, nucleophilic displacement, and electron transfer (59)]; and OH^- [proton abstraction from carboxylic acids, alcohols, and ketones to generate intense $(M - 1)^-$ ions (168)].

A trend of current research in this area is to develop an arsenal of reagent gases to produce particular reagent ions to probe structural features of an unknown molecule (101). The relative tendency to form molecular anions by resonance capture or to react with particular negatively charged reagents should vary widely with molecular structure. Thus, these techniques hold out the promise of being able to analyze target molecules selectively in the presence of lower electron affinity or less reactive interferents (cf. gas chromatography with electron capture detection) (56a).

4. Field Ionization

The technique of field ionization mass spectrometry (FIMS) provides an alternative approach to the problem of producing usably intense peaks characteristic of the intact ionized molecule. The quantitative theory of the field ionization (FI) process is complex and not fully developed (20). However, it is qualitatively readily understood. When a molecule approaches the surface of a conductor in the presence of a very high electric field (ca. 10^8 V/cm), quantum-mechanical tunneling of an electron from the molecule to the conductor can occur. Once ionized (and, if necessary, desorbed from the conductor's surface), the ion will be accelerated past the cathodic counterelectrode and into the mass analyzer portion of the mass spectrometer.

The electric field required to induce ionization and/or desorption will depend inter alia on the molecule's ionization potential, the distance the tunneling electron must traverse, and the chemical composition of the electrode's surface. However, even in the most favorable cases, the fields required are very high. This is conventionally produced by applying a positive potential of 3–10 kV to the field anode and a negative potential of several kilovolts to the cathodic counterelectrode. Thus, a potential difference of 10–12 kV exists between two electrodes that are separated by perhaps several tenths of a centimeter. Simple arithmetic indicates that if the field between the two electrodes were uniform, it would correspond to ca. 10^5 V/cm, at least several orders of magnitude below values needed for FI. The key to FIMS, then, is to alter the surface of the field electrode to produce locally much stronger fields. This is accomplished by introducing sharp discontinuities onto the surface of the field electrode. For example, the field strength F_0 at the apex of a typical metal tip whose radius of curvature is R_0 is given by (74)

$$F_0 = \frac{U}{5R_0}$$

where U is the potential difference between electrodes. Thus, if $U = 10^4$ V and

$R_0 = 2 \times 10^{-5}$ cm, the field strength at the tip's apex will be 10^8 V/cm, sufficient for FIMS. Single and multitip metal emitters, sharp metal blades (e.g., commercial razor blades), thin foils, and, especially, thin wires have been widely used (72). The ability of an emitter to produce useful ion currents can be drastically enhanced by various etching, activation, and conditioning processes. One of the goals of this activity is to grow large numbers of microneedles with small radii of curvature to increase the density of high-field-strength sites. Although these techniques have recently been reviewed, research into the development and production of new emitters continues (71, 116, 120, 151, 166).

The major advantage of FIMS over EIMS is the markedly more abundant unfragmented ions produced by this "soft" ionization technique (Fig. 13). It is believed that FI results in the transfer of only small amounts of internal excitation to the ion (< 0.5 eV) (34) and thus minimal fragmentation occurs. This can be contrasted with the ca. 5 eV excitation imparted by EI (159). The major disadvantage of FIMS is that the ionization efficiency of FI is one or two orders of magnitude lower than EI (124a). Full-scan spectra may require introduction of 100 ng of sample (143). However, since the information obtained from EI and FI experiments is often complementary, it is notable that combined EI–FI–field desorption sources are now readily available commercially.

5. Other Techniques

Although the ionization techniques just described are widely used, they often fail to produce ions characteristic of the intact molecule when the sample is nonvolatile or thermally labile. A number of techniques have been developed in an attempt to address this problem. These include field desorption (FD), ^{252}Cf plasma desorption mass spectrometry (PDMS), electrohydrodynamic ionization mass spectrometry (EHDMS), secondary ion mass spectrometry (SIMS), laser desorption mass spectrometry (LDMS), in-beam EI and CI, fast-atom bombardment (FAB), and so on. These methods are discussed briefly in a subsequent section.

C. MASS ANALYSIS

After ions are produced, the mass spectrometer must separate them according to their mass-to-charge ratio (m/z). An important measure of an instrument's ability to perform this task is its resolution R or resolving power RP; the terms are synonymous. A definition particularly favored by users of quadrupole and time-of-flight instruments is based on the width W of a peak at mass M at some arbitrary percentage of the peak's height:

$$R = \frac{M}{W} \tag{15}$$

For example, if a peak at m/z 1000 exhibited a width at 10% of peak height of 0.5 m/z, $R_{10\%} = 2000$. An alternative definition is based on the intensity of the

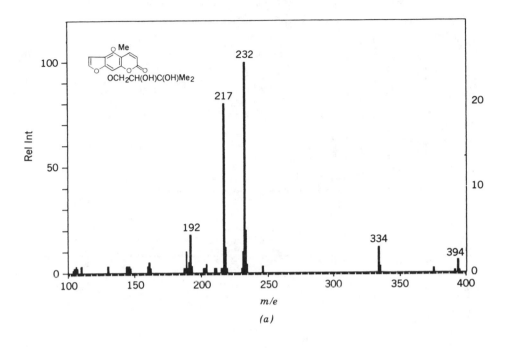

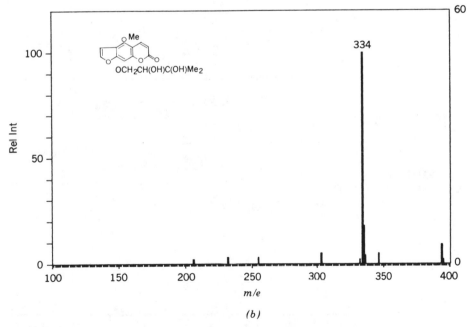

Fig. 13. The EI (*a*) and FI (*b*) spectra of coumarin byak-angelicin (70).

G. A. EADON

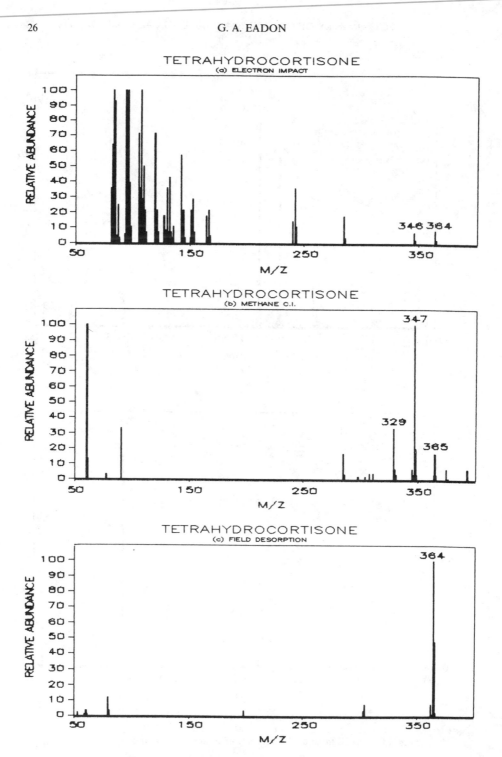

Fig. 14.   The EI (*a*), CI (*b*), and FD (*c*) mass spectra of tetrahydrocortisone (79).

signal in the valley between two peaks of equal height (Fig. 14). The resolving power of the instrument is the mass at which this valley is an arbitrary percentage (often 10%) of the height of either peak. This definition is most favored by users of magnetic sector instruments. Circuitry permits displacement of any peak by a selectable and accurately known fraction of a mass unit. Thus, if

$$R = \frac{M}{\Delta M} \tag{16}$$

is used, resolution can be determined at any mass. A third, less commonly used definition of resolution is based on the contribution one member of a doublet makes to the maximum height of the other (the "interference" or "crosstalk"). The resolution then corresponds to the mass at which the crosstalk equals some arbitrary value (often 1%) (Fig. 15). It is apparent that if the peaks in question are well shaped (Gaussian), resolving powers determined by various definitions are related by simple arithmetic factors.

Resolution is an important criterion of mass spectrometer performance for several reasons. Obviously, a low-resolution instrument ($R = 100$) would be a poor choice for studies of high-molecular-weight compounds (e.g., a steroid with

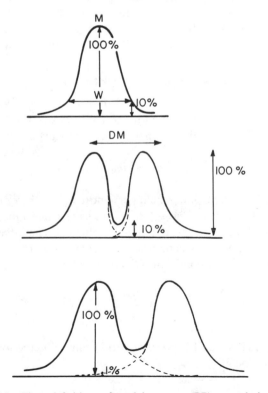

Fig. 15.   Three definitions of resolving power (RP) or resolution (R).

a molecular weight of 500 amu) since the higher mass peaks would overlap severely. More subtly, however, high resolution can permit separation and characterization of peaks of the same integral mass but of different elemental composition. As a simple example, consider the mass spectrum of 3-heptanone. At low resolution, a single peak is observed at m/z 57. At higher resolution, the m/z 57 peak consists of a pair of closely spaced peaks. The effect arises because the exact masses of most atoms are not integral. Thus, C_4H_9 has an exact mass of 57.070, while the exact mass of C_3H_5O is 57.034. If a mass spectrometer can resolve a peak into its constituents, it becomes possible to determine the exact mass of each and thus the elemental composition of each. Obviously, this considerably increases the information content of a mass spectrum. Commercial mass spectrometers are now available with resolving powers in excess of 100,000, permitting resolution of nearly all peaks likely to be encountered. It should be noted that mass measurements of sufficient accuracy to define elemental composition are possible at much lower resolutions (e.g., 3000) (11). This is satisfactory so long as the peak of interest is not an unresolved multiplet. If this is the case, the center of the peak will be shifted from its true position and an inaccurate mass measurement will result.

Although dozens of designs have been constructed to permit mass analyses of ions, this chapter will discuss only four of the five most widely used: single-focusing magnetic, double-focusing magnetic, quadrupole, and time-of-flight analyzers. The fifth popular technique (ion cyclotron resonance) is discussed in Chapter 2.

1. Single-focusing Magnetic Analyzers

Figure 16 depicts a highly simplified version of a 180° single-focusing magnetic analyzer. As the ions drift out of the ion source, they are accelerated through a large potential V. They acquire a kinetic energy ($\frac{1}{2}mv^2$) of zV:

$$zV = \frac{1}{2}mv^2 \tag{17}$$

If these ions then enter a magnetic field perpendicular to their direction of travel (perpendicular to the page in Fig. 16), they will experience a force perpendicular to their direction of travel and to the magnetic field B. The force zBv will induce an acceleration v^2/R, where R is the radius of curvature of the path imposed on the ion. Using Newton's second law, ($F = ma$) yields

$$zBv = \frac{mv^2}{R} \tag{18}$$

Equations (17) and (18) can be combined to eliminate v, the ion's velocity, and produce the equation

$$\frac{m}{z} = \frac{R^2B^2}{2V} \tag{19}$$

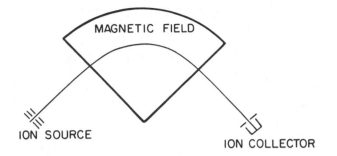

Fig. 16.   Simplified diagram of single-focusing magnetic mass spectrometer.

This equation relates the mass-to-charge ratio of an ion to three variable parameters: the radius of curvature R, the magnetic field strength B, and the accelerating potential V. Most commonly V and R are held constant while the magnetic field strength is varied (scanned). At some value of B, each m/z will travel an appropriate R to permit its impingement on the collector.

An alternative mode of operation holds R and B constant and scans V, the accelerating voltage. Again, at the appropriate value of V, a particular m/z ion will reach the collector. This technique's advantage is that it permits more rapid scanning; the rate at which a magnetic field can be changed is limited by the phenomena of magnetic hysteresis. (Nevertheless, modern magnetic instruments are capable of extremely rapid scanning, e.g., 0.2 s/decade with 0.2 s flyback time.) Accelerating voltage scanning is not a widely used technique since it produces mass-dependent effects on ion abundances and focusing. In particular, it discriminates against high-mass ions.

The maximum resolving power obtainable with a single-focusing magnetic sector instrument is limited to ca. 5000. It is worthwhile to consider why. First, as the ions enter the magnetic field, they will not be traveling in a perfectly collinear fashion. Rather, as a result of interionic repulsions and the finite width of the source slits, they will be traveling in a slightly divergent direction (Fig. 17). As these ions travel through the magnetic sector, they are brought into a partial focus at the collector. If the half-angle of deviation at the source is α, the beam width at the collector is $\alpha^2 R$; the magnetic field has produced "directional focusing." It is important to note that the focusing action is imperfect. Since it depends on α^2, it is called "first-order focusing." If the beam width varied as α^3, it would be "second order," and so on. (42a).

As a result of directional focusing, ion divergence is not the principal limitation on resolving power in these instruments. Rather, it is due to the fact that ions of a given mass are not as strictly homogeneous in their kinetic energy as Eq. (17) would indicate. Because of the translational energies possessed by the molecules before ionization, because ions are produced at points of slightly different potential within the source, and because fragmentations occur with release of varying amounts of kinetic energy, these ions are slightly inhomogeneous in kinetic energy. One strategy to minimize this inhomogeneity is to use

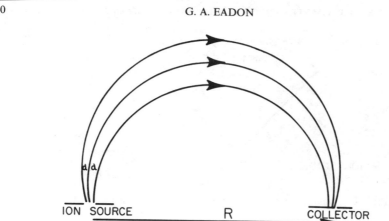

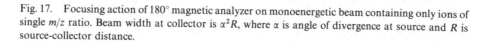

ION SOURCE   R   COLLECTOR

Fig. 17. Focusing action of 180° magnetic analyzer on monoenergetic beam containing only ions of single m/z ratio. Beam width at collector is $\alpha^2 R$, where α is angle of divergence at source and R is source-collector distance.

a very high accelerating voltage (e.g., 8000 V). Then, the fractional spread in ion beam energy will be relatively small (e.g., <0.1%). Nevertheless, because the magnetic field is not energy focusing, these inhomogeneities do produce significant variations on the point of focus of ions.

Finally, it should be mentioned that although the design depicted in Fig. 15 corresponds to a 180° deflection angle, 60° and 90° magnetic sectors are more commonly used. The magnets are lighter in weight and less costly; further, the source and collector are more readily accessible and are no longer immersed in the magnetic field. The physical principles involved in mass dispersion and focusing with wedge-shaped fields are essentially the same as those already discussed in connection with the 180° sector (14, 174).

2. Double-focusing Mass Analyzers

Double-focusing mass analyzers incorporate a device called an electric sector (Fig. 18). If positive ions are under study, a positive potential is applied to the outer plate and a negative potential to the inner plate to produce a uniform radial electric field E. An ion of charge z traveling through the electric sector will experience a force zE perpendicular to its direction of travel; it will therefore undergo a centripetal acceleration v^2/R_e. Using Newton's second law ($F = ma$) yields

$$zE = \frac{mv^2}{R_e} \tag{20}$$

Rearranging Eq. (20),

$$\tfrac{1}{2}R_e zE = \tfrac{1}{2}mv^2 \tag{21}$$

it becomes obvious that the radius of curvature of an ion depends on its kinetic

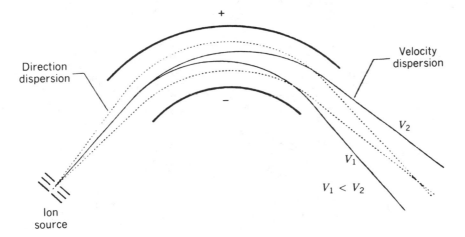

Fig. 18. Schematic drawing of electrostatic analyzer. Ions leaving source in slightly divergent directions (dotted lines) are refocused, while ions with different velocities (solid lines) are dispersed (Roboz, General Bibliography).

energy. If a slit were located appropriately near the exit of the electric sector, only ions of a specific kinetic energy would be transmitted. The electric sector would then serve as an energy filter, and if the monoenergetic beam of ions were then subjected to a magnetic field, high resolution would result. This approach is less attractive than it appears at first glance, however, since it results in a drastic decrease in sensitivity. A more sophisticated design would make better use of the direction-focusing characteristics of the electric sector.

Two such designs have found extensive commercial application. The Nier–Johnson version (Fig. 19) (102) produces second-order direction focusing and first–order velocity focusing. As ions pass through the electric sector, directional focusing and velocity (kinetic energy) dispersion occur. The geometry of the instrument is such that the magnetic field compensates for the velocity dispersion and focuses the ions at the collection slit.

The ion optics of the Nier–Johnson design result in a point of focus; this arrangement is ideally set up for electrical detection with magnetic or electrical scanning. An alternative double-focusing design, the Mattauch–Herzog geometry (Fig. 20) (121) focuses all ions onto a plane. If a photographic plate is placed in the focal plane, the magnetic field can be held constant and the entire mass spectrum recorded without scanning. Alternatively, a slit, collector plate, and electron multiplier can be installed to permit electrical recording as the magnetic field is scanned. The advantages and disadvantages of these detection techniques are discussed in a subsequent section.

Instruments with reversed Nier–Johnson geometry (i.e., with the magnet preceding the electric sector) have recently become commercially available. The applications of this arrangement are discussed in a subsequent section.

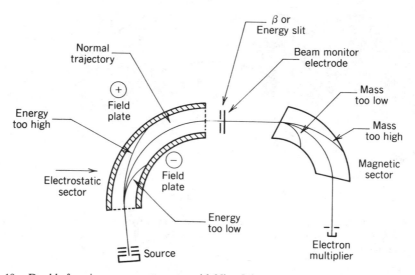

Fig. 19.   Double-focusing mass spectrometer with Nier–Johnson geometry and electronic detection (Roboz, General Bibliography).

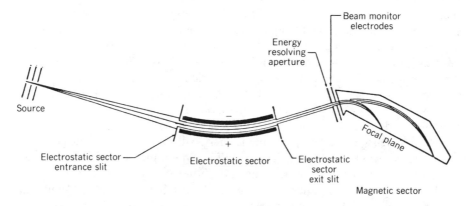

Fig. 20.   Double-focusing mass spectrometer with Mattauch–Herzog geometry and photoplate detection (Roboz, General Bibliography).

3.   Quadrupole Mass Spectrometers

The magnetic mass spectrometers already described accomplish mass analysis through the use of "static" magnetic and electric fields; these parameters are unchanged during the transmission and detection of an ion of a particular m/z. In contrast, many other mass analyzers are "dynamic"; mass separation requires the time dependence of a parameter such as magnetic or electrical field strength or ion production (Dawson, General Bibliography). The quadrupole mass spectrometer is the most widely used instrument in this category.

Unfortunately, the level of physics and mathematics required to understand the principles underlying mass separation in quadrupole mass spectrometry are far beyond those required for a magnetic instrument (50). Briefly, a quadrupole mass analyzer consists of four metal cylindrical rods (typically 10–25 cm long, 0.5–1.0 cm diameter) whose axes are located at the corners of a square and are precisely parallel (Fig. 21). If a positive dc voltage $(+V_{DC})$ is applied to one pair of diagonally opposite rods and an equal negative dc voltage $(-V_{DC})$ applied to the other pair, a time-independent electric field will be produced within the quadrupole cavity. If positively charged ions are launched along the quadrupole's long axis (conventionally, the Z axis), they will be attracted toward the negative potential rods. If the dc voltages and ion velocities are typical of those used in quadrupole mass spectrometers, no ions would survive to emerge from the opposite end of the quadrupole. All ions, regardless of their m/z, would be accelerated into the negative potential and destroyed. In contrast, suppose that a radio-frequency (rf) voltage (V_{RF}) is applied to one diagonally opposite pair of rods and the same rf voltage 180° out of phase is applied to the other pair. Under these conditions, the direction of the resulting electric field will oscillate rapidly, and there will be no persistent tendency for an ion to be attracted toward a particular rod. All ions, independent of their m/z, will have a high probability of emerging unscathed from the quadrupole.

Use of a quadrupole as a mass analyzer requires the superposition of direct and rf voltages. The potential applied to one pair of rods is $+V_{DC}+V_{RF}\cos wt$, while that applied to the other pair is $-V_{DC}+V_{RF}\cos(\pi+wt)$, where w is equal to 2π times the frequency of the RF radiation; Fig. 21 graphically depicts the net potentials applied to the two pairs of rods during transmission of a particular m/z ion. The electric fields within the quadrupole will now exhibit a complicated space and time dependence. Ion trajectories can be approximated using the Mathieu equations (128). Qualitatively, however, it is clear that the intensity and directions of the forces acting on a particle will change rapidly. Thus, at $t=0$ (Fig. 22), positively charged ions will be attracted toward rods 2 or 4. By t_2, that attraction will cease and turn into a net repulsion. An ion whose mass is too low will be accelerated rapidly toward rod 2 or 4 and might well be lost before the

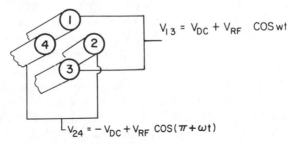

$$V_{13} = V_{DC} + V_{RF}\ \cos wt$$

$$V_{24} = -V_{DC} + V_{RF}\ \cos(\pi+\omega t)$$

Fig. 21.   Schematic diagram of quadrupole mass filter.

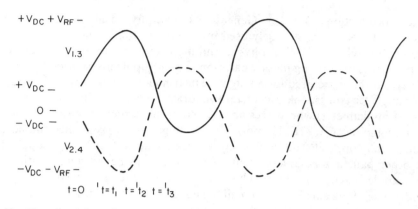

Fig. 22.   Net potentials applied to rods 1 and 3 ($V_{1,3}$, solid line) and 2 and 4 ($V_{2,4}$, dashed line) when $V_{13} = V_{DC} + V_{RF} \cos wt$ and $V_{2,4} = V_{DC} + V_{RF} \cos (\pi + wt)$.

repulsive effect of the electric field after t_2 can operate. The amplitude of oscillation of ions whose mass is too large increases steadily until, eventually, after a number of cycles, they too collide with a bar and are destroyed. A range of masses, however, undergo stable oscillations within the quadrupole cavity. Polarity reversal occurs before collision with a rod becomes unavoidable, and these ions eventually emerge from the quadrupole cavity.

The m/z that is optimally transmitted through the quadrupole depends on the frequency f and the voltage V_{RF} of the applied rf field [Eq. (22)] (50, 152):

$$\frac{m}{E} = \frac{CV_{RF}}{f^2} \tag{22}$$

Using the Mathieu equations, diagrams relating the stability of a particular m/z in the quadrupole field to these parameters and the applied DC voltage can be drawn (32, 50). A hypothetical diagram for an instrument optimized for transmission of m/z 200 is depicted in Fig. 23. As already discussed, in the absence of a dc voltage ($V_{DC}/V_{RF} = 0$) all ions are transmitted, and at very high dc voltage ($V_{DC}/V_{RF} = \infty$) no ions are transmitted. Three more interesting modes of operation are indicated by lines a, b, and c. Line a corresponds to selection of a DC voltage such that only ions with a mass very near 200 will undergo stable oscillations within the quadrupole. Operations in such a "high-resolution" mode is impractical for quadrupole. The actual amplitude of oscillation of an ion depends in addition on its entrance conditions into the quadrupole (axial displacement, angular divergence, rf phase). Under very high resolution conditions, these conditions are so demanding that only a very small fraction of ions of the "right" mass will be transmitted; the instrument's sensitivity will be insufficient for practical application. The ratio V_{DC}/V_{RF} indicated by line b corresponds to transmission of all ions between masses 199.5 and 200.5. This resolution suffices to define the mass of any ion transmitted near m/z 200 and

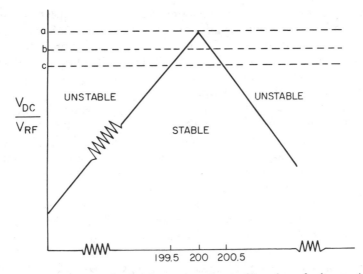

Fig. 23.   Hypothetical stability diagram for quadropole mass filter whose rf voltage and frequency have been selected to optimize transmission of m/z 200

results in less demanding entrance conditions and higher sensitivity. Of course, the voltage ratio indicated by line c would result in even greater sensitivity but would not suffice to uniquely identify the mass-to-charge ratio of collected ions. Such a mode of operation might be useful on an intermittent basis if a quadrupole were used as a chromatographic detector in a GC/MS combination.

Quadrupole mass spectrometers possess certain advantages over magnetic instruments that have led to their wide popularity (66). As Eq. (22) indicates, the m/z transmitted varies linearly with the voltage of the RF field ($m = m_0 + kV_{RF}$). This results in a simple assignment of m/z to unknown peaks since RF voltages are readily measured. In contrast, magnetic scanning results in a nonlinear mass scale so that the spacing between masses decreases proportional to the square root of the masses as mass increases. A second advantage commonly cited is that it is possible to scan quadrupole very rapidly (tenths of a second); since only electric fields are being changed, problems of magnet hysterisis do not intervene. The ability to scan rapidly is an advantage in applications such as capillary column GC/MS, where peaks elute very rapidly and more than one scan per peak is desired. However, contemporary magnetic instruments can scan at rates sufficiently rapid for most practical applications. The quadrupole's real advantage is in the technique of "selected ion monitoring," in which the instrument rapidly switches conditions to monitor only a few ions in a mass spectrum. The sensitivity of the instrument is vastly increased in this mode since much less time is spent monitoring ions of no interest.

Another advantage of quadrupole mass analyzers is that they are "continuous focusing." If an ion undergoes a collision during its passage down the quadrupole, it will be refocused and transmitted. This feature permits operation at relatively high pressure in the analyzer region of the spectrometer. Such conditions might be encountered during GC/MS and LC/MS experiments. In contrast, if an ion undergoes collision in a magnetic sector instrument after acceleration but before exiting the magnetic field, it will not be focused normally. Magnetic instruments circumvent this problem by high-speed pumping or differential pumping on the analyzer region.

On the other hand, quadrupole mass filters also possess certain disadvantages vis-à-vis magnetic mass spectrometers. Long-lived ions that fragment after leaving the source but before detection ("metastable ions") can provide much useful information about ion structure and chemistry. Such ions are readily studied in magnetic instruments, as will be discussed subsequently. However, as a result of the continuous-focusing feature just alluded to, metastable ions are not transmitted through quadrupole mass filters. A second disadvantage is that the quadrupole's practical upper limit of resolution is about 3000. Of course, the significance of this limitation will depend on the applications contemplated for the instrument.

A third disadvantage of quadrupole instruments is that they tend to discriminate against higher mass ions unless special care is taken. Ions travel down the quadrupole with rather low velocities in the forward (z) direction; typically, they have been accelerated through a potential drop of 5–30 V. These low velocities are necessitated by the requirement for repeated oscillations (e.g., 10–100) within the quadrupole region in order to produce useful mass filtering action and resolution. Poor transmission of higher masses is related to the fact that these ions travel with slower velocity in the z direction. Therefore, the fringe fields near the entrance to the quadrupole are more likely to divert these ions into a negative rod (50). Such discrimination is a significant problem for several reasons. High-mass peaks are often of special utility for identifying unknown molecules; if they are seriously attenuated, they may be missed or lost in the background level of the spectrum. Further, most reference spectra in mass spectral libraries have been obtained on magnetic instruments. Severe mass discrimination might preclude the successful matching of a reference spectrum obtained on a magnetic instrument with an unknown's spectrum from a quadrupole. Several instrumental modifications have lessened the severity of mass discrimination. Contemporary instruments now scan the accelerating voltage in parallel with the rf field. As a result, the z velocities of high- and low-mass ions are more nearly comparable. Further, focusing lenses are interposed between the ion source and the quadrupole system to permit differential focusing of ions. Then, the instrument can be "tuned" with a reference compound on a regular basis to permit close simulation of magnetic instrument spectra. Unfortunately, the tuning process can be laborious.

4. Time-of-Flight Mass Spectrometer

The time-of-flight (TOF) mass spectrometer provides another example of dynamic mass separation; in this case, analysis requires modulation of the rate of ion production. The general operating principles of the TOF mass spectrometer are rather simple. After a cluster of ions of various m/z values are accelerated through a potential V of several kilovolts, all will have essentially the same kinetic energy [Eq. (23)] if their preacceleration thermal energies are neglected:

$$zV = \tfrac{1}{2}mv^2 \qquad (23)$$

Thus, high-mass ions have smaller velocities than lower mass ions. If the ions all start down a flight tube at the same instant (Fig. 24), an ion's time of arrival at the detector will depend on its m/z. For example, an ion of m/z 500 accelerated through a potential of 2 kV will traverse a 1-m drift tube in ca. 38 μs, while an ion of m/z 501 will require 370 ns longer. Mass analysis is thus accomplished.

In the ideal case, the cluster of ions entering the accelerating region would be infinitely compact, and individual ions would have no thermal velocities. In practice, resolution is limited to ca. 600 by an inability to attain these goals. Two procedures are commonly used to produce well-defined packets of ions for acceleration. In the pulsed mode, the electron beam is switched on very briefly (ca. 1 μs) and is then switched off while the resulting ions are repelled into the accelerating region. Alternatively, in the continuous mode of operation, the electron beam remains on much longer (e.g., for 70% of the 20–100-μs TOF duty cycle). The resulting ions are stored in a "potential well" within the source. After the electron beam is pulsed off, the repeller voltage is pulsed on for a few nanoseconds to eject the stored ions into the accelerating region. Since the electron beam is operational for a higher fraction of the time in the continuous mode, this is the more sensitive procedure (128). In either mode, 10,000–50,000 packets of ions might be produced each second.

Since ions of adjacent m/z values arrive at the detector within a few nanoseconds of each other, the entire detection system must have a very rapid

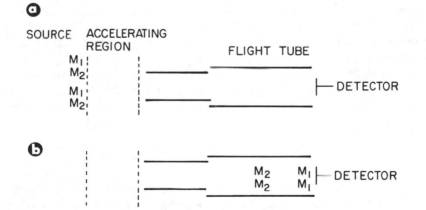

Fig. 24.   Schematic diagrams illustrating operating principle of TOF mass spectrometer. (a) Packet of m_1 and m_2 ions is in source, about to be accelerated the flight tube. (b) A few nanoseconds later M_1 ions are about to be detected, while heavier, slower M_2 ions are still in flight tube.

response time. Electron multiplier detectors (discussed subsequently) are there-fore universally used. One technique for displaying spectra synchronizes the sweep of an oscilloscope with the pulse circuit of the ion source. Up to 10,000 spectra per second can be produced and displayed using this procedure (152). This rapid-scanning feature facilitates studies of very fast gas phase reactions such as those occurring in flames, explosions, shock tubes, and so on. Alter-natively, a gating system can be used to pulse the electron multiplier's output to a short-term integrating device only during the period when a particular m/z ion is being collected. The intensity of the resulting signal will be proportional to that ion's abundance in the spectrum. By selecting various delay times, the entire mass spectrum can be scanned more slowly but with a better signal-to-noise ratio than obtained using the oscilloscopic technique (32). Alternatively, if selected ion monitoring is desired, only a few preselected masses can be monitored with high sensitivity (152).

The "reflectron" is an interesting modification of the conventional TOF instrument. An electrostatic mirror installed in the flight tube permits reflection of the ions back toward a second detector. This procedure improves resolution by providing energy focusing; the small variations in the velocity of ions of a given mass are substantially compensated for. The faster moving, higher energy ions of a given mass travel more deeply into the reflecting electric field than lower energy ions. Thus, their arrival at the second detector is somewhat delayed. This procedure increases mass resolution from 500–600 to 3000–4000 (31a, 52b).

The reflectron technique also permits direct and selective detection of neutrals formed by metastable decompositions. When a TOF instrument is operated normally (i.e., with the reflecting mirror disabled), ions and neutrals formed by postacceleration fragmentation impinge on the detector. Since these neutrals have essentially the same velocities as their precursor ions, their arrival times (and thus apparent masses) will be identical to that of the unfragmented ion. However, when the mirror is enabled, only neutrals will pass through and impinge on the detector. In some cases, the bulk of the peak intensity observed in the "normal" operating mode arises from neutral impacts (52c). Other reflectron operating modes permit identification of fragment ions formed from particular precursor ions.

The combination of thermionic ionization and the TOF mass spectrometer has recently found extensive application to the study of high-molecular-weight molecules. This work is discussed more fully in section IV.C of this chapter.

D.   DETECTION

Accuracy, sensitivity, and response time are the most important measures of an ion detector's performance. Although no detector excels at all three criteria, the electron multiplier detector represents the best compromise for routine use. Photographic detection is a "last resort" technique practical only when high-resolution data is required on a large number of peaks and only very limited amounts of sample are available. The Faraday cup finds its application in cases where the relative abundances of ions must be very accurately known (e.g., measurements of isotopic abundance). The three main detection systems are discussed in greater detail in below.

1. Faraday Cup Collector

A Faraday cup detector (Fig. 25) consists of a metal collector grounded through a 10^{10}–10^{12}-Ω resistor and is enclosed within a metal cage or cup that is open at one end (149). The ion beam enters the cage through the open end and strikes the collector electrode. The cage's function is to capture any reflected ions or ejected secondary electrons emitted from the collector. The resulting positive charge on the collector–cage assembly is neutralized by the flow of electrons from ground. Since $V=iR$, a very high resistance maximizes the voltage drop across the resistor. The voltage drop is then amplified by an electrometer or DC amplifier. If a 10^{12}-Ω resistor (the maximum practical) is used, collection of about 6000 ions per second will result in a 1-mV signal to the amplifier.

The Faraday cup detector's best feature is its accuracy. To an excellent approximation, each positively charged ion will require a single electron for its neutralization essentially independent of its chemical nature; the other detection systems discussed here exhibit significant discrimination in their responses. The Faraday cup detector exhibits low electrical noise, and its sensitivity remains constant over time. Its main disadvantage is its slow response time, due to the high-impedance amplifiers required to measure the potential across a 10^{10}–10^{12}-Ω resistor. This detector is therefore unsuitable for recording rapidly scanned mass spectra.

2. Electron Multiplier Detector

The electron multiplier lacks many of the solid virtues of the Faraday cup detector (114). Its responsiveness to an ion depends, in a poorly understood

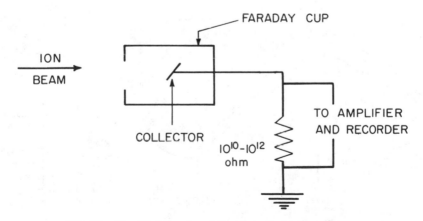

Fig. 25.   Simplified drawing of Faraday cup detection system.

fashion, on that ion's mass, energy, and molecular structure. Its gain varies over time depending on its prior use. Its response is affected by the presence of stray magnetic fields. When its output current exceeds 10^{-8} A, it becomes nonlinear.

Despite these disadvantages, the electron multiplier is certainly the most widely used detector in organic and inorganic mass spectrometry. It has extraordinary sensitivity and is capable of detecting the arrival of single ions. Further, it is capable of extremely rapid response (time constant of ca. 10^4 Hz). This feature is particularly important when rapid scanning is required to measure transient phenomena (e.g., elution of GC peaks from capillary columns or arrival of ions in a TOF mass spectrometer).

In an electrostatic electron multiplier (Fig. 26) (2), ions are accelerated by a large potential into the conversion dynode (so called because it converts an ion current into an electron current). Each collision results in the expulsion of two to three electrons. The emitted electrons are accelerated toward the second dynode. As each electron collides with the second dynode, two to three additional electrons are emitted. The process continues as the electrons cascade down the 10–20 dynodes typical of contemporary electron multipliers. The final dynode is connected to an amplifier–recorder system. The maximum gain of an electron multiplier is typically 10^5–10^8.

Figure 27 depicts a continuous-dynode electron multiplier of a design commonly used with quadrupole mass spectrometers. A tube of decreasing diameter is coated with a high-resistance tin–tin oxide coating; a large (-1.0 to -2.0 kV) potential is applied to the open end of the tube. As positively charged ions are drawn into the tube and strike the inner surface, electrons are emitted. These electrons travel toward the ground potential at the closed end of the tube; because of the curved electric field within the multiplier, the emitted electrons collide repeatedly with the walls, emitting more and more electrons. The process repeats itself many times as the electrons cascade down the tube, resulting in gains on the order of 10^5 in routine operation.

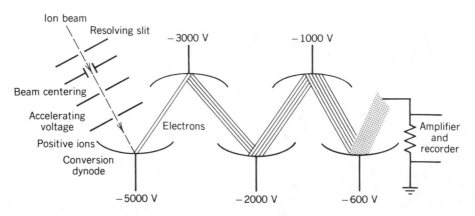

Fig. 26.   Electrostatic electron multiplier (Roboz, General Bibliography).

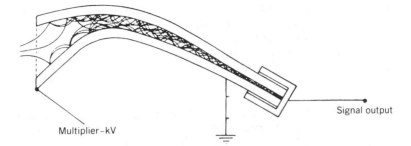

Fig. 27.   Simplified diagram of continuous-dynode electron multiplier.

The noise level ("dark current") is sufficiently low and the gain sufficiently high that either electron multiplier design is capable of detecting the arrival of a single ion. However, as discussed in the next section, considerably larger numbers of ions are required to define the relative abundance and exact mass of any ion in a mass spectrum.

3. Photographic Plates

Double-focusing mass spectrometers with Mattauch–Herzog geometry can use photograph plates as a detector–recorder system for determining mass spectra. The technique has some clear disadvantages. Development of the plates is required before the data can be used. Determination of accurate masses when the data are finally in hand requires expensive equipment. The response (blackening) produced by an ion depends upon its mass, its kinetic energy, and the quality of the photoplate used. The dynamic range of a photoplate is only about 100. Finally, a photoplate is intrinsically less sensitive than an electron multiplier; at least 1000 ions of a given mass are necessary for photoplate detection.

These disadvantages are sufficient to preclude the routine use of a photoplate for recording low-resolution mass spectra. However, the technique does find application to determining high-resolution mass spectra when sample sizes are extremely small. Suppose, for example, that a compound of interest elutes from a capillary GC column with a half-width of 5 s and one wishes to record the entire mass spectrum at 50,000 resolution. If a 5-s/decade scan rate is used, each "mass window" will be collected for ca. 10^{-4} s if electrical recording is used. On the other hand, photoplate detection permits collection of each "mass window" during the elution of the entire peak; as already discussed, all ions are focused simultaneously using the Mattauch–Herzog geometry. Thus, an ion of a given m/z will be collected ca. 50,000 times longer in such an instrument, more than compensating for the thousandfold loss in sensitivity of a photoplate versus an electron multiplier.

The sensitivity advantage of photoplate detection vanishes if one is interested in determining the exact masses of only a few ions. Here, the instrument using electrical detection can monitor only the region where a mass of interest is expected. Thus, these ions will be observed for a larger fraction of time, permitting drastically improved sensitivity. This technique is known as selected ion monitoring and is discussed subsequently.

E. METASTABLE IONS

In a double-focusing magnetic mass spectrometer (Fig. 28), ion fragmentation can occur in a number of distinct locations. Ions that are formed in the source and do not fragment further until their arrival in the third-field free region will be observed as "normal" ions in the mass spectrum. Ions that fragment during the 10^{-6}–10^{-5} s required to travel between the source and the third field-free region are called "metastable ions." Many of these ions will not be observed during the normal mode of operation. For example, fragmentation during acceleration in the first field-free region or in the electrostatic analyzer will all result in daughter ions with less kinetic energy than the normal ions. The electrostatic analyzer will efficiently filter out ions whose kinetic energy is significantly lower than normal. However, ions that fragment in the second field-free region of a double-focusing magnetic instrument (or, equivalently, in the field-free region preceding the magnet in a single-focusing magnetic instrument) will appear as diffuse, low-intensity peaks in the mass spectrum. If the mass of the precursor ion is m_1 and the mass of the daughter ion is m_2, the apparent mass of the metastable ion (usually designated at m^*) is given by (84)

$$m^* = \frac{m_2^2}{m_1}. \tag{24}$$

For example, if the m/z 100 ion fragments to generate an m/z 85 ion, The metastable ion will be centered at m/z 72.25. The arithmetic of Eq. (24) dictates that metastable peaks will usually occur at nonintegral masses, always less than either the parent or daughter ion. In a complex mass spectrum, considerable trial-and-error manipulations may be required to identify the pair of peaks responsible for a particular metastable peak. The process may sometimes be simplified by the observation that $m_2 - m_1 \approx m_1 - m^*$ if m_1 and m_2 are close. If the center of the peak can be accurately measured, published tables that list values of m^* corresponding to values of m_1 and m_2 up to 500 are useful (27).

Second field-free region metastable peaks are nearly always broader than normal peaks. This effect arises because fragmentations usually involve the

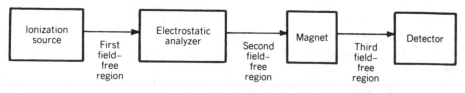

Fig. 28.   Box diagram of double-focusing mass spectrometer.

conversion of some internal excitation energy into kinetic energy. If the released kinetic energy is directed along the ion's direction of flight, the daughter ion will have a slightly greater forward velocity as it enters the magnetic field. Since the force acting on the ion in that region is proportional to the ion's velocity, such ions will be focused on the collector slit at a slightly lower field (i.e., mass) than predicted by Eq. (24). Since all orientations are equally likely, such metastable peaks are Gaussian. However, as the kinetic energy release becomes larger, certain daughter ions will be formed with sufficient velocity perpendicular to their original direction of travel that they will not be transmitted through the collector slit. The result will be a flat-topped or even dish-shaped metastable peak (Fig. 29). The width and shape of the peak are readily related to the amount of kinetic energy released during fragmentation (24).

Observation of a second field-free region metastable peak can facilitate interpretation of a mass spectrum since it can establish which of several possible precursor ions actually generated a particular daughter ion. Measurement of the kinetic energy released can also provide useful information about the mechanisms and ion structures involved in certain fragmentations (29).

1. First Field-Free Region Metastables

The first field-free region lies between the accelerating plate and the electric sector. If an ion of mass m_1 decomposes in this region, its kinetic energy (eV_1, where V_1 is the accelerating voltage) will be partitioned between the daughter ion m_2^+ and the expelled neutral according to

$$\frac{eV_2}{eV_1} = \frac{m_2}{m_1} \tag{25}$$

The electric sector, as normally operated, transmits only ions with kinetic energies near normal (eV_1); daughter ions m_2^+ will therefore be destroyed.

One technique for observing ions formed by fragmentations in the first field-free region requires location of a collector electrode and slit at the focal plane of the ions after transmission through the electric sector. Then, the potential E across the electric sector is scanned. The kinetic energy that an ion needs to successfully traverse the electric sector is linearly related to E. Thus, the kinetic energy of the daughter ion corresponding to the transition $m_1^+ \rightarrow m_2^+$ will be decreased by the factor m_2/m_1; if E is decreased proportionately from the value E_1 that transmits the main ion beam [$E = E_1(m_2/m_1)$], m_1 will be transmitted through the electric sector and collected at the auxiliary electrode and a signal will result. In practice, the electric sector voltage is usually scanned from E_1 to 0; typically, many metastable ions are observed (Fig. 30). The result is referred to as an ion kinetic energy (IKE) spectrum (26, 29). These spectra serve as "fingerprints" of a particular molecule. However, their complexity, the presence of numerous overlapping peaks, and the occasional ambiguity encountered in

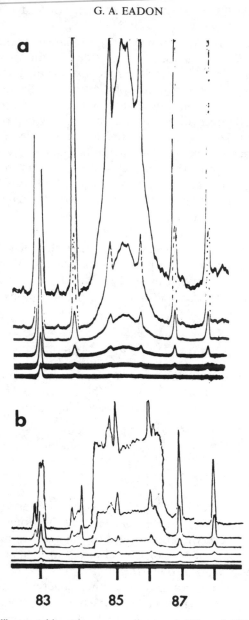

Fig. 29. (a) "Normal" metastable peak corresponding to m/z 139 → m/z 109 transition in spectrum of m-nitrophenol. (b) Flat-topped metastable peak corresponding to same transition in spectrum of o-nitrophenol (27a).

assigning a transition to a particular peak make detailed interpretation of IKE spectra a formidable task.

An alternative mode of scanning can be used to generate similar spectra. The electric sector voltage E can be kept constant and the accelerating voltage V

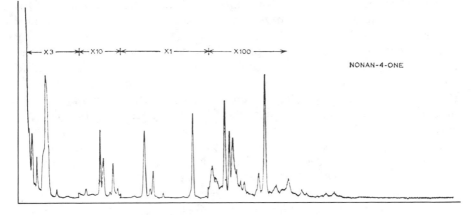

Fig. 30. Ion kinetic energy spectrum of 4-nonanone (60).

scanned toward higher values. At some value of V [specifically, at $(m_1/m_2) V_1$, where V_1 is the accelerating voltage that transmits the main beam of ions through the electric sector], the daughter ions m_2 will possess appropriate kinetic energy to be transmitted, collected, and detected at the β-slit. Operationally, this procedure is most conveniently accomplished by setting the instrument up to observe normal ions at a low accelerating voltage (e.g., 2 kV) and then scanning upward toward the maximum voltage attainable (often 8 kV). The spectra that result will be similar to IKE spectra.

A very important modification of this experiment avoids use of the auxiliary collector. The instrument is set up as before to transmit the main beam of ions at a low accelerating voltage (e.g., 2 kV). The magnet is adjusted to focus the daughter ion of interest (e.g., m_2^+) on the collector slit. Under these conditions, the magnet will transmit and focus all m_2^+ ions with kinetic energy V_1e). As already discussed, the magnet will transmit and focus only those m_2^+ ions with kinetic energy $V_1 e$. If other metastable transitions result in formation of m_2^+, they too will be detected at different values of the accelerating voltage. This technique, variously called metastable refocusing or high-voltage scanning, permits rapid determination of all ions that fragment in the first field-free region to form a particular daughter ion (13, 69, 99).

This technique has certain attractive features. The parent and daughter ions responsible for a particular peak are determined without ambiguity. Further, the procedure can be a very sensitive tool for detection of metastables. Since the main beam of ions is filtered out by the electric sector as soon as the accelerating voltage is significantly perturbed, the electron multiplier's gain can be increased to permit observation of weak transitions. One disadvantage of this procedure is the limited range over which the accelerating voltage can be varied, typically by a factor of 4. Thus, metastable transitions in which an ion's mass decreases by more than a factor of 4 cannot be observed. More perplexing, as the accelerating voltage is scanned, the "tune" of the ion source will change drastically. If the source conditions are not reoptimized during a scan, low-abundance metastable ions may be missed.

An alternative solution to the problem of determining all first field-free metastable transitions that generate a particular daughter ion uses a linked scanning technique (33, 75, 113). The daughter ion of interest is focused on the detector under normal operating conditions. The accelerating voltage is then maintained constant, while the magnetic field strength B and electric sector voltage E are scanned simultaneously. The scan is conducted in such a manner that the ratio B^2/E is kept constant. The resulting spectra (Fig. 31) exhibit high daughter ion resolution ($\Delta M/M = 500$); fairly broad peaks corresponding to the precursor ions are usually observed ($\Delta M/M = 100$) (75). This and other linked scan techniques discussed subsequently require more sophisticated instrumentation than single-parameter scans. However, the B^2/E scan has several compensating features. In contrast to the metastable refocusing technique, it can be used over a virtually unlimited mass range. Further, since the accelerating voltage is unchanged during the scan, the tune of the ion source remains constant.

The B^2/E linked scan determines all precursors of a specific daughter ion (i.e., all m_1^+'s generating a specific m_2^+). Often, it will be of interest to determine all daughter ions arising from a particular precursor ion (i.e., all m_2^+'s generated by a specific m_1^+, the mass spectrum of m_1^+). Two linked scans determine the ionic products formed by first field-free region fragmentation of a particular ion (188). In one technique, the $V^{1/2}/E$ scan, the magnetic field is held constant and the accelerating voltage V and electrostatic analyzer voltage E are scanned so that $V^{1/2}/E$ remains constant. Since V is changed during this scan, the tune of the ion source (and thus the number of m_1^+ ions emerging into the first field-free region) will vary. Further, the range of the technique is limited ($m_1/m_2 = 3$) (100, 187). An alternative linked scan holds V constant while scanning B/E so that it remains constant. This procedure does not have the disadvantages mentioned for the

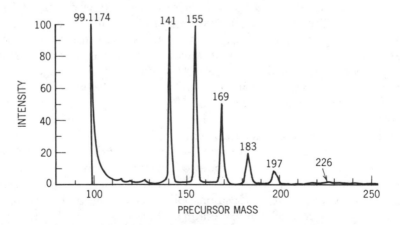

Fig. 31. The B^2/E linked scan of $C_7H_{15}^+$ ion (m/z 99) in n-hexadecane. Peaks correspond to precursor ions whose fragmentation generates $C_7H_{15}^+$ (75). Reprinted with permission from *Anal. Chem.*, **51**, 983 (1979). Copyright (1979) American Chemical Society.

$V^{1/2}/E$ scan. It exhibits low precursor ion resolution (~ 100) but high daughter ion resolution (~ 500) (75). As discussed later, techniques exist that exhibit complimentary resolution characteristics.

Finally, several linked scans permit determination of all ions undergoing a particular neutral loss (75a, 113a, 196b). Loss of a particular neutral fragment is often fairly diagnostic for the presence of a specific functional group. For example, the negative-ion isobutane CI mass spectra of carboxylic acids often exhibit abundant molecular ions with loss of carbon dioxide (44 amu). Thus, the parent ions corresponding to carboxylic acids in a complex mixture could be identified by performing a neutral loss scan for loss of 44 amu. The neutral loss scan can be performed at either constant B or constant V; in either case, the relationship between the linked variables is complex. The neutral loss scan as well as scans equivalent to the others described in this section are more conveniently and more commonly performed on quadrupole mass spectrometers as discussed in a later section.

2.   Collision-induced Dissociation

Consideration of the $k(E)$ curves that arise from analyses of the quasi-equilibrium theory equation [Eq. (26)] when plausible values of E_0, S, and γ are assumed indicates that only a narrow range of excitation energies will produce the rate constants of 10^5–10^6 s^{-1} most conducive to first and second field-free region fragmentation. Thus, metastable peaks arising from spontaneous uni-molecular fragmentation are, at best, weak. Further, as already discussed, certain sorts of fragmentations that generate abundant ions in the normal mass spectrum will fail to produce observable metastables. The technique of collision-induced dissociation (CID), also known as collisional activation (CA), can remedy both problems.

Mass spectrometers are normally designed to maintain very low pressures (e.g., 10^{-7} torr) in their flight tubes. Under these conditions, the probability of an ion in transit to the collector experiencing a collision with a neutral molecule is small. Thus, only those ions still possessing considerable excitation energy will be capable of fragmentation in the first or second field-free region. These ions will constitute only a small fraction of the ions transmitted by the mass spectrometer. However, if a low pressure of gas (e.g., 10^{-5} torr) is introduced into the second field-free region, the number of ions undergoing fragmentation will increase dramatically; this will be reflected by an increase in the number and intensity of metastable peaks in the spectrum. The effect known as CID arises from grazing collisions or near misses between the ion and a neutral molecule, which results in the conversion of 1–25 eV of ion kinetic energy into ion internal excitation (129, 157). (Ions that undergo head-on collisions will be so severely scattered that they will not be detected.)

Alternatively, similar events will ensue if the pressure in the first field-free region is increased; observation of these fragmentations will require use of a high-voltage or linked scan technique (134).

The nature of the collision gas apparently has little effect on the course of collision-induced fragmentation. The necessary pressure can be generated through use of a constant pressure leak or, more simply, by gently baking the electrostatic analyzer or by loosening an appropriate vacuum flange. There are some advantages to using a short-length differentially pumped collision chamber maintained at a fairly high pressure rather than maintaining the entire field-free region at a somewhat lower pressure (135).

Collision-induced metastable peaks are invariably more numerous and more intense than unimolecular metastable peaks. Consequently, the technique can facilitate the interpretation of mass spectra. The amount of energy transferred by the collision process is comparable to but probably slightly less than the amount transferred by 70-eV electron impact (89, 133). Since this will usually be considerably more energy than the average internal excitation of ions surviving to reach the field-free regions, the energy content of an ion after collision will be essentially independent of the mechanism of its formation (105, 131). Thus, this technique can be used to demonstrate that the structures of ions resulting from different fragmentations are the same or different. Finally, the technique finds considerable application in MS/MS studies as discussed subsequently.

III.   THEORY

A qualitative understanding of the theory and energetics involved in ion production and fragmentation is essential to understanding the influence of molecular structure, ionization technique, and ion lifetime on mass spectral behavior. There are several important differences between ion chemistry at the low pressures and short lifetimes typical of mass spectrometry and the solution and gas phase chemistries more commonly encountered. Solution chemists can use a single number (the temperature) to define both the average electronic, vibrational and rotational energies of an ensemble of molecules and also the probability of a particular energy level. The internal energies of molecules in solution follow the well-known Maxwell–Boltzmann distribution because frequent intermolecular collisions provide a mechanism for energy redistribution and reequipartition. No such mechanism exists for ions formed in an electron impact source. Ion–molecule and ion–ion collisions are extremely improbable during typical ion residence times (10^{-6} s) and at typical source pressures (10^{-5} torr). Thus, an individual ion will retain its internal energy content (unless fragmentation or photoemission occurs) from its genesis to its collection.

The internal energy content of a particular ion is the summation of the energy content of the neutral molecule before ionization and the energy imparted during the ionization process. The latter is difficult to estimate.

The probability of transferring a particular amount of energy during electron-impact-induced ionization depends on the kinetic energy of the impacting electron and the probability of the transition (the Franck–Condon factor) connecting the neutral's electronic, vibrational, and rotational states with the corresponding state of appropriate energy in the ion.

The energy distribution $[P(E)]$ resulting from electron-impact-induced ionization can be estimated by consideration of the photoelectron spectra obtained from photoionization experiments. This technique involves expulsion of the outer electrons from sample molecules as a result of ultraviolet irradiation. A monoenergetic beam of photons (obtained by filtering the light emitted when a dc or microwave discharge occurs through an inert gas) is allowed to impinge on the sample molecules, and the kinetic energies of the expelled electrons are measured. The difference between the kinetic energies of the expelled electrons and the energy of the incident photon is the energy absorbed by the molecule during the ionization process. For a polyatomic molecule, a number of broad peaks will be observed (Fig. 32) (117). Each peak corresponds to formation of a different electronic state (expulsion of an electron from a different orbital). The peaks are broad because of incompletely resolved vibrational and rotational fine structure. The intensity at a particular energy is proportional to the number of ions formed with the deposition of that amount of energy. Thus, Fig. 32 accurately depicts the energy transferred to benzene when photoionization with 20.2-eV photons occurs. After convolution with the thermal energy present before ionization (136), the actual energy distribution of the ethylene molecular ion at the instant of ionization would be obtained. It is notable that a slightly different distribution would result from photoionization by photons of different energy (10); greater differences would be observed if 70-eV electrons were the ionizing agent (137, 138).

As molecules become more complex, more bands appear in their photoelectron spectra and the bands become broader (Fig. 32b). For qualitative purposes, an energy distribution function such as that depicted in Fig. 31c is often assumed for intermediate or large organic molecules. The notable features of Fig. 31c are an absence of ions below the ionization potential, and the prevalence of ions only a few electron volts above the ionization potential (41). Only a small fraction of ions are produced with energies more than 10 eV above the ionization potential.

At the instant of their formation, molecular ions exist in a variety of electronic, vibrational, and rotational states; it is a postulate of theoretical mass spectrometry that a "quasi-equilibrium" among the various energy states is established before fragmentation occurs. Thus, the behavior of an ion will depend only on its structure and its total energy content. It will be independent of the mode of the ion's formation or the details of the ionization process involved in its genesis.

If the internal energy distribution of a population of ions is known and the quasi-equilibrium postulate is valid, knowledge of the energy-dependent rate constants $k_i(E)$ for a particular set of reactions will permit prediction of mass spectral behavior. A number of fairly sophisticated approaches have been made to this goal (156). However, a qualitative appreciation of the factors influencing these rate constants and their effect on the appearance of a mass spectrum can be gained from the original, simplified equation (158)

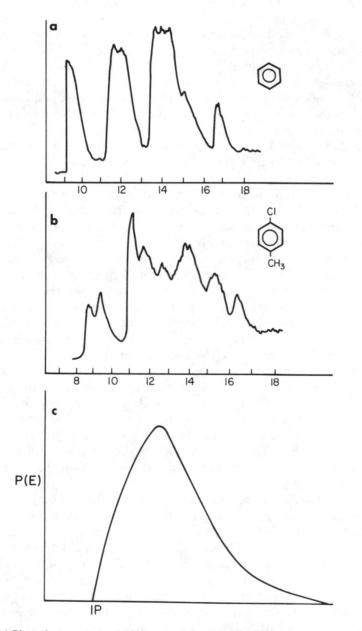

Fig. 32.   (*a*) Photoelectron spectrum of benzene obtained by irradiation with 21.2-eV photons (11).
(*b*) Photoelectron spectrum of *p*-chlorotoluene obtained by irradiation with 21.2-eV photons (11).
(*c*) Typical energy distribution function [$P(E)$] useful for qualitative arguments.

$$k = v \left(\frac{E - E_0}{E} \right)^{s-1} \tag{26}$$

The rate constant k for a fragmentation of an ion of internal energy E depends on the "effective" number of oscillators in the ion(s) and the energy of activation E_0 and frequency factor v of the fragmentation.

Qualitatively, it can be seen that Eq. (26) can lead to some useful predictions and is intuitively plausible. When $E \gg E_0$, the term in parentheses tends toward unity. Since its value is insensitive to variations in E_0, the relative rates of competing reactions from a high-energy ion will depend primarily on their frequency factors. On the other hand, when $E \approx E_0$, the term in parentheses will be dominant since it is small and is raised to a high power. The relative rates of competing reactions from a low-energy ion will usually depend most strongly on their energies of activation E_0. The physical significance of the $s-1$ term is that the probability of localizing a large amount of energy in a single vibration (a necessary prerequisite for bond cleavage) will depend on the number of oscillatory modes competing for that energy. That probability will be less for a large molecule with many vibrational degrees of freedom (large s). Theoretically, a molecule with n atoms has $3n-6$ modes of oscillation. In practice, much smaller values of s produce better agreement between calculated and experimental ion behavior. The frequency factor v depends on the "tightness" or "looseness" of the reaction's transition state. Reactions with rigid geometrical requirements (e.g., rearrangements, reactions requiring transfer of an atom) will have small frequency factors, while those with more flexibility (e.g., simple bond cleavage) will often approach the stretching frequency of the bond being cleaved.

If values of v, E_0, and s are assumed for a particular fragmentation of an ion, plots of k versus E can be produced. Figure 33 represents such a plot for the hypothetical situation in which the molecular ion $[A-B-C-D]^{+\cdot}$ can fragment only to generate the products $[A-B]^{+\cdot}$ and CD. The parameters assumed are $v = 10^{13} \text{ s}^{-1}$, $E_0 = 2$ eV, and $s = 10$. These values are typical ones for a reaction involving simple cleavage of a single bond in a 30-atom molecule.

To understand the significance of Fig. 33, it is necessary to discuss the time scale of events occurring in a typical magnetic sector mass spectrometer. A typical ion may spend 1×10^{-6}–5×10^{-6} s in the mass spectrometer's source and may require 10×10^{-6}–30×10^{-6} s to travel to the detector (25, 88). The "ordinary" ions observed in a mass spectrum have all been generated within the source and have not fragmented further during mass analyses. Thus, if a fragmentation is to produce an intense product ion, its rate constant must be sufficiently large to permit extensive reaction within the source. To simplify the following discussion, it will be assumed that for ion $ABCD^{+\cdot}$ the source residence time is 1×10^{-6} s and that 10×10^{-6} s are required for $ABCD^{+\cdot}$ to reach the detector. Therefore, unless the rate constant for formation of $A-B^{+\cdot}$ equals or exceeds 10^6 s^{-1} the intensity of the $A-B^{+\cdot}$ peak will be low.

Inspection of Fig. 33 indicates that at least 2.4 eV excitation energy is required to produce a rate constant for formation of AB^+ of 10^6 s^{-1} or greater. Ions with

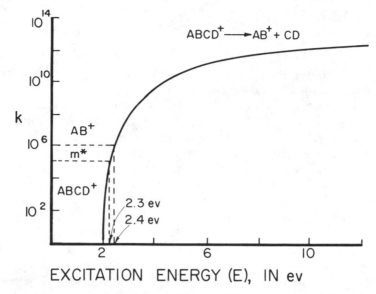

Fig. 33.   Plot of rate constant k vs. excitation energy E for reaction $A-B-C-D^+ \rightarrow A-B^+ + CD$, where $V = 10^{13}\ s^{-1}$, $E_0 = 2\ eV$, and $s = 10$.

less excitation energy are unlikely to fragment in the source and will not therefore generate intense $A-B^{+\cdot}$ ions in the mass spectrum. On the other hand, detection of the $[ABCD]^{+\cdot}$ ion requires that the ion remains intact until its mass analysis is complete. If it is assumed that 10×10^{-6} s are required to complete this process, ions with rates of fragmentation of $10^{+5}\ s^{-1}$ or less are likely to be observed as $A-B-C-D^{+\cdot}$ ions. Figure 33 indicates that this corresponds to ions with an excitation energy of 2.3 eV. Ions with rates of fragmentation between 10^5 and $10^6\ s^{-1}$ are likely to fragment in transit between the source and the detector. These "metastable" ions are of considerable interest and are discussed elsewhere in this chapter.

   Figure 34 represents another hypothetical situation. Now it is assumed that ion $[A-B-C-D]^{+\cdot}$ can fragment only to generate the products $[A-D]^{+\cdot}$ and $B=C$, with $v = 10^9\ s^{-1}$, $E_0 = 1\ eV$, and as before, $s = 10$. The lower frequency factor and activation energy values assumed here are characteristic of reactions proceeding through "tight" or geometrically constrained transition states. Note that the indicated reaction can only occur in geometries in which groups A and D approach one another within bonding distances. If the same source residence and transit times are assumed, Fig. 34 indicates that an excitation energy of at least 1.9 eV is required before formation of AD^+ in the source becomes probable and that ions with excitation energies of 0–1.6 eV are likely to survive intact until their detection. Ions with excitation energies between 1.6 and 1.9 eV are likely to fragment after leaving the source but before detection; they are the source of metastable peaks.

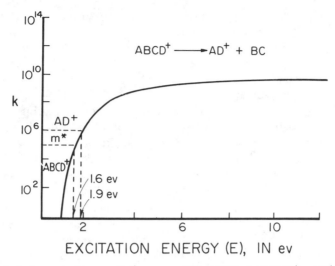

Fig. 34.   Plot of rate constant k vs. excitation energy E for reaction $A–B–C–D^+ \rightarrow AD^+ + BC$, where $V = 10^9$ s^{-1}, $E_0 = 1$ eV, and as before, $s = 10$.

Comparison of these two reactions can illustrate some important generalizations. The "energy window" for metastable ions is invariably small. The fact that such metastable ions are rather homogeneous in energy and lower in average energy than the general population of ions fragmenting in the source facilitates structural and mechanistic studies. Their reactions will often be qualitatively and quantitatively different from those of source fragmenting ions. Specifically, the energy window for high-frequency-factor, high-activation-energy processes (typical of simple cleavages) is smaller than that for low-frequency-factor, low-activation-energy processes (typical of rearrangement reactions). Thus, the latter processes are more prone to the generation of abundant metastable ions.

The preceding discussion demonstrates that knowledge of E_0, the excitation energy of a reaction, has practical utility. The appearance potential (AP) of a fragment ion is defined as the minimum energy required to produce it in detectable amounts in the mass spectrometer's source. To a crude approximation, $AP - IP = E_0$, where IP is ionization potential. In fact, $AP - IP$ will always exceed E_0 due to a phenomenon known as "kinetic shift." If an ion is to form in detectable numbers within the source, its precursor must possess sufficient internal energy to permit its formation with a rate constant of 10^5–10^6 s^{-1}. In the two reactions just discussed, the kinetic shifts are ca. 0.3–0.4 and 0.6–0.9 eV, respectively. In general, larger kinetic shifts are observed in fragmentations whose k-versus-E curves are rising slowly near E_0. Inspection of Equation 26 indicates that this corresponds to larger E_0's, smaller v's, and larger values of s.

Next, consider a slightly more complex kinetic situation in which the two reactions $A–B–C–D^{+\cdot} \rightarrow A–B^{+\cdot} + C–D$ and $A–B–C–D^{+\cdot} \rightarrow A–D^{+\cdot} + B=C)$ are occurring in competition with one another (Fig. 34). At low excitation energies

(1–2.3 eV), reaction 2 will proceed more rapidly than reaction 1; under these conditions, the $[(E - E_0)/E]^{s-1}$ term controls Eq. (26) and low-activation-energy fragmentations (e.g., rearrangements) predominate. At high excitation energies, reaction 1 is faster; at large values of E, $[(E - E_0)/E]$ approaches unity whatever the value of E_0, and high-frequency-factor fragmentations (e.g., simple cleavages) become dominant.

Inspection of Fig. 35 indicates an additional problem in equating $AP - IP$ with E_0, the energy of activation of a reaction. Reaction 1 has a rate constant of 10^5 at an ion excitation energy of 2–3 eV. If no other fragmentations were possible, A–B$^+$ might be detectable in the source if a reasonable number of A–B–C–D$^{+\cdot}$ ions had such excitation. However, among such ions reaction 2 will proceed ca. 60 times more rapidly than reaction 1; nearly all the A–B–C–D$^{+\cdot}$ ions with excitation energies near 2.3 eV will instead undergo reaction 2. Only at significantly higher values of excitation energy will reaction 1 become sufficiently competitive with reaction 2 to produce detectable product ions. This phenomenon, the "competitive shift" will raise the observed appearance potential whenever the reaction under study is not the one of lowest E_0. If the energy distribution $[P(E)]$ after ionization and the rates $[k(E)]$ of all fragmentations were accurately known, it would in principle be possible to quantitatively predict the ensuing mass spectral behavior. This discussion will assume several hypothetical $P(E)$ functions and consider only the two hypothetical reactions already discussed in an attempt to qualitatively explain the effect of ionizing conditions on mass spectral behavior.

Figure 36 depicts a hypothetical energy distribution resulting from ionization of A–B–C–D with 70-eV electrons. Consideration of Fig. 35 indicates that ions with excitation energies of 0–1.6 eV will largely survive intact until detection; these ions will be detected as molecular ions, A–B–C–D$^{+\cdot}$. Ions with excitation

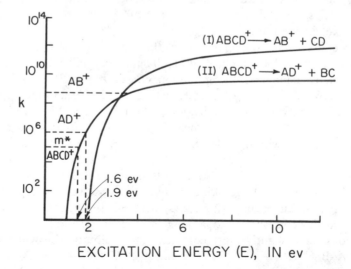

Fig. 35. Plot of ratio constant k_i and k_{ii} for two fragmentations proceeding in competition: $ABCD^+ \rightarrow AD^+ + BC$.

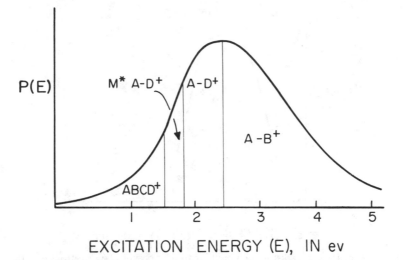

Fig. 36.   Hypothetical energy distribution resulting from impact of 70-eV electrons on ABCD. It is assumed that molecular ion can only undergo two fragmentations depicted in Fig. 34.

energies between 1.6 and 1.9 eV are likely to form A–D$^{+\cdot}$ in transit to the detector. These fragmentations will generate metastable peaks. Ions with excitation energies between 1.9 and 2.3 eV are likely to fragment in the source to generate normal A–D$^{+\cdot}$ ions. Finally, A–B^{+} ions will be formed in the source from A–B–C–D$^{+\cdot}$ ions with excitation energies greater than 2.3 eV.

If a different $P(E)$ function is assumed, the relative proportions of A–B–C–D$^{+\cdot}$, A–B^{+}, and A–D$^{+\cdot}$ ions in the mass spectrum will change. For example, if ionization is accomplished with an electron beam of lower kinetic energy, the average excitation energy will decrease (Fig. 37). The relative amount of A–B–C–D$^{+\cdot}$ detected will increase; low-activation-energy processes (e.g., rearrangement reactions such as formation of A–D$^{+\cdot}$) will become predominant over high-activation-energy processes (e.g., sample cleavages such as formation of A–B^{+}). Such behavior is observed experimentally (192). Ionization techniques such as CI and FI transfer so little excitation energy that any fragmentation of the parent ion is energetically unfavorable.

IV.   APPLICATIONS

A.   INTERPRETATION OF MASS SPECTRA

1.   Manual Interpretation

The theoretical number of possible organic compounds is unlimited; millions have already been described in *Chemical Abstracts*. Usually, then, attempts to deduce an unknown's structure based only on its low-resolution mass spectrum will fail. If information about the sample's origin and probable composition is

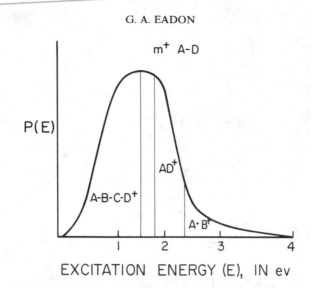

Fig. 37.   Hypothetical energy distribution resulting from impact of 12-eV electrons on ABCD. It is assumed that molecular ion can only undergo two fragmentations depicted in Fig. 34.

available, if other physical and spectroscopic techniques can be brought to bear, and if computer assistance is available, the likelihood of a successful identification will be markedly increased. Probably more directly relevant to the subject of this chapter, certain mass spectroscopic techniques can vastly facilitate interpretation of a mass spectrum.

A number of excellent books have been written that discuss the characteristic fragmentations of various classes of organic compounds and the art of interpreting mass spectra (27a, 37, McLafferty, General Bibliography). McLafferty's treatment (General Bibliography) is especially recommended as a suitable self-study text for persons interested in manual interpretation of low-resolution mass spectra. The treatment of these topics in this chapter will obviously be more superficial.

The identification of a compound based on its mass spectrum can be described as based on five not necessarily sequential steps (Table 3). Determination of the unknown's molecular weight and then its elemental composition will vastly limit the number of possible compounds. Perhaps the key step in identification is deciding the functional group(s) present in the unknown. Then if the characteristic fragmentations typical of that functional group can be ascertained, it may be possible to postulate plausible structure(s) for the unknown. Finally, the spectra of the postulated compounds should be compared to that of the unknown to confirm structural assignments. Each of these steps is discussed in more detail below.

a.   MOLECULAR WEIGHT

Most chemists would probably agree that the single most useful datum to be obtained from a mass spectrum is a compound's elemental composition. The first

TABLE 3
Manual Interpretation of a Mass Spectrum

Goal	Procedure
1. Deduce unknown's molecular weight	Tentatively locate molecular or quasi-molecular ion. Confirm assignment based on plausibility of forming daughter ions. When necessary, instrumental techniques can simplify this assignment.
2. Deduce unknown's elemental composition	Straightforward when high-resolution data are available. Can often be accomplished with low-resolution spectrum by consideration of isotopic abundance. "Nitrogen rule" is sometimes useful, as is knowledge of compound's origin, its physical and chemical properties, and its behavior when examined by other spectroscopic techniques.
3. Deduce unknown's functional groups	Clues include general appearance of spectrum, peaks corresponding to loss of certain stable neutral fragments or formation of certain characteristic ions; presence of ion series characteristic of a particular structure. Use of "rings-plus-double-bond rule" may exclude certain possibilities.
4. Tentatively deduce unknown's structure	Can be accomplished based on generalized knowledge of principles governing mass spectral fragmentation; may require reference to specific discussions of mass spectral behavior of individual functional groups.
5. Confirm conclusions	Where possible, unknown's spectrum should be compared to that of authentic compound. Reference spectra are most unambiguously generated under same instrumental conditions as unknown spectra. Often, however, adequate spectra can be located in libraries by manual or computer-aided searching.

step in deducing this information is to identify the molecular ion peak in the mass spectrum since that establishes the compound's molecular weight.

A molecule that survives intact from ionization to collection will generate a molecular ion peak (M^+). Spectra that exhibit molecular ions usually contain multiple peaks that fulfill this definition. The effect arises because many common elements exist as more than one isotope. For example, the mass spectrum of bromoethane (Fig. 38) exhibits significant peaks at m/z 108–111. The peak at m/z 108 corresponds to $^{12}CH_3\ ^{12}CH_2-^{79}Br$. The m/z 109 peak corresponds to both $^{13}CH_3\ ^{12}CH_2-^{79}Br$ and $^{12}CH_3\ ^{13}CH_2-^{79}Br$. Since about 1.1% of all carbon atoms are ^{13}C (Table 4), the m/z 109 peak is about 2.2% as intense as the m/z 108 peak. [The natural abundance of ^2H is sufficiently low (0.015%) that its effects can be ignored in these calculations.] The m/z 110 peak corresponds principally to $^{12}CH_3\ ^{12}CH_2-^{81}Br$; since 49.5% of bromine atoms are ^{81}Br, the m/z 110 peak is nearly as intense as m/z 108 (98%). Finally, the m/z 111 peak is generated from bromoethane molecules containing one ^{13}C and ^{81}Br. All of these peaks correspond to unfragmented ions. By convention, however, mass spectroscopists refer to the peak that contains only the most abundant isotope of each element as

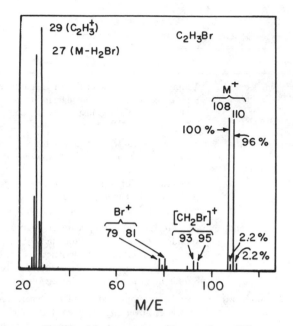

Fig. 38.   Mass spectrum of bromoethane (37).

the molecular ion peak (e.g., m/z 108 in this case). It may be useful to remember that for all common elements the most abundant isotope is the lightest one.

Identification of the molecular ion peak is straightforward if the peak is reasonably intense and the spectrum is that of a pure compound; after allowance for isotopic impurities, the molecular ion is the peak of highest m/z in the spectrum. Misidentification can arise from at least two sources. Many compounds do not generate usefully intense molecular ion peaks after EI-induced ionization. The actual molecular ion peak may then be "lost" in the noise or background and a daughter ion might appear to be the highest m/z peak. Alternatively, if the unknown is impure or if residual compounds are present in the spectrometer when the unknown is introduced, a peak generated by the impurity could be identified as the molecular ion peak.

The most straightforward solutions to these problems are instrumental. For example, CI or FI produces markedly more abundant molecular ions or quasi-molecular ions than are observed after EI (cf. Sections II.B.2 and II.B.4). In modern instruments equipped with dual or combined sources, changeover from EI to CI or FI can be rapid, even automatic. An alternative, though often less satisfactory, procedure for enhancing the abundance of molecular ion peaks relative to fragment peaks is simply to decrease the ionizing voltage from the conventional 70 eV. As the ionizing voltage is decreased to a value near the ionization potential, the average excitation imparted will decrease; a higher percentage of the ions produced will survive intact to be detected as molecular ions. A difficulty with this technique is that the probability of ionization decreases sharply as the ionizing voltage approaches the threshold. Consequently, the intensities of all peaks will be markedly decreased (although most dramatically so for fragment ions). Thus, the technique may be inappropriate in sample limited situations. Further, even at low ionizing voltages, some compounds do not generate significant molecular ion peaks.

In situations in which these instrumental techniques are not available, certain criteria can be used to disqualify a particular ion from consideration as the molecular ion. The molecular ion must contain all elements present in any fragment peak. For example, the number of chlorine or bromine atoms contained by a molecular ion or fragment can usually be deduced by consideration of peak patterns. The isotopic abundances listed for these elements in Table 4 are such that they are readily detected. If one fragment ion contains a single bromine atom and another a single chlorine atom, the molecular ion must obviously contain at least one chlorine and one bromine atom.

The molecular ion must also be able to generate all daughter ions by the loss of logical neutral fragments. Certain parent–daughter ion relationships are arithmetically implausible. For example, loss of 4–14 or 21–25 amu from a molecular ion to generate a daughter peak is quite unlikely. More plausibly, the two peaks are formed from two different compounds or from a common molecular ion peak that has not been observed. On the other hand, intense daughter peaks corresponding to loss of stable neutrals (e.g., H_2O, HX, CO, CH_3) from the candidate molecular ion lend support to this assignment.

Virtually all stable molecules are even-electron species. Molecular ions, formed by loss of a single electron, are therefore odd-electron species. The "nitrogen rule" postulates that all odd-electron species that contain no or an even number of nitrogen atoms will appear at an even m/z value. Conversely, odd-electron species with an odd number of nitrogen atoms must appear at an

TABLE 4
Natural Abundances of Isotopes of Common Elements

Element	Mass	Relative Abundance (%)	Mass	Relative Abundance (%)	Mass	Relative Abundance (%)
Hydrogen	1	100	2	0.015	—	[a]
Fluorine	19	100	—	[a]	—	[a]
Phosphorus	31	100	—	[a]	—	[a]
Iodine	127	100	—	[a]	—	[a]
Carbon	12	100	13	1.1	—	[a]
Nitrogen	14	100	15	0.37	—	[a]
Oxygen	16	100	17	0.04	18	0.20
Silicon	28	100	29	5.10	30	3.4
Sulfur	32	100	33	0.8	34	4.4
Chlorine	35	100	—	[a]	37	32.5
Bromine	79	100	—	[a]	81	98

[a] Natural abundances less than 0.2%

odd m/z value. These axioms can exclude certain ions from consideration as the molecular ion provided that one knows the number of nitrogen atoms to be expected in the molecular ion. Occasionally, the sample's origin, chemical, or spectroscopic properties will permit the appropriate deduction. Alternatively, high-resolution mass spectrometry can be used to demonstrate the number of nitrogen atoms in a candidate peak. As already discussed, that number must be appropriate for an odd-electron species.

b.   ELEMENTAL COMPOSITION

Identification of the molecular ion immediately establishes the molecular weight of the unknown. Almost invariably, numerous elemental compositions will correspond to a particular molecular weight. The most satisfactory procedure for determining a peak's elemental composition is the use of high-resolution mass spectrometry. If that approach is precluded, the relative intensities of isotope peaks can assist deducing elemental composition.

Fluorine, phosphorus, and iodine are the only monoisotopic elements commonly found in organic molecules. The other common elements exist in at least two isotopic forms. Isotope peaks can unambiguously demonstrate the presence or absence of the elements chlorine or bromine. These elements exist in two abundant isotopic forms separated by 2 amu. The patterns resulting from single or multiple chlorine or bromine atoms are sufficiently striking that the number and identity of these atoms are usually readily apparent. Sulfur and silicon can often be identified by reference to the relative abundances of isotope peaks. These elements also possess significant isotopes 2 amu heavier than the most abundant one (Table 4). In these cases, however, the heavy isotopes are markedly less abundant than those of chlorine or bromine, complicating their identification.

Isotope peaks are often insufficient to completely define the elemental composition of an unknown. However, in most cases, their intensities will drastically limit the number of elemental compositions possible. For example, consider the spectrum depicted in Fig. 39 (actually that of the steroid 5α-pregnane) to be of an unknown. The molecular ion appears at m/z 288. The peaks observed at m/z 289 (23% as intense as the m/z 288 peak), 290 (2.5%), and 291 (0.2%) must be attributed to the presence of various heavy isotopes in the sample. The presence of certain elements can be readily excluded by reference to the intensity of the M + 2 peak (m/z 290). Reference to Table 4 confirms that even a single atom of Si, S, Cl, or Br per molecule would produce an m/z 290 peak significantly larger than actually observed. The only common elements that exist in appropriate isotopic forms to contribute significantly to the M + 1 peak (m/z 289) are carbon and nitrogen. If a reasonable experimental error is assumed for the measurement of the intensity of m/z 289 (e.g., 23 ± 3%), the total of carbon and nitrogen atoms must equal at least 18. The presence of iodine (monoisotopic, mass 127) can therefore also be excluded since even 18 nitrogen atoms combined with an iodine atom weigh more than 288 amu. A further refinement of the possible elemental composition of m/z 288 can be made through use of the

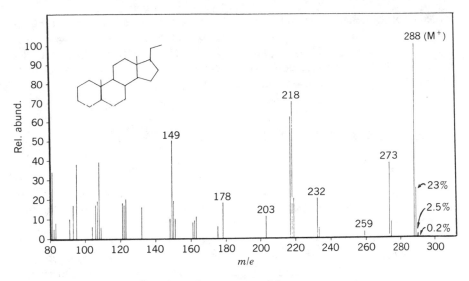

Fig. 39. Mass spectrum of 5α-pregnane.

nitrogen rule. Since the molecular ion appears at an even m/z value, it must contain zero or an even number of nitrogen atoms. However, even within these constraints a variety of formulas containing C, H, N, O, and F can be concocted consistent with the observed isotope peaks, the nitrogen rule, the valencies of the elements and the requirement that the sum of masses total 288 amu. Of course, in most real analytical situations, additional information could be brought to bear to exclude many of these possibilities.

Several caveats should be borne in mind when using isotope peaks to deduce elemental compositions. When mass spectra are determined under conditions conducive to ion–molecule reactions (e.g., elevated source pressures or lengthened ion residence times in the source), artifact peaks may be observed in the molecular ion region. Many ion–molecule reactions generate M + 1 ions. Thus, the intensity of the M + 1 ion is probably best thought of as providing an upper limit on the number of C, N, and Si atoms present. Similar caution should be exercised in interpreting weak M + 2 peaks since even a minor impurity could significantly enhance the intensity of such a peak. Finally, the relative intensities of weak peaks can be distorted by ion statistical effects, especially when spectra are rapidly scanned or sample size is very limited.

c. IDENTIFICATION OF FUNCTIONAL GROUPS AND INTERPRETATION
    OF FRAGMENTATIONS

Determination of the functional group(s) of an unknown molecular will simplify interpretation of its mass spectral fragmentations. A compound's elemental composition usually limits the possible functional groups in the unknown. The rings-plus-double-bond rule is helpful in this regard. For a molecule of the general formula $C_xH_yN_zO_n$, the total of rings plus π bonds equals $x-\frac{1}{2}y+\frac{1}{2}z+1$. Thus, for example, the number of rings plus double bonds in benzamide (III) ($C_7H_7NO_2$) equals $7-\frac{1}{2}(7)+\frac{1}{2}(1)+1$, or 5. The total of 5 represents the three double bonds of the benzene ring, the double bond of the carbonyl group, and the six-membered ring:

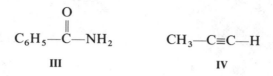

III   IV

Triple bonds contain two π bonds; application of the rings-plus-double-bond formula to 1-propyne (IV) produces the result that two double bonds plus rings are present.

The rings-plus-double-bond formula is based on the valencies of the elements. Thus, if elements other than C, H, N, or O are present, they can be counted as the element of the appropriate valence. For example, Si counts as C, halogens as H, P as N, and S as O. Certain elements (e.g., P and S) can assume other valencies. The rings-plus-double-bond rule will not reveal the presence of multiple bonds to elements in their higher valence states. For example, structure V has the empirical formula $C_{11}ClH_{24}O_4PSi$:

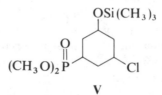

V

If P is treated as N, Cl as H, and Si as C, the rings-plus-double-bond formula gives $12-\frac{1}{2}(25)+\frac{1}{2}(1)+1$, or 1 (corresponding to the six-membered ring). The P–O double bond was not detected because its formation required pentavalent phosphorus. It should also be noted that the rings-plus-double-bond rule can be applied to fragment ions. However, in certain cases (i.e., when the ion is an even-electron species), the calculated value will be incremented by 0.5.

In many cases, however, a molecule's elemental composition will still permit multiple possibilities for its functionality. Sometimes, a closer inspection of the mass spectrum will resolve these ambiguities. For example, the general appearance of a mass spectrum can provide some useful clues about molecular structure and functionality. Polynuclear aromatic compounds have relatively low ionization potentials (resulting from their extensively delocalized π electron system) and highly fused structures. Other things being equal, molecules with low ionization potentials tend to exhibit more abundant molecular ions. Further, molecules whose fragmentation requires cleavage of multiple bonds tend to exhibit relatively abundant molecular ions. For both reasons, polynuclear aromatics usually exhibit intense molecular ion peaks and weak fragmentation

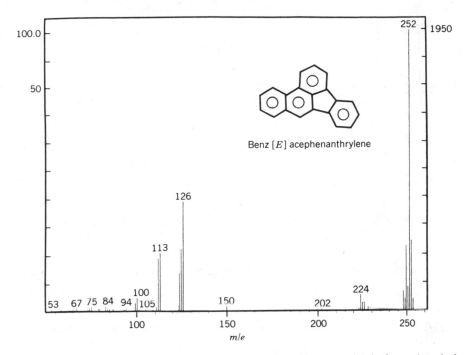

Benz [E] acephenanthrylene

Fig. 40.   Mass spectrum of bene(e)acephenanthrylene exhibiting intense molecular ion peak typical of polynuclear aromatics. Peaks near m/z 126 correspond to doubly charged ions.

peaks (Fig. 40). Conversely, a linear alkane of similar molecular weight will have a higher ionization potential and will contain numerous carbon–carbon bonds whose cleavage will generate fragment ions. Since these bonds are all readily cleaved, the resulting mass spectrum exhibits regularly spaced (14-amu or 1-CH_2 unit) clusters of peaks of gradually increasing intensity as the fragment's molecular weight decreases (Fig. 41). Between these extremes, the general appearance of a spectrum can provide useful information about an unknown's structure. For example, molecules that exhibit only one or a few intense fragments usually contain structural features that labilize particular bonds toward cleavage. Similarly, the intensity of the molecular ion peak can provide useful clues about an unknown's structure and functionality. McLafferty has produced a useful tabulation of the molecular ion abundances characteristic of particular structural types and functionalities (General Bibliography). Of course, even two molecules with identical functional groups can produce molecular ions of widely differing abundances if only one structure has features that introduce a low activation energy–high frequency factor decomposition.

Certain high mass peaks in a mass spectrum can also assist in determining an unknown's functional groups. For example, alcohols usually exhibit $M^{+\cdot}$ — — 18 peaks (H_2O), acetates show $M^{+\cdot} - 60$ peaks (CH_3CO_2H), alkyl and aryl halides show $M^{+\cdot} - X$ and/or $M^{+\cdot} - HX$ peaks, and trimethysilyl ethers exhibit intense $M^{+\cdot} - 15$ (CH_3) and $M^{+\cdot} - 90$ peaks [$(CH_3)_3SiOH$].

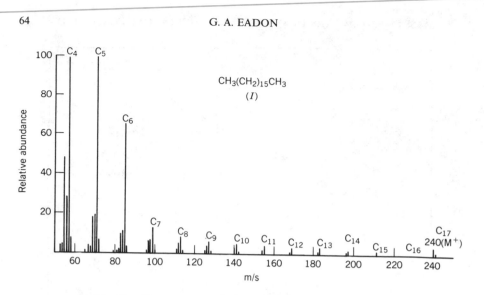

Fig. 41.   Mass spectrum of *n*-heptadecane, normal alkane (37).

Usually, the number of possible precursor peaks for a low mass peak is much larger than for a high mass peak. Further, low-mass ions are often formed by complex rearrangement reactions. Thus, when interpreting mass spectra, greater attention is usually given to high-mass peaks than to low-mass peaks. However, certain low-mass peaks are fairly diagnostic of particular structural subunits or functional groups. The m/z 77 ion often indicates the presence of a C_6H_5 group in the parent molecule. An intense m/z 91 ($C_7H_7^+$) usually signals the presence of a benzyl group. The benzoyl group

can be relied upon to generate an intense m/z 105 peak. Of course, other origins are possible for these low-mass ions, and these interpretations must be considered preliminary.

It is often possible to deduce the functional group(s) present in an unknown after marshaling all available physical, chemical, and spectroscopic evidence. If so, the next step is to attempt detailed interpretation of the mass spectrum to further define the unknown's structure. On the other hand, if several possibilities remain for the unknown's functionality, careful consideration of the observed fragmentations may permit a unique choice. In either case, a working knowledge of the "ground rules" governing mass spectral fragmentation will be necessary. The following discussion is intended to illustrate the principles that permit rationalizing and predicting mass spectral behavior. The references cited earlier contain much more exhaustive listings of the fragmentations characteristic of particular functionalities and might profitably be consulted at this stage of the interpretation.

Two fundamental principles suffice to rationalize most EI-induced fragmentations. The first postulates that the most abundant ions will be formed in fragmentations proceeding from the most stable electronic state. The most stable electronic state will correspond to that in which the "missing" electron has been lost from the highest occupied molecular orbital. For example, the most stable state of tertiary amine **VI** corresponds to structure **a** in which the lost electron is removed from a nonbonding orbital on nitrogen rather than a structure such as **b** wherein a strongly bound σ-electron has been lost:

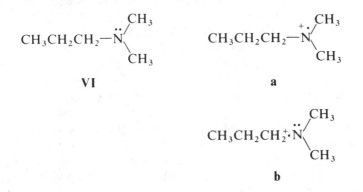

Generally, σ-electrons will be most strongly held and structures analogous to **b** will be important only in ions that contain no multiple bonds or heteroatoms with nonbonding electrons. Nonbonding electrons are usually most easily removed. Thus, structures analogous to **a** are important in the fragmentation of most heteroatom-containing species. As a general rule, the ionization potentials of nonbonding electrons on heteroatoms decrease as one proceeds from right to left and from top to bottom in the periodic table. Thus, a molecule containing a nitrogen and an oxygen atom in similar environments should exhibit fragmentations derived largely from ions in which the charge is localized in the nitrogen; similarly, if other things are equal, a sulfur atom will more effectively localize charge than an oxygen. Electrons in π orbitals typically exhibit ionization potentials intermediate between those of σ and nonbonding electrons. Ionization potentials for electrons in aromatic or highly conjugated systems may, however, sometimes be lower than those of nonbonding electrons. Ionization of an alkene will generate ions such as **c**:

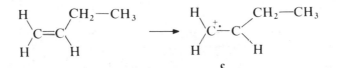

It is unarguably true that this postulate permits rationalization of large amounts of empirical data. It remains controversial, however, whether the concept of charge localization is an accurate description of physical reality (21).

A reasonable description of the ionization–fragmentation processes might begin with more or less indiscriminate removal of various bonding and nonbonding electrons by 70-eV electrons. Numerous electronic states will then be rapidly interconverting with one another, especially if the expelled electron was strongly held. According to Quasiequilibrium Theory (QET), fragmentation requires localizing a number of vibrational quanta of energy into a particular bond. This is unlikely to occur from a high-energy electronic state (e.g., E_2 in Fig. 42) since it will possess less vibrational excitation energy than the most stable electronic state (e.g., E_1). Very few fragmentations have been shown to occur from excited electronic states.

The second important postulate is that mass spectral fragmentations take place by processes that are chemically plausible. Students of mass spectrometric behavior have long noted many similarities between EI-induced fragmentation and conventional pyrolytic, photolytic, and radiolytic reactions. The fundamental origin of this similarity is that the driving forces of these apparently disparate reactions are similar.

Molecular ions are charged odd-electron species; they combine the reactivities of a cation and a radical. The individual steps of fragmentation reactions therefore usually involve "pairing" of odd electrons or shifting of the positive charge to a more stable location. The overall result is usually to form relatively stable, neutral and ionic products. Since these fragmentations are "energetically uphill," the transition state for fragmentation will closely resemble the products [Hammond's postulate (76)], and therefore the energy of activation for a fragmentation will be strongly dependent on the stabilities of its products. The

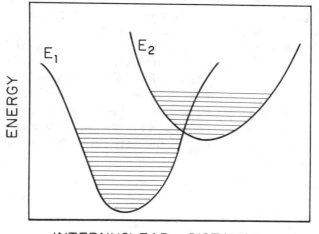

Fig. 42. Interconversion between E_1 (ground electronic state of M^+) and E_2 (excited electronic state) will occur rapidly if sufficient internal excitation is present. However, when ion is in state E_1, it will possess much greater vibrational excitation energy than when in state E_2. Therefore, fragmentation from E_1 state is most probable.

following discussion illustrates these principles and presents a few of the most important and diagnostically useful mass spectral fragmentations.

(1) α-Cleavage

As already discussed, the most favorable site for charge localization in ions containing a heteroatom is often in a nonbonding orbital of the heteroatom. Cleavage of a bond α to the heteroatom is then a favorable process [Eq. (27)]:

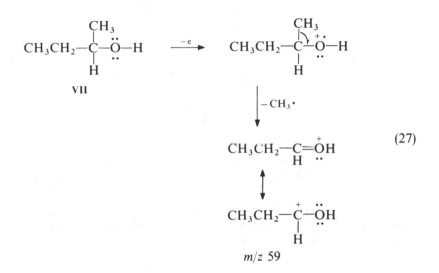

$$m/z \ 59 \tag{27}$$

An alkyl radical is expelled as the neutral particle, and the positive charge of the product ion is delocalized through resonance. The energy consumed in cleaving the carbon–carbon bond is partially compensated for by formation of a carbon–oxygen π bond. The cleavage of Eq. (27) has been depicted as a homolysis of the α C–CH$_3$ bond by use of the "fish hook" convention. A single barbed arrow (fish hook) indicates migration of a single electron; conventional double-barbed arrows are reserved for situations in which electron pair migration is to be indicated.

Note that 2-butanol (VII) can also undergo loss of H and CH$_3$CH$_2$ by strictly analogous fragmentations to generate m/z 73 and 45. The α-cleavage reaction indicated is closely analogous to α-cleavage reactions observed in the spectra of other heteroatoms containing species (e.g., N, S, hal).

A related α-cleavage reaction is observed when the charge is localized on a heteroatom that is part of a multiple bond. For example, the most stable electronic state of 2-hexanone (VIII) corresponds to charge localization in a nonbonding orbital of the oxygen atom:

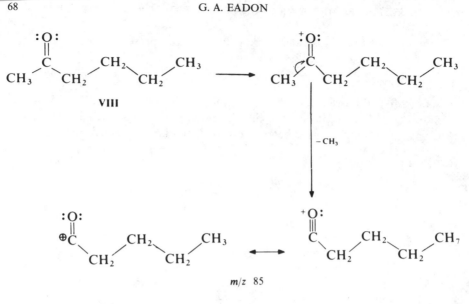

VIII

$$\tag{28}$$

Its mass spectrum exhibits major peaks corresponding to cleavage of either α carbon–carbon bond. The driving force for this reaction is again formation of an ionic species whose positive charge is resonance delocalized. Similar processes occur in the spectra of other carbonyl-containing compounds (e.g., esters, amides, acids) and compounds containing other multiply bound heteroatoms (e.g., oximes, imines, sulfoxides).

(2) Allylic Cleavage

Ionization of the π-electrons in a multiple bond (usually favorable only for bonds that do not contain heteroatoms) labilizes the allylic bonds Eq. (29). Homolytic cleavage generates a resonance-stabilized allylic carbonium ion and an alkyl radical:

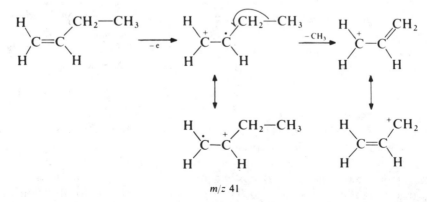

$$\tag{29}$$

An interesting variation of this reaction is observed in cyclohexene derivatives:

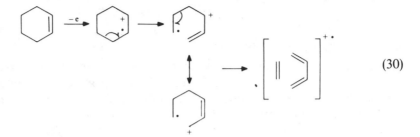

$$(30)$$

Here, cleavage of the first allylic bond does not produce a fragment ion. However, cleavage of another bond results in formation of either an ionized alkene and a neutral diene or a neutral alkene and an ionized diene. The charge distribution between the resulting products will be in accord with Stevenson's rule (175): the positive charge will remain on the fragment of lower ionization potential. Unless the alkene portion is substituted with a strongly charge stabilizing group (e.g., C_6H_5), the charged species will usually be predominantly the diene. This reaction (the "retro Diels–Alder reaction") can facilitate deduction of the substitution pattern on cyclohexene derivatives.

Benzylic cleavage [Eq. (31)] is closely related to allylic cleavage and has a similar driving force. In at least some cases it has been demonstrated that the initially formed benzylic carbonium ion is in equilibrium with the more stable symmetrical tropylium ion:

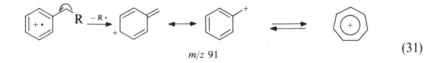

m/z 91

$$(31)$$

(3) Heterolytic Cleavages

Another group of reactions proceed via heterolytic cleavage of a single bond; their driving force is usually the expulsion of a stable neutral species.

For example, alkyl halides fragment with heterolysis of the C X bond to generate a carbonium ion and a halide radical:

$$
CH_3-\overset{\displaystyle CH_3}{\underset{\displaystyle CH_3}{C}}-\overset{..}{\underset{..}{I}}{}^{\cdot+} \longrightarrow CH_3-\overset{\displaystyle CH_3}{\underset{\displaystyle CH_3}{C}}{}^{+} \quad + \quad \overset{..}{\underset{..}{I}}{\cdot}
$$

$$(32)$$

The acylium ion [commonly formed by α-cleavage of a carbonyl-containing ion, Eq. (28)] can similarly expel a stable molecule of carbon monoxide to generate a carbonium ion:

$$R-\overset{\overset{\overset{\cdot\cdot}{O^+}}{\lVert\rVert}}{C} \longrightarrow R^+ + :C\equiv O: \qquad (33)$$

(4) Rearrangements Proceeding from Odd-Electron Ions

Rearrangement reactions proceed with the transfer of one or more atoms (usually hydrogen atoms) to or from the site of charge localization prior to fragmentation. Since such transfers can only occur from specific molecular conformations, their frequency factors are necessarily lower than those of simple cleavages. The driving force for their occurrence is that they usually result in the elimination of a very stable neutral species and formation of a stable ion; as a consequence, these reactions have lower activation energies than simple cleavages.

The McLafferty rearrangement is certainly the most thoroughly studied mass spectral fragmentation (106). It occurs quite generally in molecules that contain multiple bonds (e.g., ketones, esters, oximes, nitriles, alkenes, alkynes) and an appropriately located hydrogen. The initial step of the reaction involves transfer of a γ-hydrogen atom to the site of charge localization through a cyclic six-membered transition state [Eq. (34)]. Cleavage of the β carbon–carbon bond then results in formation of a stable alkene and a resonance-stabilized cation radical. It is notable that a closely analogous reaction (the Norrish Type II cleavage) can be induced photochemically:

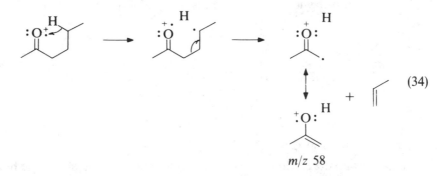

$$m/z \ 58 \qquad\qquad (34)$$

Another interesting group of rearrangement reactions involve dehydration and dehydrohalogenation (107). Although cursory inspection of the mass spectral behavior of alcohols or alkyl halides might suggest that these reactions are closely analogous to their solution chemistry analogs, deuterium labeling experiments have demonstrated that the mass spectral reactions are mechanistically unique. For example, thermal or acid-catalyzed dehydration of alcohols is exclusively a 1,2-elimination. The EI-induced reaction proceeds with predominant 1,3- and 1,4-elimination:

$$C_5H_{10}^{+\cdot} \quad m/z \ 70$$
$$+$$
$$H_2O \qquad\qquad (35)$$

The mass spectral reaction must be unimolecular; perforce, the eliminated hydrogen must be transferred directly to the hydroxyl group prior to carbon–oxygen bond cleavage. The tendency toward 1,3- and 1,4-elimination probably reflects the entropic factors favoring five- and six-membered transition states over the four-membered transition state necessary for 1,2-elimination.

Rearrangement reactions (e.g., the McLafferty rearrangement) can provide great assistance in deducing molecular structure. Often, rearrangement ions can be readily identified. Molecular ions are nearly always odd-electron species. Any fragmentation of an ion containing an odd number of electrons must produce an ion or a neutral product with an odd number of electrons. If the fragmentation is a simple cleavage involving breakage of a single bond, the ionic product will be an even electron and the neutral product an odd electron [cf. Eqs. (27)–(29), (31), (32)]. On the other hand, if an odd-electron ion undergoes a fragmentation requiring cleavage of two bonds (typical of many rearrangement reactions), an even-electron neutral and an odd-electron ion will be formed [cf. Eqs. (30), (34), and (35)]. According to the nitrogen rule already discussed, odd-electron ions containing no or an even number of nitrogen atoms must appear at even m/z values. The corresponding even-electron ions will appear at odd m/z values. When interpreting a mass spectrum, it is often useful to use these arithmetic relationships to note those ions whose genesis involves cleavage of multiple bonds.

d. Use of Reference Spectra

After considering all physical, chemical, and spectroscopic data relevant to an unknown, it will often be possible to narrow the possible structures to one or a few likely candidates. In this case, it will often be desirable to consult one of the reference libraries of mass spectra for confirmation. The two largest compilations of complete spectra are the Wiley Registry of Mass Spectral Data (173) and the EPA/NIH collection distributed by the National Bureau of Standard's National Standard Reference Data System (142a,178a). Since both are printed versions of databases now principally used for computerized searching, they are discussed briefly in the following section. Not uncommonly, however, the candidate molecules will not be listed in any reference compilation. Confirmation then requires determining the mass spectra of authentic samples of these compounds. Although this procedure is laborious, it does permit a more unambiguous identification. The relative intensities of peaks in a mass spectrum can vary considerably depending on the conditions and instrumentation used during its determination.

If, on the other hand, attempts to interpret the mass spectrum have been unsuccessful, these reference spectra compilations can provide useful inspiration. The "Eight Peak Index of Mass Spectra" consists of three tables in three volumes (62) and contains 31, 101 spectra. The most useful lists compounds according to the m/z value that is largest, second largest, and so on, in their mass spectra. The *Compilation of Mass Spectral Data*, or 10-peak index (42) also has a similar list and extensive cross indexing. Since the 10 peak compilation (42) contains data on only ca. 10,000 compounds, the eight-peak index (62) is the more valuable compilation.

Even if the unknown's spectra is not contained in these references, they can provide useful information about the functional groups and structural features that generate ion series similar to those found in the unknown's spectrum.

B. COMPUTER-ASSISTED INTERPRETATION

Manual interpretation of mass spectra is, at best, an intellectually demanding and time-consuming task. A GC/MS or LC/MS might easily produce the spectra of hundreds of compounds in a day's analyses of complex environmental or biological samples. Few mass spectroscopists would care to tackle the interpretation of so many spectra without computer assistance.

Even in cases where only one or a few spectra require interpretation, computer assistance can be invaluable. Not uncommonly, either it will result in a direct identification or it will suggest a compound sufficiently similar in structure to simplify identification of an unknown that did not yield to manual interpretation. Since the subject of computer-assisted interpretation of mass spectra has been extensively reviewed (83, 118c, 132, 153, 167a, 169), the subject will be discussed here only briefly.

Numerous algorithms have been developed to rapidly and accurately accomplish the difficult task of extracting the "best" match to an unknown spectrum from a large library (118c, 181). The substantial information content (i.e., many peaks with widely varying intensities) in most spectra presents a complication. Extensive disk storage and long searches would be necessary if retrieval from a large library was based on evaluation of complete peak mass and intensity data. Additionally, the matching process is made more challenging by the sometimes substantially different spectra observed when a pure compound is analyzed under different experimental conditions (e.g., inlet technique, source temperature) or on different instruments and by the frequent presence of extraneous peaks due to impurities in the library spectrum or the unknown spectrum. The two most widely used searching systems, the Probability Based Matching System [PBM, developed at Cornell (111, 130b)] and the Biemann system [developed at MIT (9, 83)], approach these problems differently.

The PBM system abbreviates spectra by selecting the 15–25 most significant peaks. Significance is determined by peak intensity (since strong peaks are less common in mass spectra than weak ones) and mass. Peaks at masses that occur less frequently in the library database are obviously more useful in distinguishing among spectra than peaks at commonly occurring masses. Thus, for example, high-mass peaks, which are less common and have more structural significance, tend to be retained in preference to low-mass peaks (135).

A second interesting feature of the PBM system is the use of "reverse searching." The searching process begins by selecting peaks from the reference spectrum; then the program determines whether these peaks are also present in the unknown's spectrum. This procedure will ignore any "extra" peaks present in the unknown's spectrum due to background or impurities.

Obviously, the speed with which this process can be performed will depend on the power of the computer used, and the accuracy of the proposed identification will depend on the structure of the unknown. However, a 1983 report claimed a search time of 10 s per unknown while a ca. 80,000-spectra library was searched in "backround" mode (i.e., GC/MS data were simultaneously being acquired) (130c). In another study of 392 unknown compounds whose reference spectra were contained in a ca. 80,000-spectra library, the unknown spectrum, determined under different conditions, was correctly matched with the reference spectrum in 74% of the trials (130d).

The Biemann system differs from PBM both in its approach to spectrum abbreviation and in its search strategy. The two most intense peaks in each 14-amu interval of a mass spectrum are retained. A similarity index is then calculated between the unknown spectrum and the reference spectra based on the relative intensities of peaks present in both as well as the relative intensities of peaks present only in one. The similarity index thus measures the closeness of the match between two spectra and can be related to the probability that the identification is correct.

The utility of any system for matching unknown and library spectra obviously depends on the size and quality of the library database. An industry standard nine-track magnetic tape available from NBS contains the mass spectra of about 44,000 compounds. Associated with each compound is a Chemical Abstracts Service Reference Number (which, inter alia, avoids inclusion of multiple spectra per compound), compound name and all available synonyms, a classification flag (e.g., alkaloid, pesticide, steroid), the compound's molecular weight and molecular formula, the spectrum's quality index (calculated by computer), instrument operating parameters, and contributor. Many instrument manufacturers provide this database and a spectrum matching and retrieval program as part of the mass spectrometer's data-processing system. Alternatively, the database and associated software can be accessed through several commercial time-sharing systems. Most conveniently, the database and necessary searching software are available on floppy disks for installation on IBM PCs and compatible computers. A printed version of the database is expected to be available soon (178a).

Computers can be used to facilitate identification of unknowns in a conceptually different fashion. They can be asked to deduce structural features or the entire structure based on the compound's mass spectrum. A number of different approaches have been employed to accomplish this goal with some success. At Stanford, the artificial-intelligence-based DENDRAL technique has been used. The unknown spectrum is used to deduce "substructures" present in the molecule. A structure generation program then produces a series of candidate structures for the molecule. Then, since the computer has been programmed with the "rules" governing mass spectrometric fragmentation, it predicts the mass spectrum of each candidate structure. Finally, the predicted spectra are compared to the unknown's spectrum and ranked on the basis of their similarity (39b, 118c, 169, 170, 170a).

In an alternative approach, the computer teaches itself to recognize compound types or substructures by applying pattern recognition techniques to reference spectra that do or do not contain those features (97). However, the most generally available (through commercial time-sharing systems) interpretive procedure is the Self-Training Interpretive and Retrieval System (STIRS) (112, 130e, 135). The STIRS system uses 15 classes of mass spectral data to predict structural features present in the unknown. For example, for one of the classes, the eight most intense peaks between m/z 47 and 102 in the unknown's spectrum are matched to the corresponding data in ca. 30,000 reference spectra. The computer selects the closest matches for each class of data, permitting inferences about structural similarities between the matches and the unknown. The interpretive procedures are much more expensive than simple retrieval procedures. However, the interpretive procedures can elucidate molecular structure even when the unknown's spectrum does not appear in the reference library. Thus, the usual procedure is to first attempt identification using a "matching" algorithm, proceeding to an interpretive algorithm only as a last resort.

It should be emphasized that the ideal combination for structure elucidation by mass spectrometry would be a skilled interpreter armed with a powerful computer search system. Together the two would successfully identify many more compounds than either could manage separately. Further, the interpreter would likely "screen out" many false-positive identifications that appear plausible to the computer.

C. QUANTITATION BY MASS SPECTROMETRY

Many examples of quantitation by mass spectrometry make use of the gas chromatographic inlet system. In fact, the quantitation techniques used in GC/MS and conventional GC are broadly similar. In the "external standard technique," the signal produced by injection of aliquots containing known amounts of the analyte of interest is compared to the signal produced by the analyte in the sample matrix. The signal examined by the mass spectroscopist is most commonly the integrated abundance of one (or more) ions whose m/z is characteristic of the analyte. Alternatively, the "internal standard technique" involves use of a reference compound. Typically, solutions containing a constant, known amount of the internal standard and varying, known amounts of the analyte are injected onto the GC/MS. The integrated area of one or more characteristic ions of the internal standard and one or more characteristic ions of the analyte are then used to calculate response factors. The solution to be quantitated is then spiked with the standard concentration of the internal standard and injected onto the GC/MS. The ratio of the abundance of the characteristic ion of the internal standard to the abundance of the characteristic ion of the analyte combined with the predetermined response factor permits calculation of the analyte's concentration. The advantage of the internal standard technique is that it will at least partially compensate for variations in the volume of solution spiked, volume injected, changes in the responsiveness of the GC/MS system, and matrix effects.

Within the broad categories of the external and internal standard techniques, there are enormous variations possible. Spectra can be recorded using full-scan techniques and the characteristic ion's abundances extracted, most conveniently under computer control. Alternatively, selected ion-monitoring procedures can be used; the mass spectrometer monitors the abundances of only the ions of interest, resulting in drastically (e.g., thousandfold) improved sensitivity. The disadvantage of selected ion monitoring, of course, is the possibility of mis-identification since only limited spectral information is obtained. Spectra can be recorded at high or low resolution. Obviously, high-resolution spectra or selected ion abundances are more useful than low-resolution data. However, obtaining high-resolution data requires more sophisticated instrumentation and may result in poorer sensitivity. When the internal standard technique is used, the optimal internal standard is usually an analog labeled with a stable isotope other than deuterium (e.g., ^{13}C). Then it can reasonably be assumed that the self-correcting features of the internal standard technique will be optimized since adsorption effects, extraction efficiencies, derivatization yields, and so on, are unlikely to exhibit significant isotope effects, and the "carrier effect" (which postulates that losses of, e.g., a trace-level analyte through adsorption are minimized by the presence of an excess of labeled analog) will be strongest. Deuterium-labeled analogs are usually more readily available and are therefore widely used, although it should be recognized that significant isotope effects and retention time differences may be observed. In some cases, appropriately labeled analogs are not available and structural analogs must be used as the internal standard.

In order to clarify these concepts and to begin to illustrate the diversity of quantitation techniques available to the mass spectroscopist, two very different analyses appropriate to environmental samples will be discussed. More extensive compilations of applications of quantitative mass spectrometry to biomedical problems have recently appeared (52, 53, 118b, 141), and a general review of quantitation techniques is available (Millard, General Bibliography).

1. Priority Pollutant Analysis

The Federal Water Pollution Control Act of 1972 (Clean Water Act) and the Clean Water Act of 1977 required that the U.S. Environmental Protection Agency (EPA) establish a program to limit discharge of "toxic pollutants" into the nation's waterways. An analytical scheme was therefore developed and promulgated to identify and quantitate 114 organic chemicals (the organic priority pollutants) in industrial wastewater. The analysis was accomplished using gas chromatography combined with low-resolution scanning mass spec-trometry and a computerized data acquisition and handling system. Since its promulgation in 1978, the analytical methodology has become increasingly sophisticated. Capillary gas chromatography has replaced packed-column gas chromatography, and advanced quality control procedures have been mandated. Further, in its most current form [U.S. EPA Method 8270 (180a)] the range of sample types has been expanded to include groundwater, soil, sludge, sediment, and solid waste.

In its various perturbations, the priority pollutant analysis is by far the most frequently performed analytical procedure that utilizes a GC/MS unit. Numerous commercial firms perform the analysis, some on a 24-h/day basis, with sufficient automation and efficiency to permit pricing at under $500 per sample for many matrices. An outline of the procedure is presented here because it illustrates the performance and sophistication routinely attainable in GC/MS laboratories of widely differing abilities.

For purposes of this analysis, the 114 organic priority pollutants fall into two classes. The 30 volatile organics (boiling point $<150°C$) are vaporized by bubbling an inert gas through the sample. The gas flows through a cartridge packed with Tenax (a porous polymeric material based on 2,6-diphenyl-p-phenylene oxide), which effectively traps out the organic volatiles. After a few minutes, trapping is complete and the cartridge is backflushed with flash heating into the inlet system of the GC/MS unit. Since quantitation and quality control procedures used during the volatile analysis are conceptually similar to those used during the more elaborate semivolatile analysis, the volatile procedure will not be discussed further here.

Daily, prior to analysis for the semivolatile organic priority pollutants, a GC/MS tuning standard containing three difficult-to-chromatograph compounds (4,4'-DDT, pentachlorophenol, and benzidine) and decafluorotriphenyl-phosphine (DFTPP) each at 50 ng/μl must be injected. The first three compounds are used to verify injection port inertness and chromatographic performance of the capillary GC column. The DFTPP is used to verify that the mass spectrometer has been properly tuned. The various quantitative tuning criteria that must be attained are designed to assure that a compound's mass spectrum recorded on different days or on different instruments will be similar. This will improve the likelihood of a successful computer match between a compound's recorded and library spectra.

Another quality control procedure requires spiking each sample matrix with appropriate "surrogate standards." Surrogate standards are compounds whose chemical and physical properties are at least broadly representative of the analytes sought but that are unlikely to be naturally present in the sample. Recommended compounds include phenol-d_6, 2-fluorophenol, 2,4,6-tribromo-phenol, nitrobenzene-d_5, 2-fluorobiphenyl, and p-terphenyl-d_4. After sample extraction, clean-up, and concentration, the recoveries of the surrogate standards are determined along with the concentrations of any analytes found. Unusual recoveries of one or more surrogate standards raise suspicions about the validity of analytical results for priority pollutants with similar physical and chemical properties.

Quantitation of analytes and surrogate standards is accomplished by the internal standard technique. Response factors are determined by injecting solutions containing known concentrations of an internal standard (often 20 μg/ml of d_{10}-anthracene) and one or more of the analytes. The response factor (RF) is then computed according to

$$RF = \frac{A_s C_{is}}{A_{is} C_s} \tag{36}$$

where A_s and A_{is} are, respectively, the areas (or heights) of the peaks corresponding to the characteristic ions of the analyte and the internal standard, and C_s and C_{is} are their concentrations. An example of a response factor determination for the m/z 275 ion of hexachlorobutadiene versus the m/z 188 ion of d_{10}-anthracene appears in Fig. 43. A solution containing 27 μg/ml of hexachlorobutadiene and 20 μg/ml of d_{10}-anthracene was injected onto a computer-controlled GC/MS unit. During the run, a complete low-resolution mass spectrum was recorded every three seconds. At the conclusion of the run, the computer can produce extracted ion current profiles (EICPs), which plot the abundance of selected ions versus the scan number. In Fig. 44, the total ion chromatogram (corresponding to a plot of the sum of the intensity of peaks at all m/z values) and the EICPs of m/z 225 and 188 are depicted. The computer-generated areas of m/z 225 and 188 are 13,202 and 55,943 counts, respectively. Therefore, the response factor can be calculated as

$$RF = \frac{(13,202)(20)}{(55,943)(27)} = 0.17 \tag{37}$$

Since it is at least possible that a response factor might vary significantly with large variations in analyte concentration, the response factor must be determined over the range of concentrations likely to be encountered in samples.

After a satisfactory set of response factors has been determined, the sample itself can be quantitated. Just prior to injecting the extract suspected to contain one or more priority pollutants, it is spiked with the internal standard (d_{10}-anthracene at 20 μg/ml in this case). Figure 45 represents the total ion chromatogram (TIC) resulting from capillary column analysis of a sample containing all semivolatile priority pollutants. Full-scan mass spectra are recorded every 3 s during the chromatographic run. Criteria for qualitative identification of an analyte include appropriate retention time, maximization of key ions' intensities within one spectrum of each other, and the ratios of peak heights of the key ions agreeing within $\pm 20\%$ with the corresponding ratios in the spectrum of the authentic compound. Quantification is accomplished by comparing the integrated intensity of the characteristic ion of the internal standard with the integrated intensity of the charactcristic ion of the analyte. For example, Fig. 44 depicts analysis of an extract containing hexachlorobutadiene and d_{10}-anthracene at 20 μg/ml. The area of the m/z 225 cm of hexachlorobutadiene is 59,710, counts and the area of the m/z 188 ion of d_{10}-anthracene is 51,625 counts.

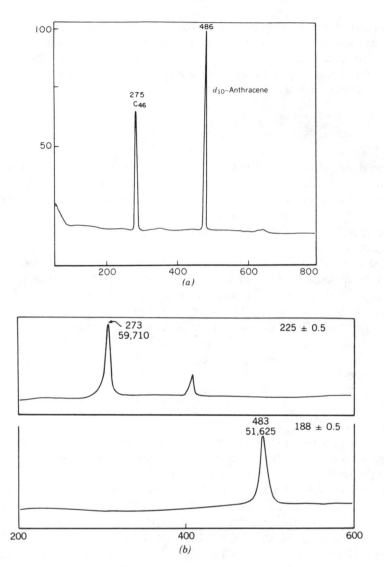

Fig. 43. (a) Total ion chromatogram produced by injection of standard solution containing hexachlorobutadiene at 27 μg/ml and d_{10}-anthracene at 20 μg/ml. (b) The EICP of m/z 188 (characteristic ion of d_{10}-anthracene) and m/z 225 (characteristic ion of hexachlorobutadiene). Numbers printed above peaks correspond to scan number in which ion abundance was maximized and total area under peak.

Therefore,.using the predetermined response factor, the concentration of hexachlorobutadiene in the extract is readily calculated as

$$C_s = \frac{(59,710)(20)}{(0.17)(51,625)} = 136 \ \mu g/ml \tag{38}$$

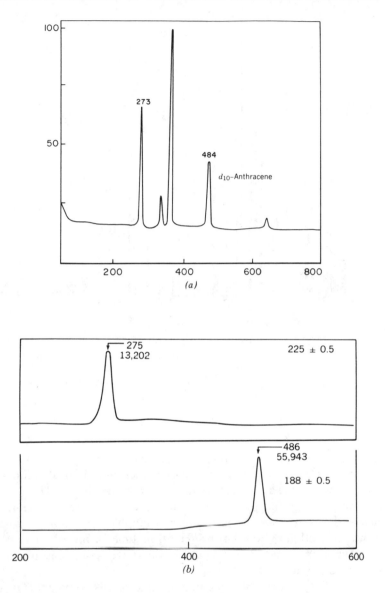

Fig. 44. (a) Total ion chromatogram produced by injection of unknown solution spiked with d_{10}-anthracene at 20 μg/ml. (b) The EICP of m/z 188 and 225.

The EPA has estimated the detection limit for most base-neutral analytes at 10 μg/liter; experience has suggested that for many analytes and matrices much lower detection limits are possible.

The advantages of quantitation by internal standard have already been alluded to; matrix effects, uncertainties in extract volumes and volumes injected,

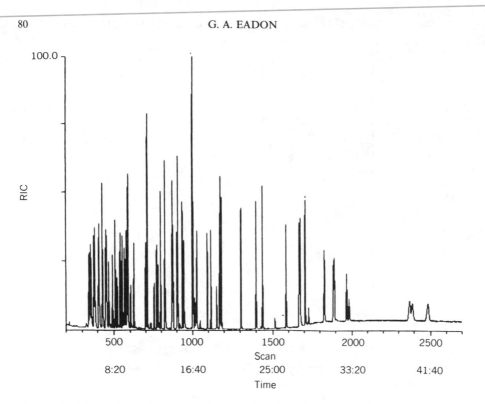

Fig. 45. Total ion chromatogram resulting from injection of extract containing all semivolatile priority pollutants.

and variations in instrument response are at least partially compensated for. These compensatory effects are more reliable when the physical and chemical properties of the analyte and internal standard are more similar. Thus, the most advanced versions of the priority pollutant method currently use multiple internal standards so that the gas chromatographic retention times of any analyte will be within $\pm 20\%$ of the retention time of an internal standard (Table 5). The ultimate internal standard is, of course, a stable-isotope-labeled analog of the analyte of interest.

A feature of the priority pollutant analysis that merits comment is its use of full-scan rather than selected ion monitoring (SIM) techniques. Obviously, SIM would, in favorable cases, give orders-of-magnitude improvements in detection limits. Collection of full-scan data, on the other hand, gives more unambiguous identification of the target analytes. Further, the 130 priority pollutants represent a small fraction of the 50,000 or so compounds in the NIH/EPA library of mass spectra or the millions of organic compounds described in *Chemical Abstracts*. By recording full-scan mass spectra, the analyst or the data system can frequently provide qualitative identification for peaks that do not correspond to priority pollutants.

TABLE 5
Semivolatile Internal Standards with Corresponding Analytes Assigned for Quantitation

1,4-Dichlorobenzene-d_4	Naphthalene-d_8	Acenaphthene-d_{10}
Aniline	Acetophenone	Acenaphthene
Benzyl alcohol	Benzoic acid	Acenaphthylene
Bis(2-chloroethyl)ether	Bis(2-chloroethoxy)methane	1-Chloronaphthalene
Bis(2-chloroisopropyl)ether	4-Chloroaniline	2-Chloronaphthalene
2-Chlorophenol phenyl ether	4-Chloro-3-methylphenol	4-Chlorophenyl phenyl ether
1,3-Dichlorobenzene	2,4-Dichlorophenol	
1,4-Dichlorobenzene	2,6-Dichlorophenol	Dibenzofuran
1,2-Dichlorobenzene	α,α-Dimethylphenethylamine	Diethyl phthalate
Ethyl methanesulfonate		Dimethyl phthalate
2-Fluorophenol (surr.)	2,4-Dimethylphenol	2,4-Dinitrophenol
Hexachloroethane	Hexachlorobutadiene	2,4-Dinitrotoluene
Methyl methanesulfonate	Isophorone	2,6-Dinitrotoluene
2-Methylphenol	2-Methylnaphthalene	Fluorene
4-Methylphenol	Naphthalene	2-Fluorobiphenyl (surr.)
N-Nitrosodimethylamine	Nitrobenzene	
N-Nitroso-di-n-propylamine	Nitrobenzene-d_8 (surr.)	Hexachlorocyclopentadiene
Phenol	2-Nitrophenol	
Phenol-d_6 (surr.)	N-Nitroso-di-n-butylamine	1-Naphthylamine
2-Picoline	N-Nitrosopiperidine	2-Naphthylamine
	1,2,4-Trichlorobenzene	2-Nitroaniline
		3-Nitroaniline
		4-Nitroaniline
		4-Nitrophenol
Perylene-d_{12}	Phenanthrene-d_{10}	Pentachlorobenzene
		1,2,4,5-Tetrachlorobenzene
Benzo(b)fluoranthene	4-Aminobiphenyl	2,3,4,6-Tetrachlorophenol
	Anthracene	
Benzo(k)fluoranthene	4-Bromophenyl phenyl ether	2,4,6-Tribromophenol (surr.)
	Di-n-butyl phthalate	
Benzo(g, h, i)perylene	4,6-Dinitro-2-methylphenol	2,4,6-Trichlorophenol
	Diphenylamine	
Benzo(a)pyrene	1,2-Diphenylhydrazine	2,4,5-Trichlorophenol
Dibenz(a, j)acridine	Fluoranthene	
Dibenz(a, h)anthracene	Hexachlorobenzene	Chrysene-d_{12}
	N-Nitrosodiphenylamine	
7,12-Dimethylbenz(a)anthracene	Pentachlorophenol	
	Pentachloronitrobenzene	
Di-n-octylphthalate	Phenacetin	Benzidine
Indeno(1,2,3-cd)pyrene	Phenanthrene	Benzo(a)anthracene
	Pronamide	Bis(2-ethylhexyl)phthalate
3-Methylcholanthrene		Butylbenzylphthalate
		Chrysene
		3,3'-Dichlorobenzidine
		p-Dimethylaminoazobenzene
		Pyrene
		Terphenyl-d_{14} (surr.)

2. Analysis for Polychlorinated Dibenzodioxins and Polychlorinated Dibenzofurans

The priority pollutant analysis is a routine procedure capable of reliably quantitating over 100 analytes with widely differing physical and chemical properties at concentrations of a few parts per billion in many matrices. Analysis of environmental samples for polychlorinated dibenzodioxins (PCDDs) and polychlorinated dibenzofurans (PCDFs) presents a very different set of problems. The extraordinary toxicity exhibited by certain of these compounds necessitates extremely low detection limits. For example, the EPA has recommended a maximum total concentration of TCDD in ambient air of about 0.1 fg/m^3 (10^{-13} g/m^3) (180b), and plausible arguments have been advanced for the desirability of a 1-ppt (parts per trillion) detection limit in tissue analysis (39). The latter requirement demands the ability to detect subpicogram ($< 10^{-12}$-g) amounts of these compounds when injected into a GC/MS unit. In contrast, the instrumental detection limit demanded by the priority pollutant analysis is typically greater than 10 ng ($> 10^{-8}$ g).

A second complication inherent in the PCDD/PCDF analysis is the large number of structurally homologous compounds that are typically present and can interfere with the quantitation of a particular isomer of interest. For example, the 22 isomers of tetrachlorodibenzodioxin differ by factors in excess of 10^3 in their biological potency (61a). For many purposes, then, it is important to conduct the analysis in an "isomer-specific" manner so that the peaks corresponding to the most potent isomers (usually those substituted at at least the 2, 3, 7, and 8 positions) are chromatographically distinct from isomeric interferences. Qualitative identification of a particular peak, demonstration of its isomeric purity, and quantitation are further complicated by the fact that authentic samples of all 75 chlorinated dibenzodioxins and all 135 chlorinated dibenzofurans are not readily available.

Because identification of PCDDs or PCDFs in an environmental or tissue sample can have significant political, medical, or biological implications, the specificity of mass spectral detection is a virtual necessity. A broad range of mass spectrometric techniques have been applied to this problem (48,180c). For example, some procedures include use of oxygen as reagent gas in a negative-ion CI procedure (92), use of atmospheric pressure ionization mass spectrometry (144), use of metastable transitions associated with molecular ion fragmentation (39), a GC/GC/MS technique in which a packed column was interfaced with a capillary column prior to mass spectrometric analysis (114b), use of GC/(hybrid)MS/MS with selected reaction monitoring (180c), as well as more conventional low-resolution and high-resolution techniques. The EPA procedure 8290 (180b) will be discussed here as an example of a "state-of-the-art" high-resolution GC–high-resolution MS method intended for broad implementation.

The methodology is intended to be applicable to such diverse matrices as soil, sediment, fly ash, water, sludge (including paper pulp) still bottoms, fuel oil, chemical reactor residue, fish tissue, or human adipose tissue. Detection limits for 2,3,7,8-TCDD range from 2.5 ppq (parts per quadrillion) to 12.5 ppt depending on matrix and sample size. Each sample is spiked with nine fully ^{13}C-labeled PCDDs and PCDFs containing from four to eight chlorines for use as standards. Since the data of greatest interest resulting from the analysis are the concentrations of the biologically potent 2,3,7,8-substituted congeners, the standards used all have this substitution pattern. The distinction between standard usage in this procedure versus the priority pollutant procedure is significant. Since the internal standard is added to the sample matrix *before* extraction, clean-up, and concentration in the PCDD/PCDF analysis, the internal standard technique will automatically correct for any analyte losses during these steps provided the chemical and physical behavior of the internal standards and the corresponding analyte are very similar. This condition should be well fulfilled since the analytes and internal standard are often isotopic analogs.

Extraction of the spiked sample is accomplished using various techniques (e.g., Soxhlet, liquid–liquid, or Dean–Stark extraction); clean-up is also matrix specific but typically involves one or more adsorptive column chromatography steps followed by chromatography on Celite-adsorbed carbon. (Carbon has an unusually strong affinity for planar aromatic molecules, thus allowing easy and effective separation of PCDDs and PCDFs from molecules not possessing these characteristics.) After the volume of the cleaned extract has been reduced to 1 ml, it is spiked with 10–50 μl of a tridecane solution containing two fully ^{13}C-labeled recovery standards (e.g., ^{13}C$_{12}$-1,2,3,4-TCDD and ^{13}C$_{12}$-1,2,3,7,8,9-HxCDD). The role of the recovery standards is most easily understood by considering a case where equal amounts of internal standard and recovery standard are added to the sample. The extract ultimately injected onto the GC/MS unit will contain fewer picograms of internal standard than of recovery standard since losses of internal standard inevitably occur during sample extraction, clean-up, and concentration. The ratio of the amounts of the two types of standards thus gives a quantitative measure of the extent of such losses and serves as an important indicator of the quality of the analysis.

The instrumental requirements to perform Method 8290 are demanding. The mass spectrometer must be capable of performing selected ion monitoring at a resolving power of at least 10,000 (10% valley definition) on nine ions and a PFK lock mass with a cycle time of 1 s or less. The gas chromatograph is fitted with a 60M DB-5 column; this column possesses the useful ability to separate the compounds of interest by chlorine number, thus limiting the number of ions to be monitored at a particular time. For example, Table 6 lists the ions monitored during elution of the tetrachloro dibenzofurans and dibenzodioxins.

A "column performance check solution" is injected regularly in order to verify the system's chromatographic performance and to establish the time "windows" during which each chlorine group elutes. This solution contains 100 pg/μl ^{13}C$_{12}$-2,3,7,8-TCDD, native 2,3,7,8-TCDD, and the four TCDD isomers whose retention times on DB-5 are nearest that of 2,3,7,8-TCDD. Successful completion of this check requires separation of 2,3,7,8-TCDD from any other isomer with a valley of 25% or less (Fig. 46). In addition, this solution contains similar

TABLE 6
Ions Monitored for HRGC/HRMS analysis of PCDD/PCDFs[a]

Accurate[b] Mass	Ion ID	Elemental Composition	Analyte
303.9016	M	$C_{12}H_4 {}^{35}Cl_4O$	TCDF
305.8987	M+2	$C_{12}H_4 {}^{35}Cl_3 {}^{37}ClO$	TCDF
315.9419	M	${}^{13}C_{12}H_4 {}^{35}Cl_4O$	TCDF (S)
317.9389	M+2	${}^{13}C_{12}H_4 {}^{35}Cl_3 {}^{37}ClO$	TCDF (S)
319.8965	M	$C_{12}H_4 {}^{35}Cl_4O_2$	TCDD
321.8936	M+2	$C_{12}H_4 {}^{35}Cl_3 {}^{37}ClO_2$	TCDD
331.9368	M	${}^{13}C_{12}H_4 {}^{35}Cl_4O_2$	TCDD (S)
333.9339	M+2	${}^{13}C_{12}H_4 {}^{35}Cl_3 {}^{37}ClO_2$	TCDD (S)
375.8364	M+2	$C_{12}H_4 {}^{35}Cl_6O$	HxCDPE
[354.9792]	Lock	C_9F_{13}	PFK

[a] Abbreviations: HRGC, high-resolution gas chromatography; HRMS, high-resolution mass spectrometry; S, internal/recovery standard.

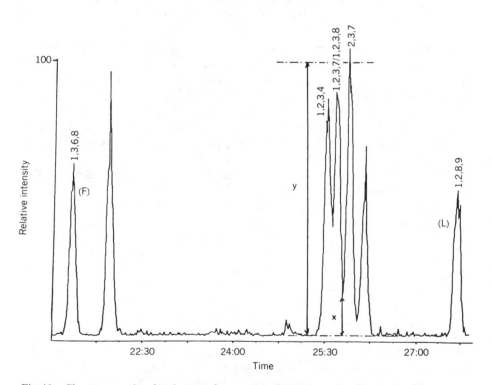

Fig. 46. Chromatography of "column performance check solution" containing 100 pg/μl of various TCDD isomers. Data collected in SIM mode at 10,000 resolution.

concentrations of the first and last eluting PCDD and PCDF isomer of each chlorination number. Thus, analysis of this solution defines the time windows when each chlorination group elutes. Defining the elution windows improves instrumental sensitivity since, for example, ions characteristic of pentachloro compounds need not be monitored during the tetrachloro window. As Table 6 demonstrates, the analysis nevertheless requires monitoring at least nine ions simultaneously. For example, in the tetrachloro window, monitored ions include the molecular ions and the $M + 2$ ions of the unlabeled TCDFs, of the $^{13}C_{12}$-labeled TCDFs, of the unlabeled TCDDs, and of the $^{13}C_{12}$-labeled TCDDs. Two ions characteristic of each compound are acquired because an important criterion for identification is the presence of the correct intensity ratio between the M^+ and $M + 2$ ions. In addition, a PFK lock mass is monitored to compensate for the possibility of the mass spectrometer's magnetic field "drifting" away from the value that is optimal for focusing the target ions.

The wide range of masses that must ultimately be acquired even for a tetrachloro through octachloro analysis (303.9016, the M^+ of TCDF, through 469.7779, the $M + 2$ of OCDD) introduces some complications into the analysis. As discussed earlier in this chapter, technical difficulties preclude performing ion-monitoring experiments by manipulation of magnetic field strength. Instead, the magnetic field strength is held constant during a particular time window, and the accelerating voltage and electric sector voltage are varied to permit the necessary rapid cycle times. However, large changes in these voltages detune the source and result in large decreases in resolution and sensitivity; thus there is a practical limit to the mass range that can be monitored at a given magnet setting. For the analysis under discussion here, this limitation means that if the full range of furans and dioxins is to be acquired in a single chromatographic run, the mass spectrometer's magnetic field must be "jumped" to a higher value one or more times during a chromatographic run. Before the new set of ions can be acquired, the magnetic field must stabilize at the new strength, the lock mass must be reacquired, and the new set of masses to be monitored must be established. Often, these processes must be accomplished quickly. For example, about 15 s typically elapses between elution of the last tetrachlorinated congener and the first pentachlorinated congener. Obviously, this demands sophisticated electronics and a powerful data system.

Quantitation of analytes is accomplished by the internal standard technique and requires determination of the relative response factors between each ^{13}C internal standard and the analytes quantitated from that internal standard [e.g., the relative response factor (RRF) $^{13}C_{12}$–TCDD and $^{12}C_{12}$–TCDD is obtained by dividing counts per picogram 2378-TCDD at 319.8965 by counts per picogram $^{13}C_{12}$–2378-TCDD at 331.9368]. To that end, seven high-resolution concentration calibration solutions containing the ^{13}C-labeled internal standards at constant concentration and 17 unlabeled analytes at concentrations ranging over two orders of magnitude (Table 7) are injected. The resulting data are used to calculate relative response factors for the unlabeled analytes and, by extension, for all homologs of that chlorination group.

G. A. EADON

TABLE 7
High-Resolution Concentration Calibration Solutions

Compound	Concentration (pg/μl)						
	7	6	5	4	3	2	1
Unlabeled Analytes							
2,3,7,8-TCDD	200	100	50	25	10	5	2.5
2,3,7,8-TCDF	200	100	50	25	10	5	2.5
1,2,3,7,8-PeCDD	200	100	50	25	10	5	2.5
1,2,3,7,8-PeCDF	200	100	50	25	10	5	2.5
2,3,4,7,8-PeCDF	200	100	50	25	10	5	2.5
1,2,3,4,7,8-HxCDD	500	250	125	62.5	25	12.5	6.25
1,2,3,6,7,8-HxCDD	500	250	125	62.5	25	12.5	6.25
1,2,3,7,8,9-HxCDD	500	250	125	62.5	25	12.5	6.25
1,2,3,4,7,8-HxCDF	500	250	125	62.5	25	12.5	6.25
1,2,3,6,7,8-HxCDF	500	250	125	62.5	25	12.5	6.25
1,2,3,7,8,9-HxCDF	500	250	125	62.5	25	12.5	6.25
2,3,4,6,7,8-HxCDF	500	250	125	62.5	25	12.5	6.25
1,2,3,4,6,7,8-HpCDD	500	250	125	62.5	25	12.5	6.25
1,2,3,4,6,7,8-HpCDF	500	250	125	62.5	25	12.5	6.25
1,2,3,4,7,8,9-HpCDF	500	250	125	62.5	25	12.5	6.25
OCDD	1000	500	250	125	50	25	12.5
OCDF	1000	500	250	125	50	25	12.5
Internal Standards							
$^{13}C_{12}$-2,3,7,8-TCDD	50	50	50	50	50	50	50
$^{13}C_{12}$-2,3,7,8-TCDF	50	50	50	50	50	50	50
$^{13}C_{12}$-1,2,3,7,8-PeCDD	50	50	50	50	50	50	50
$^{13}C_{12}$-1,2,3,7,8-PeCDF	50	50	50	50	50	50	50
$^{13}C_{12}$-1,2,3,6,7,8-HxCDD	125	125	125	125	125	125	125
$^{13}C_{12}$-1,2,3,4,7,8-HxCDF	125	125	125	125	125	125	125
$^{13}C_{12}$-1,2,3,4,6,7,8-HpCDD	125	125	125	125	125	125	125
$^{13}C_{12}$-1,2,3,4,6,7,8-HpCDF	125	125	125	125	125	125	125
$^{13}C_{12}$-OCDD	250	250	250	250	250	250	250
Recovery Standards							
$^{13}C_{12}$-1,2,3,4-TCDD[a]	50	50	50	50	50	50	50
$^{13}C_{12}$-1,2,3,7,8,9-HxCDD[b]	125	125	125	125	125	125	125

[a] Used for recovery determinations of TCDD, TCDF, PeCDD, and PeCDF internal standards.
[b] Used for recovery determinations of HxCDD, HxCDF, HpCDD, HpCDF, and OCDD internal standards.

After these preliminary steps are completed, the extract to be quantitated is injected and the necessary mass spectral data acquired. Calculation is based on the formula

$$C_x = \frac{A_x Q_{is}}{A_{is} W [\overline{RRF(n)}]} \tag{39}$$

where

C_x = concentration of unlabeled PCDD/PCDF congeners (or group of coeluting isomers within homologous series) (pg/g)

A_x = sum of integrated ion abundances of quantification ions (Table 6) for unlabeled PCDDs/PCDFs

A_{is} = sum of integrated ion abundances of quantification ions (Table 6) for labeled internal standards

Q_{is} = quantity of internal standard added to sample before extraction (pg)

W = weight of sample (solid or liquid) (g)

$\overline{RRF}(n)$ = calculated mean relative response factor for analyte

A typical fly ash extract, for example, will produce dozens of peaks in the ion chromatogram of each mass monitored. After each peak's area is integrated and its retention time established, the peak must be evaluated to determine whether it represents a PCDD or PCDF. The evaluation focuses on agreement between the ratio of areas observed for the molecular ion peak versus the M + 2 peak and the theoretical ratio. If this criterion is met, it must be determined whether the observed retention time is characteristic of a 2,3,7,8-substituted isomer. This is usually accomplished by calculating the ratio of the retention time of the peak in question to the retention time of the nearest eluting internal standard. This ratio can then be compared to the corresponding ratio observed during a standard injection.

If attention is focused only on the tetrachloro through octachlorodibenzo-dioxins and dibenzofurans in a single sample, seventeen 2,3,7,8-substituted isomers will be reported as will 10 "total" chlorination groups (e.g., total TCDF). If it is noted that problems as simple as a misintegration of the area under a peak in one of the ion chromatograms, or the presence of an interference at one or more ions can lead to an incorrect decision about whether a peak represents a PCDD or PCDF, and that there may be more than 100 peak pairs where such problems will influence the final data reported, it is obvious that successful execution of this analysis is challenging. Its advancement by a government agency as a method to be performed in commercial and government laboratories testifies to the advances made in mass spectrometers and data systems in recent years.

D. MASS SPECTROMETRY OF NONVOLATILE AND THERMALLY UNSTABLE COMPOUNDS

Most of the ionization techniques discussed in Section II.A have a common feature: the sample must be introduced into the ionization region as gaseous molecules. This constraint places important limitations on the volatility and thermal stability required to perform the mass spectral experiment. If a molecule is so involatile or thermally labile that it cannot be vaporized without extensive decomposition, even a very mild ionization technique such as FI will at best generate ions characteristic of the decomposition products rather than the intact molecule. It is not surprising, therefore, that Beckey's demonstration (15) of a new technique that permits ionization and mass analysis of adsorbed substances without prior volatization sparked considerable interest. The new technique was called field desorption mass spectrometry (FDMS).

FIMS and FDMS differ principally in the procedures used to admit the sample to the ionizing region. In FDMS, the sample is applied directly to the cathodic electrode as a solution either by dipping (15), as an aerosol (164), or by use of a microliter syringe (17, 115). After evaporation of the solvent, the electrode is inserted into the field desorption source (usually through a vacuum lock), mounted, and carefully aligned. After application of a ca. 10-kV potential between the anodic and cathodic electrode, the emitting (anodic) electrode is heated, usually by passage of a small electric current. Spectra are generally recorded at the "best anode temperature" (BAT), defined as the emitter temperature that generates maximum molecular ion currents and minimum fragmentation (191). This temperature (often expressed in milliamperes) may range from subambient to 300°C for organic and biological molecules (163). Since current variations of only a few percent can substantially alter the duration and intensity of a spectrum, determination of the BAT can be tedious. More recently, a laser has been used to heat the emitter electrode (165). Advantages claimed for the use of a laser include greater sensitivity and longer emitter lifetimes.

Several distinct mechanisms may contribute to the FD ionization process. The conventional explanation for FDMS is similar to that already discussed for FIMS. Ionization involves electric-field-induced removal of a valence electron from the sample molecule. Heating of the emitter electrode serves to facilitate desorption of the field-induced ions and to permit more rapid migration of neutrals from their site of deposition to the high-field regions where ionization occurs. Experimental results suggestive of a field-independent ionization and desorption process have also been obtained (85, 86). The conventional explanation is more generally accepted (38, 49) for neutral species. However, the observation of intact quaternary ammonium ions upon intense heating of quaternary ammonium salts in a quadrupole mass spectrometer (in the absence of an electron beam and any strong electric fields) shows that thermal volatilization of ionic species is at least a contributing process in the FD experiment for ionic compounds (114a, 177a).

A simplistic comparison of FIMS and FDMS suggests that in FIMS the molecules are ionized while adsorbed on the electrode's surface. Actually, however, it is likely that the genesis of many FIMS ions involves initial adsorption of the molecule onto the emitting electrode's surface followed by FDMS. Thus, the aforementioned controversy is also relevant to the mechanism of FIMS.

The principal advantage of FDMS over all techniques discussed so far is its ability to produce ions characteristic of intact, unrearranged molecules even when these molecules are nonvolatile or thermally labile (Fig. 14). Disadvantages of the technique include its lower sensitivity versus electron impact mass spectrometry (EIMS), the high degree of skill involved in obtaining reproducible results (163), the difficulties encountered in using the technique for quantitative work (19) and the sometimes transient spectra produced. Recently, other techniques have been developed that do not suffer from some of these handicaps and show great promise for the study of nonvolatile and thermally unstable molecules.

Direct insertion probes have long been used to obtain mass spectra of compounds of moderate volatility and lability. The sample is coated onto a glass or ceramic tip or packed into a glass capillary and then volatilized very near an entrance port of an EI or CI source and allowed to diffuse into the region of ionization. This procedure works satisfactorily only for molecules sufficiently stable and volatile to produce a vapor pressure of $10^{-6}-10^{-7}$ torr without thermal decomposition.

Three distinct improvements can be made in the operation of direct insertion probes to extend their utility. First, a fairly minor redesign of the probe geometry can permit (12, 52, 155) sample volatilization within the ion source, typically 2–3 mm from the electron beam (77, 150) source. This technique, known as "in-beam" ionization, can be used in either an EI or CI source and itself considerably enhances the range of molecular stability and volatility capable of mass spectral analysis. The mechanism of in-beam CI and EI is not well established. The consensus of opinion, however, is that ionization occurs predominantly in the gas phase and not on the probe's surface (44). It is notable that when using an EI source, protonated molecular ions (MH^+) often predominate immediately after sample introduction. Later, the spectrum resembles a conventional EI spectrum. These changes are consistent with changes in the pressure of neutrals within the ion source. A surge of pressure probably occurs immediately after sample introduction resulting in CI-like conditions.

Other modifications of the conventional solid probe technique are designed to increase a sample's volatility by weakening the forces binding sample molecules to each other and to the probe's surface. A significant improvement in volatility can be produced by simply constructing the probe tip of Teflon rather than glass. For example, the energy of activation for vaporization of the tripeptide thyrotropin-releasing hormone (TRH) from glass has been estimated at 60 kcal/M while the corresponding fugure from Teflon is 30 kcal/M (22). Other workers have used the microneedle wire emitters developed for FDMS as inert surfaces in EI or CI sources (93, 153a, 172). A related approach to enhancing volatility is to attempt to disperse sample molecules over a large surface area, minimizing intermolecular interactions. Thus, samples are applied as dilute solutions and the solvent evaporated. It is unclear whether the enhanced volatility that this procedure produces can be entirely ascribed to this effect (23).

By themselves, these modifications of the conventional direct insertion technique of sample introduction extend the range of molecules capable of undergoing mass spectral analysis. However, even greater versatility can be obtained by taking advantage of what might at first glance appear to be a paradox; thermally unstable molecules are often best studied at high temperatures. For example, the spectrum of TRH exhibits a parent ion at m/z 363 and a peak corresponding to thermal decomposition at m/z 235. At 160°C, the artifact peak predominates 3:1. At 215°C, the parent peak is predominant by 2:1 (23). The origin of this effect can be explained if it is assumed that the rate of evaporation of the thermal artifact is determined by the energy of activation for its formation and that the energy of activation for evaporation of the intact molecule is higher. The rate of the higher energy of activation process will increase most rapidly with increasing temperature. At sufficiently high temperatures, vaporization of the intact molecule will be faster than decomposition (3, 22a) (Fig. 47). Thus, when an unstable molecule's mass spectrum is sought, it is advantageous to use rapid heating rates. A typical procedure is to use a copper-tipped probe where temperature is increased at 10°C/s. Under these conditions, the sample will be consumed quickly so rapid scanning (2–3 s) is desirable. Daves et al. (3) attempt to minimize even further the time during which the sample is at temperatures favoring decomposition over volatization; their "flash desorption technique" uses drastically more rapid sample heating (25–1000°C in ca. 0.2 s). Photoplate detection must be used with this technique. Even more rapid heating can be induced by short pulses of laser light (45, 45a, 145). Brief (40-ns) repeated pulses produce localized "hot spots" that desorb sample molecules and cool before substantial pyrolysis can occur. This principle can be used on bulk samples in more or less conventional instrumentation. Alternatively, a com-

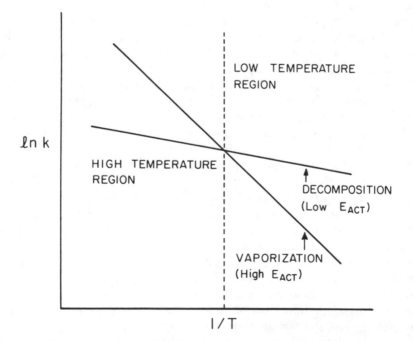

Fig. 47.   Generalized diagram illustrating variation in rate constants for decomposition and volatization with temperature (3). Reprinted with permission from *J. Am. Chem. Soc.*, **100**, 1974 (1978). Copyright, 1978, American Chemical Society.

mercial instrument for laser microprobe analysis can characterize the surface composition of very small areas (180e).

Direct comparison of the relative merits of these techniques is difficult since they have rarely all been applied to the same compounds. However, steroidal glycosides do provide some interesting examples. Androsterone glucuronide, when examined using conventional direct insertion probe techniques, does not exhibit peaks attributable to vaporization of the intact molecule. The in-beam CI spectrum does (Fig. 48); significant $MH^+ - H_2O$ and $MH^+ - 2H_2O$ peaks are observed, although MH^+ ions are absent (45). However, the laser desorption spectrum (Fig. 49) does exhibit a significant MH^+ peak (45). Balanced against the apparently better results obtained with laser desorption is the instrumental simplicity of the in-beam and flash desorption technique. It is likely that modified probes permitting use of the latter techniques on conventional instrumentation will soon be commercially available.

Californium-252 plasma desorption mass spectrometry, though requiring more exotic instrumentation than the rapid heating techniques discussed, utilizes

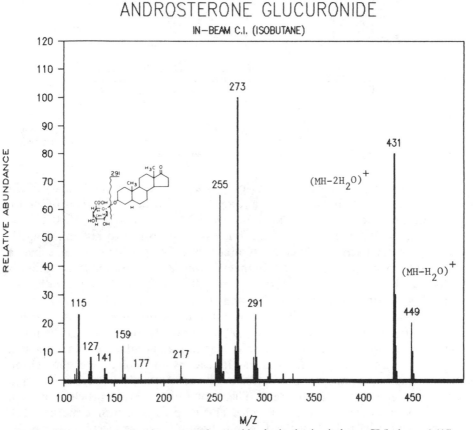

Fig. 48.   Mass spectrum of androsterone glucuronide obtained using in-beam CI (isobutane) (45). Reprinted with permission from *Anal. Chem.*, **52**, 1767 (1980). Copyright, 1980, American Chemical Society.

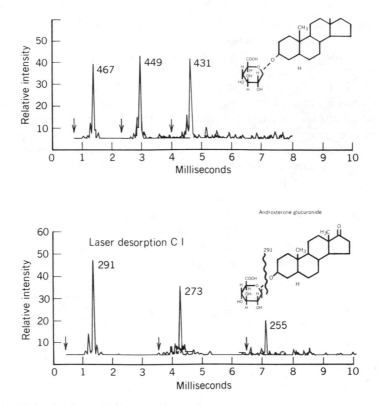

Fig. 49. Molecular ion and fragmentation region mass spectra of androsterone glucuronide obtained using laser desorption. Arrows indicate laser firing time (45). Reprinted with permission from *Anal. Chem.* **52**, 1767 (1980). Copyright, 1980, American Chemical Society.

even more intense local heating for desorption of high-molecular-weight or thermally labile molecules (125a) and appears to be the method of choice for ionizing molecules with molecular weights in excess of 10,000 (177b). Californium-252 undergoes spontaneous fission to form two high-mass, high-kinetic-energy particles travelling in almost exactly opposite directions. These ions are capable of penetrating thin metallic foils and depositing large amounts of thermal energy during their transit time (10^{-12} s). Thus, if sample molecules are thinly coated over a nickel foil and the foil is mounted in a TOF mass spectrometer near a ^{252}Cf source, each collision will produce an 80-Å-diameter plasma region in the sample ($T \sim 10,000$ K). (Actually, the details of the ionization process remain unresolved. For a discussion and leading references, cf. Ref. 39a). Vaporized, highly excited dimers can react by ion–molecule reaction or ion pair formation to produce positive and negative ions capable of acceleration and detection in the TOF spectrometer. The second high-energy particle emitted in the opposite direction can be detected and used as a zero-time marker. The technique has been used for the determination of the mass spectra of thermally labile molecules (126) (Fig. 50) and of very high molecular weight compounds

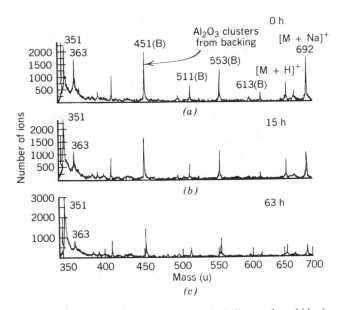

Fig. 50. Californium 252 PDMS positive-ion spectra of 1,2-diester of arachidonic acid, molecule that is unstable at room temperature. Ion at m/z 692 corresponds to the $[M + Na]^+$ adduct and decays with half-life of ca. 20 h (125a).

[e.g., a deoxydodecanucleotide exhibited an $M^+ + Na$ ion at m/z 6301 ± 3, corresponding to $C_{279} H_{249} O_{82} N_{45} P_{11} Cl_1$ Na (137a) and the molecular ion of the protein porcine trypsin was detected at 23,463 (177c)].

During plasma desorption, mega-electron-volt ions pass through and perpendicular to the plane of surface-adsorbed sample molecules. A range of procedures uses a beam of kilo-electron-volt particles directed at a grazing angle to the plane of the adsorbed sample. Discussion of these techniques is complicated by the currently ongoing controversy over nomenclature. In secondary ion mass spectrometry (SIMS) as conventionally performed, the incident particles are ions, usually Cs^+ or K^+ (49), and the sample is an adsorbed solid. This SIMS technique has a history of successful applications to the study of solid inorganic compounds but has been less successful when applied to organic solids of surfaces. The yield of secondary organic ions is sufficiently low that a high flux of primary particles is needed to produce a useful mass spectrum. Under these conditions, the adsorbed solid is quickly damaged, resulting in a transient or altered spectrum.

Fast-atom bombardment (FAB) differs from conventional SIMS in two major ways (13a). Atoms (often Ar or Xe), not ions, are the incident particle in this technique. Although the details of the ionization processes resulting from kilo-electron-volt particle impact remain a subject of research (55a, 118b), it is an empirical fact that the charge state of the incident particle is not the most important distinction between FAB and SIMS (114d). Use of neutral particles does simplify the problem of directing a particle beam at a target within the high-field source region of the spectrometer. However, experiments have clearly demonstrated that this problem can be surmounted and that mass spectra produced by xenon atoms and cesium ion bombardment are virtually identical (1a). Since the cesium ion gun does not produce significant neutral atom emission, its use permits maintenance of lower source pressures when compared to atom guns. Also, since ions are more readily focused than neutrals, the ion beam can be localized more tightly on the sample, producing a more intense spectra. Thus, many observers believe that, on balance, ions are superior to atoms as the incident particle. Instead, the innovation that made FAB applicable to a broad class of medium- and high-molecular-weight solid organic compounds was the use of a low-vapor-pressure viscous liquid (commonly glycerol) to dissolve the sample molecules (13a, 55a). Advantages resulting from the liquid matrix include more intense molecular ions, more persistent spectra, and greater sensitivity (55a, 114c). The more intense molecular ion (i.e., less extensive fragmentation) observed with liquid matrices is explained by postulating incident-particle-induced ejection of large clusters containing analyte and multiple-solvent molecules. Excess energy present in the cluster can be dissipated by loss of solvent molecules, producing a "cool" analyte ion with low probability for further fragmentation (38b). The enhanced yield of ions frequently observed when the sample is dissolved in a liquid matrix is believed to arise from surface tension effects. Molecules that contain both highly polar or ionic groups (hydrophyllic group) and hydrophobic groups can behave as effective surfactants, lowering surface tension of the liquid matrix and thus providing a driving force for the migration of the molecule toward the surface. For example, the prototypical surfactant cetyltrimethylammonium bromide (IX) forms a monolayer on the surface of a glycerol drop at a concentration of 5×10^{-4} mol/liter (13a):

$$CH_3(CH_2)_{15} \overset{\oplus}{N}(CH_3)_3 \overset{\ominus}{Br}$$

IX

Obviously, when sputtering occurs, molecules at the surface are most readily desorbed, so this effect markedly enhances the technique's sensitivity; a useful spectrum of cetyltrimethylammonium bromide can be obtained by placing 10 μl of glycerol containing 100 pmol of the analyte on a probe tip. The more persistent spectra often observed from liquid matrices result from the fact that the droplet serves as a sample reservoir. As the analyte is sputtered from the droplet surface, new analyte molecules from the bulk solution replace them as a result of surface tension effects, convection, and diffusion (55a, 114c).

The root of the SIMS/FAB nomenclature problem lies in the fact that the acronym FAB emphasizes the nature of the incident particle (generally considered to be of secondary importance) and not the presence of a liquid matrix. Thus, bombarding a sample in a liquid matrix with ions might be called "liquid

SIMS," while "FAB" might be used to describe an identical experiment using atoms. This sharp distinction in nomenclature between two experiments that are fundamentally similar is considered objectionable by many workers in the field. An influential review has proposed discarding the term *FAB* and instead using terms such as *liquid SIMS* or *solid SIMS* as appropriate (38c).

Although liquid matrix FAB/SIMS can be applied to a broad range of organic molecules including completely nonionic compounds, its greatest utility is for the measurement of spectra of higher molecular weight compounds with very polar or preionized groups since surface tension effects are maximized for such compounds and since such compounds are intractable by most ionization techniques. In particular, the technique has been successfully applied to many classes of biologically derived molecules since these attributes are characteristic of, for example, polynucleotides, peptides, oligosaccharides, and glycolipids (52a). Some illustrative examples chosen arbitrarily from the near explosion of recent papers featuring the technique include use of negative-ion FAB to sequence an octanucleotide (M–H ion at m/z 2408) (74b), the molecular weight determination of the protein proinsulin (M + H observed at m/z 9390) (13b), and the structure elucidation (in combination with other spectroscopic techniques) of a permethylated glycophospholipid containing 25 sugar units (M + Na ion at m/z 6184) (52a).

E. MIXTURE ANALYSES BY MASS SPECTROMETRY/MASS SPECTROMETRY

Identification or quantitation of a particular component of a complex mixture by conventional mass spectrometry without separation into components is a formidable task. Each component generates its usual characteristic ions. However, in a mixture it is often unclear which ions are associated with a particular component. Further, even if a compound is known to be present, quantitation will be complicated by the possibility that other compounds may also generate the characteristic ions used in quantitation.

The most common solution to the problem of mixture analysis is to interface the mass spectrometer with a device such as a gas chromatograph or liquid chromatograph capable of chromatographically separating the mixture's components and sequentially admitting them into the source of the mass spectrometer. Mass spectrometry/mass spectrometry (MS/MS), though still in its infancy, shows considerable promise as an alternative or complimentary solution.

The conceptually most straightforward and most frequently used mode of MS/MS operation is the daughter scan, which determines all daughter ions arising from a particular parent ion (102a). This is accomplished by using the first mass analyzer to selectively transmit the parent ion of interest into the region where collision-induced dissociation occurs and then scanning the second mass analyzer to permit transmission of different m/z daughter ions. An obvious application of this mode of operation is the confirmation of the structure of a particular ion in the spectrum of a complex mixture.

The parent scan, conversely, determines all parent ions that fragment to generate a specific daughter ion (102a). Much of the utility of this scan results from the tendency of all members of certain classes of compounds to generate a common characteristic ion. For example, a characteristic fragmentation of the $(M + H)^+$ ion of most phthalates generates a peak at m/z 149 (92a, 102a). Thus, performing a parent scan on a mixture's 149 ion will identify potential phthalate molecular ions.

The neutral loss scan involves scanning both mass analyzers simultaneously so that only ions arising from loss of a particular neutral mass are ultimately transmitted. The utility of this scan is largely derived from the fact that certain classes of compounds fragment with loss of a particular neutral species. For example, the molecular ions of chlorinated dibenzodioxins and dibenzofurans fragment with loss of 63 amu (COCl). A neutral loss scan of a mixture for 63 amu would reveal any parent ions potentially in this class of compounds (180c, 180d).

A disadvantage of MS/MS techniques is loss of sensitivity as additional analyzers are interposed between the ion source and the detector. Analyzer transmission efficiencies and ion yields from collision-induced dissociations are often low. The technique of selected reaction monitoring (SRM) is designed to compensate for this problem. In this mode, the analyzers are not scanned. Instead, they are "jumped" among the settings that transmit the few parent–daughter combinations of greatest interest. Since a particular ion or process will be observed far longer in this mode than in a scanning mode, sensitivity can be substantially increased. In a conventional operated mass spectrometer, there is a closely analogous relationship between scanning and ion-monitoring procedures.

A broad range of techniques and instrumentation have been used to generate MS/MS data. Approaches include use of a conventional geometry double-focusing instrument [i.e., an electrostatic analyzer (E) followed by a magnetic analyzer (B)], use of a reversed-geometry double-focusing instrument (i.e., a BE design), as well as use of instruments containing three or more electrostatic, magnetic, and/or quadrupole (Q) analyzers in various combinations (71a). In general, as the instrumentation becomes more elaborate, resolution improves and artifacts become less prevalent. Obviously, however, the conventional two-analyzer instruments are most widely available and have found many interesting applications.

1. MS/MS on Two-Analyzer Instruments

Conventionally, double-focusing mass spectrometers have been constructed so that the electrostatic analyzer precedes the magnet (e.g., Fig. 17). However, the alternative arrangement (Fig. 51) has a major advantage; it simplifies the MS/MS technique considerably. Thus, the magnet can be adjusted to transmit only an ion of a particular mass (e.g., m_1^+). These ions will undergo metastable or collision-induced dissociation in the field-free region between the magnet and the electrostatic analyzer. If m_1^+ fragments to generate m_2^+, its kinetic energy will be

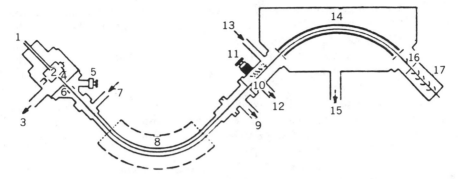

Fig. 51. Ion optical arrangement used in MIKE spectrometer. Note that magnetic sector (8) precedes electrostatic sector (14): 1, sample introduction; 2, ion source; 3, source diffusion pump line; 4, source slit; 5, source isolation valve; 6, y, z deflectors; 7,13, collision gas inlet lines; 8, magnetic sector; 9,12,15, ion pump lines; 10, intermediate slit; 11, intermediate electron multiplier; 14, electrostatic sector; 16, final collector slit; 17, electron multiplier (28). Reprinted with permission from *Anal. Chem.*, **45**, 1023A (1974). Copyright, 1974, American Chemical Society.

decreased by a factor of m_2/m_1. Thus, if the electrostatic analyzer's voltage is maintained at its normal value (E_1), only those ions that have not fragmented will be transmitted to the detector. However, if the electrostatic analyzer's voltage is set at $E_1(m_2/m_1)$, the m_2^+ ions that formed in the first field-free region will be transmitted. In general, as E is swept from E_1 to 0, all daughter ions formed from m_1^+ in the first field-free region will be successively collected. The result is referred to as a mass-analyzed ion kinetic energy (MIKE) spectrum. However, since an ion's kinetic energy is directly related to its mass, the MIKE spectrum is equivalent to a mass spectrum. The structure of the m_1^+ ion can then be confirmed by comparison of its MIKE spectrum to that obtained from an authentic sample. Spectra produced via the MIKE technique usually exhibit precursor ion resolution near 500 (1). However, daughter ion resolution is markedly lower due to collisional scattering and the conversion of internal energy to kinetic energy during fragmentation. Additionally, artifact peaks (arising from second field-free region fragmentations) are sometimes observed (30a, 144a), and scan modes requiring rapid adjustment of the magnetic field are difficult (33a). Reversed-geometry double-focusing instruments capable of performing MIKE techniques are commercially available.

The utility of the MIKE technique in the analysis of natural products is well illustrated in a now classic series of papers related to cocaine (**IX**) and cinnamoyl cocaine (**X**). These compounds are present at the percentage level in the leaves, twigs, and berries of the coca plant. Cooks et al. crushed small (milligram) portions of the plant material of interest under liquid nitrogen and introduced the resulting powder into a CI source via a direct probe (109). Chemical ionization was utilized in order to minimize the number of fragment ions produced from each compound. Nevertheless, the resulting spectrum was sufficiently complex that the alkaloids in the crude material could not be characterized (196). However, if the magnet of the MIKE spectrometer is adjusted to transmit only m/z 304 (the protonated molecular ion of cocaine, **IX**) and the ESA is scanned to detect the CA-induced decompositions of m/z 304, the resulting MIKE spectrum is clearly characteristic of cocaine (Fig. 52). If, on the other hand, a similar experiment is performed on the m/z 330 ion [the $(M + H)^+$ ion of cinnamoyl cocaine], a different but characteristic spectrum is observed (Fig. 53) (108, 109, 196).

Greater sensitivity (although, obviously, less specificity) can be obtained by continuously monitoring one or a few characteristic fragmentations of a particular ion. For example, inspection of Fig. 52 demonstrates that the favored CA-induced fragmentation of protonated cocaine (m/z 304) is loss of benzoic acid to generate m/z 182. If the magnet is adjusted to transmit only m/z 304 and the electrostatic analyzer (ESA) voltage is adjusted to 182/304 of the value that transmits the main beam, only m/z 304 ions that fragment in the first field-free region to generate m/z 182 will be detected. The integrated area of the electron multiplier's output while the sample is being desorbed from the probe is

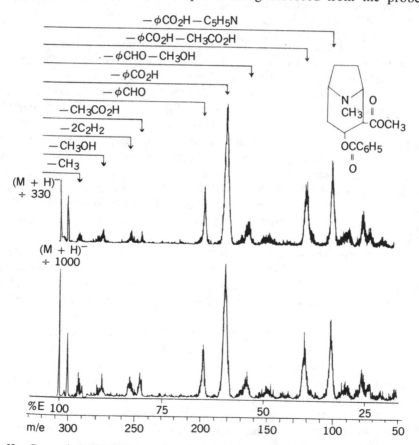

Fig. 52. Comparison of MIKE spectrum of m/z 304 from coca leaves with corresponding MIKE spectrum from authentic cocaine (109). Reprinted with permission from *Anal. Chem.*, **50**, 81A (1978). Copyright, 1978, American Chemical Society.

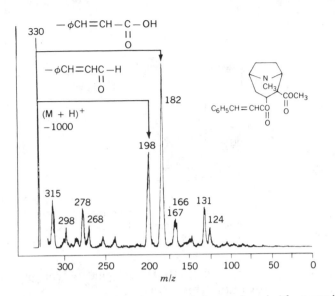

Fig. 53.   A MIKE spectrum of m/z 330 ion (protonated cinnamoyl cocaine) from coca leaves. Note similarities of fragmentation reactions exhibited here and in Fig. 52 (108). Reprinted with permission from *Anal. Chem.*, **50**, 2017 (1978). Copyright, 1978, American Chemical Society.

proportional to the amount of cocaine in the sample. If an appropriate calibration curve has been constructed for the fragmentation of interest from authentic cocaine, reasonably accurate quantitation is possible at the 1-ng level ($\pm 30\%$); more accurate quantitation could probably be obtained if a deuterated analog of cocaine were used as an internal standard (108). Cooks et al. have demonstrated that concentrations of cocaine and cinnamoyl cocaine can be mapped over very small areas (1 mm^3) of various plant tissues. Since little preliminary sample handling is required, large numbers of samples can be processed. This technique may have applications in chemotaxonomy, plant physiology, and forensic chemistry (196).

Some interesting applications of the MIKE procedure to biochemical analysis take advantage of the unusual selectivity exhibited when negative chemical ionization (NCI) techniques are used (122). Since the tendency of a molecule to undergo chloride ion attachment ($(M-Cl)^-$ ion formation) or proton abstraction [$(M-H)^-$ ion formation] varies markedly with structure, NCI can result in considerable simplification of a mixture's mass spectrum. Nevertheless, the mass spectrum of a complex mixture will often not be readily interpretable. For example, Fig. 54a exhibits the m/z 200–235 region in the NCI mass spectrum of a urine sample. However, a MIKE technique permits detection and quantification of individual compounds even in such a morass of peaks. For example, glucose in an isobutene–CH_2Cl_2 NCI mixture generates an intense m/z 215 ion $(M+Cl)^-$ that undergoes CA-induced loss of hydrogen chloride to form m/z 179. If the fragmentation m/z 215 $\rightarrow$ m/z 179 is monitored during the volatization of 1 μl of urine in an NCI source, clear evidence of the presence of glucose can be obtained.

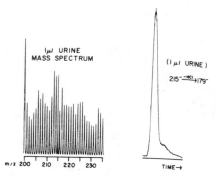

Fig. 54.   (a) Negative CI mass spectra (isobutane–CH_2Cl_2) of 1 μl urine. (b) Negative CI MIKE single-reaction monitoring of the m/z 214 → m/z 179 fragmentation of glucose. Peak corresponds to ca. 100 ng of glucose (122). Reprinted with permission from *J. Am. Chem. Soc.*, **100**, 6045 (1978). Copyright, 1978, American Chemical Society.

The indicated trace corresponds to 100 ng of glucose, roughly the expected amount in 1 μl of normal urine (Fig. 54b) (122).

Analysis of hair samples for drugs of abuse can circumvent the short half-lives exhibited by many of these compounds in blood or urine. Morphine ($C_{17}H_{19}NO_3$, m/z 285), the principal metabolite of heroin, can be detected in crude hair extracts using the MIKE technique (152a). Desorption of a 1-μg hair extract using a direct insertion probe and EI-induced ionization produces a hopelessly complex spectrum (Fig. 55a). However, the CID MIKE spectrum of the m/z 285 ion in a hair sample known to be positive (Fig. 55b) closely resembles the CID MIKE spectrum (Fig. 55c) obtained by spiking a blank hair sample with 5 fg of morphine.

As discussed earlier, MS/MS data can also be obtained from either a conventional (EB) or reversed-geometry instrument by appropriate scans and linked scans of V (accelerating voltage), E, and B. The "constant B/E scan" (which detects daughter ions from a common precursor) and the "constant B^2/E scan" (which detects precursor ions to a common daughter) are most widely used. The resolutions obtainable from conventional geometry scans are complimentary to those obtained using the MIKE technique on a reverse-geometry instrument. Both B/E and B^2/E linked scans produce spectra exhibiting high daughter ion resolution ($M/\Delta M > 500$) but low precursor ion resolution ($M/\Delta M \sim 100$) (75). Because of the low precursor ion resolution, artifact peaks due to transmission of daughter ions arising from precursor ions of adjacent mass will often be observed. Further, in contrast to MIKE spectra, the narrow daughter ion peaks do not provide information about kinetic energy release during fragmentation (104). The important advantage of the linked scan technique is that it utilizes the widely available conventional geometry double-focusing mass spectrometer with only minor modifications.

The linked B/E scan has been used to demonstrate the presence of O-methyl abscisic acid (O–Me–ABA) in methylated and partially purified whole-plant

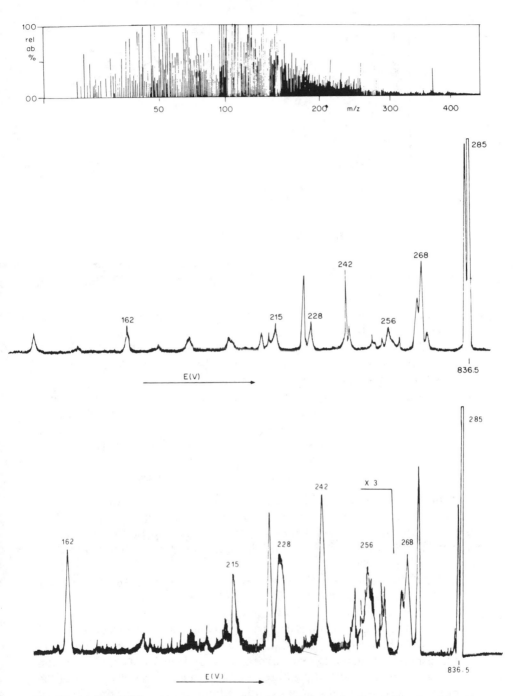

Fig. 55.   (a) A 70-eV EI mass spectrum of positive hair sample extract directly introduced into ion source. (b) A CID MIKE spectra of m/z 285 ions from 1-μg sample of positive hair extract. (c) A CID MIKE spectra of m/z 285 ions from 1-μg sample of positive hair extract (152a).

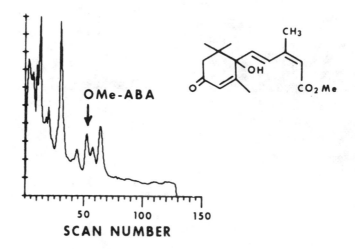

Fig. 56. A TIC profile obtained during GC/MS analyses of methylated sugar cane extract (75). Reprinted with permission from *Anal. Chem.*, **51**, 983 (1979). Copyright, 1979, American Chemical Society.

sugar cane extract. Figure 56 shows the TIC profile obtained by GC/MS analysis of the extract; O–Me–ABA obviously constitutes only a small part of the complex mixture. The same mixture introduced via direct probe with isobutane CI and helium collisional activation gave the linked scan spectrum for m/z 279 (the M + 1 ion of O–Me–ABA) in Fig. 57, which depicts the reference spectrum obtained for authentic O–Me–ABA under identical conditions. The presence of m/z 261, 247, 229, and 205 in the extract's spectrum confirms the presence of O–Me–ABA. The additional peaks at m/z 149 and 167 arise by collision-induced decomposition of m/z 279 ions from dibutyl and dioctylphthalate (75).

2. MS/MS with Three- and Four-Analyzer Instruments

A remarkable variety of three- and four-analyzer instruments have been produced for application to MS/MS. The triple quadrupole and the four-analyzer dual high-resolution mass spectrometer represent the extremes of the commercial MS/MS market (38a). The full range of instrumentation has recently been reviewed (32, 38a).

The triple quadrupole is the most widely used instrument designed expressly for MS/MS studies (Fig. 58 (102a, 193–195). The first analyzer functions as the parent ion selector. It transmits only the ion of interest to the second quadrupole. The second quadrupole functions as a reaction chamber. Since only rf fields are applied to these quadrupoles, all ions, even those undergoing collision, are likely to be transmitted. If unimolecular dissociations are to be studied, the reaction chamber is maintained at low pressure. If more extensive fragmentation of the parent ion is desired, a collision gas can be admitted to that region. The third quadrupole functions as a conventional mass analyzer, determining the m/z values of all parent and daughter ions transmitted through the second quadrupole.

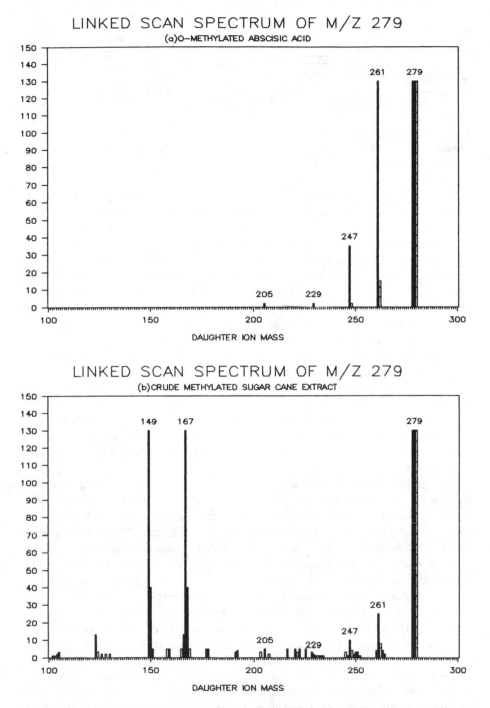

Fig. 57.   Linked-scan CA mass spectrum of (a) m/z 279 in methylated sugar cane extract and (b) m/z 279 ion from authentic O-Me-ABA (75). Reprinted with permission from *Anal. Chem.*, **51**, 983 (1979). Copyright, 1979, American Chemical Society.

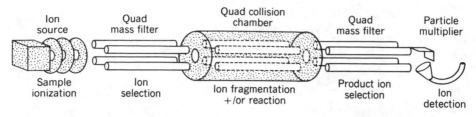

The triple quadrupole will exhibit lower precursor and daughter ion resolution and mass range (ca. 2000–3000) than the double-focusing instruments described in this section. However, the triple quadrupole has a compensating feature. As a result of the relatively low kinetic energies of ions in a quadrupole (< 30 eV) and the continuous-focusing action of a quadrupole, 50–75% of the daughter ions formed in the central quadrupole will be transmitted at a collision gas pressure of 2×10^{-4} torr. Under these conditions, 15–65% of the precursor ions will be induced to fragment. Thus, secondary ion yields of 10–75% are achieved rather than $< 1\%$ efficiencies realized in conventional CA studies. As a result, the instrument is extremely sensitive when operated in the MS/MS mode. The detection limit for the transition $C_6H_5NO_2 \rightarrow C_6H_5^+$ was reported at 120 fg (193). In comparison, Cooks reports a 10-pg detection limit for the $140^+ \rightarrow 123^+$ transition of protonated nitrophenol produced by CI (110, 193) in a magnetic instrument.

Other advantages of the triple quadrupole include more rapid "jumping" of the masses monitored, low initial and maintenance costs, ease of implementing all scan modes, and ease of computer interfacing, all as compared to magnetic instruments.

The quantitation of ergot alkaloids in plasma is a challenging problem because of the low therapeutic dose used. A triple-quadrupole MS/MS procedure has been applied to approximately 1000 samples with sensitivity and precision superior to more conventional methods (78a). Ergotamine, widely used in the abortive treatment of vascular headaches, exhibits an intense fragment peak at m/z 314 in the negative CI mode; the m/z 314 peak undergoes CID dissociation to form m/z 243. A structurally similar internal standard exhibits analogous peaks at m/z 308 and 209. A 1-ml blood sample is spiked with 100 pg internal standard; after a simple extractive work-up, a portion of the extract is applied to the direct insertion probe of a triple-quadrupole mass spectrometer operated in the negative-ion CI mode. The first quadrupole was adjusted to transmit m/z 314 and 308, CID occurs in the second quadrupole, and the third transmits only m/z 243 and 209. The resulting selected reaction chromatograms (Fig. 59) are collected on the manufacturer's supplied data system, and ergotamine concentration is calculated automatically by the internal standard technique. A detection limit of 2 pg/ml is reported.

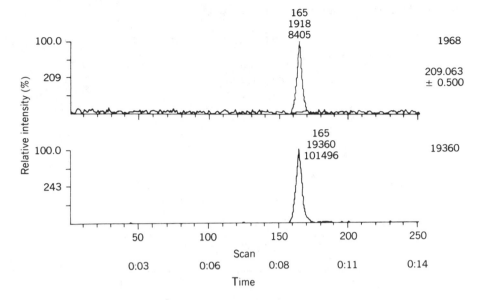

Fig. 59.   (a) Chromatogram generated by monitoring CID-induced transition m/z 314 → m/z 243. fragmentation characteristic of ergotamine, in plasma sample containing 20 pg/ml ergotamine. Triple quadrupole is operated in negative-ion CI mode and sample is introduced from direct insertion probe. (b) Simultaneously generated chromatogram generated by monitoring transition m/z 308 → m/z 209, fragmentation characteristic of ergotamine analog added as internal standard (78a).

Another commercially available triple-analyzer MS/MS unit uses the $E–B–E$ geometry. The first two analyzers function as a high-resolution mass spectrometer. Collision-induced dissociation occurs between the two electrostatic analyzers, and the products are separated by the final electrostatic analyzer. This arrangement allows high-resolution (up to 100,000) separation of ions before CID; however, because of the peak broadening the results from energy release during CID, the final electrostatic analyzer exhibits a mass resolution of about 100 (74a). This instrument has been applied to the analysis of fatty acid mixtures. For example, a mixture of 22 bacterial fatty acids with various structural features (e.g., branching, unsaturation, cyclopropane rings, epoxy groups) was analyzed by FAB MS/MS. Carboxylate anions are readily desorbed by FAB without significant additional fragmentation. Nevertheless, when a complex mixture of acids is FAB desorbed from a probe, an uninterpretable spectrum results (Fig. 60a). However, as Figs. 60b, c demonstrate, if the first two analyzers are adjusted to transmit particular M–H ions, the resulting CID spectra are structurally diagnostic.

Several commercial four-sector instruments exist. A $B_1 E_1 – E_2 B_2$ instrument has been used to determine the structures of acylcarnitines (**X**) in urine obtained from children with certain metabolic disorders:

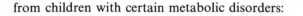

$$R'-\underset{\underset{O}{\|}}{C}-O-\overset{\displaystyle CH_2COOR}{\underset{\displaystyle CH_2\overset{\oplus}{N}CH_3}{CH}}\qquad\begin{array}{l}R'=n\text{-Bu}\\[2ex]R'=i\text{-Bu}\end{array}$$

X

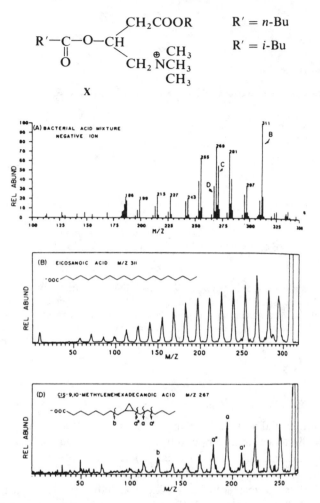

Fig. 60.   (*a*) Mass spectrum of negative ions produced by FAB desorption of 22-component mixture of bacterial acids. (*b*) A CID MIKE negative-ion spectrum of m/z 267 ion (M – H ion of *cis*-9,10-methylenehexadecanoic acid) of mixture. (*c*) A CID MIKE negative-ion spectrum of m/z 311 ion (M – H ion of eicosanoic acid) of mixture (180g). Reprinted with permission from *Anal. Chem.*, **58**, 2429 (1986). Copyright, 1986, American Chemical Society.

Defining the structure of the acyl group, central to understanding the metabolic disorder, is difficult by conventional means. Fast-atom bombardment of the acylcarnitines generates abundant protonated molecular ions. Interestingly, if B_1E_1 is adjusted to transmit the M + H ions, CID dissociation and scanning of E_2B_2 generates spectra diagnostic of the various isomeric acyl groups. For example, Fig. 61 demonstrates discrimination between the isomeric *n*-butyl and isobutyl alkyl groups in authentic samples and confirms the presence of the *n*-

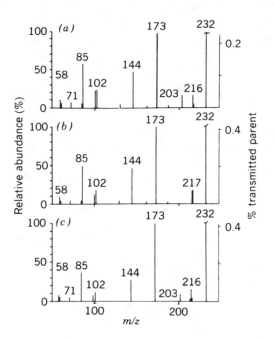

Fig. 61. (a) Daughter ion spectra of protonated n-butylcarnitine (M + H ion at m/z 232). (b) Daughter ion spectrum of isomeric protonated isobutylcarnitine m/z 232. (c) Daughter ion spectrum of m/z 232 in untreated urine sample (70a). Reprinted with permission from *Anal. Chem.*, **58**, 2801 (1986). Copyright, 1986, American Chemical Society.

butyl derivative in an untreated urine sample (70a). In these experiments, $B_1 E_1$ was operated at a resolution of about 1000, and $E_2 B_2$ at about 300; sample sizes typically corresponded to about a microgram of acylcarnitine. Interestingly, an earlier attempt to distinguish isomeric acylcarnitines by linked scanning of first field-free region fragmentation was unsuccessful, a result attributed in part to the poorer signal-to-noise ratios observed using that technique (70a, 141a). Nevertheless, as the authors of these studies note, the complexity and cost of the four-sector instrument make its routine use for mixture analysis somewhat impractical.

GENERAL BIBLIOGRAPHY

Beckey, H. D., *Principles of Field Ionization and Field Desorption Mass Spectrometry*, Pergamon Press Oxford, 1977.

Chemical Abstracts Selects—Mass Spectrometry, Chemical Abstracts Service, American Chemical Society, Columbus, Ohio.

Chapman, J. R., *Practical Organic Mass Spectrometry*, Wiley, Chichester, 1985.

Dawson, P. H., ed., *Quadrupole Mass Spectrometry and Its Application*, Elsevier, Amsterdam, 1976.

Harrison, A., *Chemical Ionization Mass Spectrometry*, CRC Press, Boca Raton, Fl., 1983.

Gaskell, S. J., ed., *Mass Spectrometry in Biomedical Research*, Wiley, Chichester, 1986.

Karasek, F., ed., *Mass Spectrometry in Environmental Sciences*, Plenum Press, New York., 1985.

Mass Spectrometry Bulletin, Mass Spectrometry Data Centre, Aldermaston, England.

McLafferty, F. W., *Interpretation of Mass Spectra*, University Science Books, Mill Valley, Calif., 1980.

McLafferty, F. W., ed., *Tandem Mass Spectrometry*, Wiley, New York, 1983.

McNeal, C. J., ed., *Mass Spectrometry in the Analysis of Large Molecules*, Wiley, Chichester, 1986.

Message, G. M., *Practical Aspects of Gas Chromatography/Mass Spectrometry*, Wiley, New York, 1984.

Millard, B. J., *Quantitative Mass Spectrometry*, Heyden, London, 1978.

Roboz, J., in *Physical Methods of Chemistry*, Vol. 3, Pt. A, (B. W. Rossiter and John F. Hamilton, Eds.) Wiley, New York, 1987, Chapter 5.

Rose, M. E., and Johnstone, R. A. W., *Mass Spectrometry for Chemists and Biochemists*, Cambridge University Press, New York, 1982.

Specialist Periodical Reports in Mass Spectrometry, Royal Society of Chemistry, Burlington House, London.

Waller, G. R., and Dermer, O. C. (eds.), *Biochemical Applications of Mass Spectrometry*, First Supplemental Volume, Wiley, New York, 1980.

Watson, J. T., *Introduction to Mass Spectrometry*, Raven Press, New York, 1985.

REFERENCES*

1.   Abbott, S. J., S. R. Jones, S. A. Weinman, F. M. Bockhoff, F. W. McLafferty and J. R. Knowles, *J. Am. Chem. Soc.*, **101**, 4323 (1979).

1a. Aberth, W., K. M. Straub, and A. L. Burlingame, *Anal. Chem.*, **54**, 2029 (1982).

2.   Allen, J., *Rev. Sci. Instrum.* **18**, 739 (1947).

3.   Anderson, W. R., Jr., W. Frick, and G. D. Daves, Jr., *J. Am. Chem. Soc.*, **100**, 1974 (1978).

4.   Ardenne, M. V., *Kernenergie*, **1**, 1029 (1958).

5.   Ardenne, M. V., and K. Sternfelder, *Kernenergie*, **3**, 717 (1960).

6.   Arpino, P. J., and G. Guiochon, *Anal. Chem.*, **51**, 682 (1979).

7.   Aston, F. W., *Philos. Mag. VI*, **38**, 707 (1919).

8.   Aston, F. W., *Proc. Roy. Soc. London, Ser. A*, **116**, 487 (1927).

9.   Aston, F. W., *Proc. Roy. Soc. London, Ser. A*, **163**, 391 (1937).

10. Baker, A. D., C. Baker, C. R. Brundle, and D. W. Turner, *Int. J. Mass Spectrom. Ion Phys.*, **1**, 285 (1968).

11. Baker, A. D., D. P. May, and D. W. Turner, *J. Chem. Soc. (B)*, **22** (1968).

12. Baldwin, M. A., and F. W. McLafferty, *Org. Mass Spectrom.*, **7**, 1353 (1973).

13. Barber, M., and R. M. Elliott, "12th Annual Conference on Mass Spectrometry and Allied Topics," Montreal, ASTM Committee E-14, 1964.

13a. Barber, M., R. S. Bordoli, G. J. Elliott, R. D. Sedgwick, and A. N. Tyler, *Anal. Chem.*, **54**, 645A (1982).

13b. Barber, M., R. S. Bordoli, G. J. Elliott, N. J. Horoch, and B. N. Green, *Biochem. Biophys. Res. Commun.* **110**, 753 (1983).

14. Barber, N. F., *Proc. Leeds Phil. Soc. Sci. Sect.* 2, 427 (1933).

15. Beckey, H. D., *Int. J. Mass Spectrom. Ion Phys.*, **2**, 500 (1969).

---

* Some references listed are not cited in text.

16. Beckey, H. D., *J. Phys. E: Sci. Instrum*, **12**, 72 (1979).
17. Beckey, H. D., A. Herndricks, and H. U. Winkler, *Int. J. Mass Spectrom. Ion Phys.*, **3**, 9 (1970).
18. Beckey, H. D., and F. J. Comes, in A. L. Burlingame, Ed., *Advances in Analytical Chemistry and Instrumentation*, Vol. 8, Wiley-Interscience, New York, 1970.
19. Beckey, H. D., and H. R. Schulten, in C. Merritt, Jr. and C. N. McEwen, Eds., *Field Ionization and Field Desorption Mass Spectrometry*, Marcel Dekker, New York, 1979.
20. Beckey, H. D., K. Levsen, F. W. Röllgen, and H. R. Schulten, *Surf. Sci.*, **70**, 235 (1978).
21. Bentley, T. W., in *Mass Spectrometry*, Vol. 4, Specialist Periodical Reports, Chemical Society, London, Chapter 2 (1977).
22. Beuhler, R. J., E. O. Flanigan, L. J. Greene, and L. Friedman, *Biochem. Biophys. Res. Commun.* **46**, 1082 (1972).
22a. Beuhler, R. J., and L. Friedman, *Int. J. Mass Spectrom. Ion Proc.*, **78**, 1 (1987).
23. Beuhler, R. J., E. Flanigan, L. J. Greene, and L. Friedman, *J. Am. Chem. Soc.*, **96**, 3990 (1974).
23a. Beynon, J. H., *Biomed. Mass Spectrom.*, **8**, 381 (1981).
24. Beyon, J. H., and A. E. Fontaine, *Z. Naturforsch.* **22a**, 334 (1967).
25. Beynon, J. H., J. A. Hopkinson, and G. R. Lester, *Int. J. Mass Spectrom. Ion Phys.*, **1**, 343 (1968).
26. Beynon, J. H., J. W. Amy, and W. E. Baitinger, *Chem. Commun.*, 1969, 723.
27. Beynon, J. H., R. A. Saunders, and A. E. Williams, *Table of Meta-Stable Transitions for Use in Mass Spectrometry*, Elsevier, 1965.
27a. Beynon, J. H., R. A. Saunders and A. E. Williams, *The Mass Spectra of Organic Molecules*, Elsevier, Amsterdam, 1968.
28. Beynon, J. H., R. G. Cooks, J. W. Amy, W. E. Baitinger, and T. Y. Ridley, *Anal Chem.*, **45**, 1023A (1973).
29. Beynon, J. H., R. M. Caprioli, W. E. Baitinger, and J. W. Amy, *Int. J. Mass Spectrom. Ion Phys.*, **3**, 313 (1969).
30. Beynon, J. H., and R. P. Morgan, *Int. J. Mass Spectrom. Ion Phys.*, **27**, 1 (1978).
30a. Bilton, J. N., N. Kyriakidis, and E. S. Waight, *Org. Mass Spectrom.*, **13**, 489 (1978).
31. Biros, F. J., *Anal. Chem.*, **42**, 537 (1970).
31a. Boese, U., H. J. Neusser, R. Weinkauf, and E. V. Schlag, *J. Phys. Chem.*, **86**, 4857 (1982).
32. Bonelli, E. J., M. S. Story, and J. B. Knight, *Dynamic Mass Spectrom.*, **2**, 177 (1971).
32a. Borman, S. A., *Anal. Chem.*, **58**, 406A (1986).
32b. Borman, S., *Anal. Chem.*, **59**, 769A (1987).
33. Boyd, R. K., and J. H. Beynon, *Org. Mass Spectrom.*, **12**, 163 (1977).
33a. Boyd, R. K., *Spectrosc. Int. J.*, **1**, 169 (1982).
34. Brand, W., and K. Levsen, *Int. J. Mass Spectrom. Ion Phys.*, **28**, 203 (1978).
35. Brewer, A. K., and V. H. Dibeler, *J. Res. Nat. Bur. Stand.*, **35**, 125 (1945).
35a. Brooks, C. J. W. (Ed.) *Gas Chromatography–Mass Spectrometry Abstracts. PRM Science and Technology Agency*, London.
36. Budzikiewicz, H., in C. Merritt, Jr. and C. N. McEwen, Eds., *Mass Spectrometry, Part A*, Marcel Dekker, New York, 1979, Chapter 1.
37. Budzekiewicz, H., C. Djerassi, and D. H. Williams, *Mass Spectrometry of Organic Compounds*, Holden-Day, San Francisco, 1967, Chapter 1.
38. Burlingame, A. L., T. A. Baille, P. J. Derrick, and O. S. Chizhov, *Anal. Chem.*, **52**, 214R (1980).
38a. Busch, K. L., and R. G. Cooksin, in F. W. McLafferty, Ed., *Tandem Mass Spectrometry*, Wiley, New York, 1983, p. 11.
38b. Busch, K. L., B. H. Hsu, Y.-X. Xie, and R. G. Cooks, *Anal. Chem.*, **55**, 1157 (1983).

38c. Burlingame, A. L., T. A. Baille, and P. J. Derrick, *Anal. Chem.*, **58**, 165R (1986).

39. Cairns, T., L. Fishbein, and R. K. Mitchum, *Biomed. Mass Spectrom.*, **7**, 484 (1980).

39a. Chain, B. T., W. C. Agosta, and F. H. Field, *Int. J. Mass Spectrom. Ion Phys.*, **39**, 33a (1981).

39b. Carhart, R. E., D. H. Smith, N. A. B. Gray, J. G. Nourse, and C. Djerassi, *J. Org. Chem.*, **000**, 1708 (1981).

40. Chapman, J. R., G. A. Warburton, P. A. Ryan, and D. Hazelby, *Biomed. Mass Spectrom.*, **7**, 597 (1980).

41. Chupka, W. A., and M. Kaminsky, *J. Chem. Phys.*, **35**, 1991 (1961).

42. Cornu, A., and R. Massot, *Compilation of Mass Spectral Data*, Heyden & Sons, 1966, 1967, 1971, London.

42a. Cooks, R. G., J. H. Beynon, R. M. Caprioli and G. R. Lester, *Metastable Ions*, Elsevier, Amsterdam, 1973.

43. Costa, J. L., *Ann. Phys.*, **4**, 425 (1925).

44. Cotter, R. J., *Anal. Chem.*, **52**, 1589A (1980).

45. Cotter, R. J., *Anal. Chem.*, **52**, 1767 (1980).

45a. Cotter, R. J., *Anal. Chem.*, **56**, 485A (1984).

46. Cotter, R. J., and C. Fenselau, *Biomed. Mass Spectrom.*, **13**, 642 (1978).

46a. Covey, T. R., E. D. Lee, A. P. Bruins, and J. D. Henion, *Anal. Chem.*, **58**, 1451A (1986).

47. "Compilation of Mass Spectral Data." Centre d 'Etudes Nucleaires de Grenoble, France.

48. Crummett, W. B., *Toxicol. Environ. Chem. Revs.*, **3**, 61 (1979).

49. Daves, G. D., Jr., *Accts. Chem. Res.*, **12**, 359 (1979).

50. Dawson, P. H., *Mass Spectrom. Rev.*, **5**, 1 (1986).

51. Day, R. J., S. E. Unger, and R. G. Cooks, *Anal. Chem.*, **52**, 557A (1980).

52. Dell, A., D. H. Williams, H. R. Morris, G. A. Smith, J. Feeney, and G. C. K. Roberts, *J. Am. Chem. Soc.*, **97**, 2497 (1975).

52a. Dell, A., and G. W. Taylor, *Mass Spectrom. Rev.* **3**, 357 (1984).

52b. Della Negra, S., and Y. Lebeyec, *Int. J. Mass Spectrom. Ion Pro.*, **61**, 21 (1984).

52c. Della Negra, S., and Y. Lebeyec, Institut de Physique Nucle'oire, Universite' Paris Sud, Report No. IPNO-DRE-85-01, 1985.

53. de Leenhier, A. P., Ed., *Quantitative Mass Spectrometry in Life Sciences II*, Elsevier, Amsterdam, 1978.

54. de Leenhier, A. P., and R. R. Roncucci, Eds., *Quantitative Mass Spectrometry in Life Sciences*, Elsevier, Amsterdam, 1977.

55. Dempster, A. J., *Phys. Rev.*, **18**, 415 (1921).

55a. DePauw, E., *Mass Spectrom. Rev.*, **5**, 191 (1986).

56. Dempster, A. J., *Phys. Rev.*, **11**, 316 (1918).

56a. Dougherty, R. C., *Anal. Chem.*, **53**, 625A (1981).

57. Dougherty, R. C., J. Dalton, and F. J. Biros, *Org. Mass Spectrom.*, **6**, 1171 (1972).

58. Dougherty, R. C., J. D. Roberts, and F. J. Biros, *Anal. Chem.*, **47**, 54 (1975).

59. Dzidic, I., D. I. Carroll, R. N. Stillwell, and E. C. Horning, *Anal. Chem.*, **47**, 49 (1975).

60. Eadon, G., C. Djerassi, J. H. Beynon, and R. M. Caprioli, *Org. Mass Spectrom.*, **5**, 917 (1971).

61. Eadon, G., and D. Mammato, *J. Org. Chem.*, **40**, 1784 (1975).

61a. Eadon, G., L. Kaminsky, J. Silkworth, K. Aldous, D. Hilker, P. O'Keefe, R. Smith, J. Gierthy, J. Hawley, N. Kim, and A. DeCaprio, *Environ. Health Perspect.*, **70**, 221 (1986).

62. "Eight-Peak Index of Mass Spectra." Mass Spectrometry Data Centre, AWRE, Aldermaston, Berkshire, England, 1975.

63. Emolf, N., and M. S. B. Munson, *Int. J. Mass Spectrom. Ion Phys.*, **9**, 141 (1972).

64. Fales, H. M., G. W. A. Milne, and R. S. Nicholson, *Anal. Chem.*, **43**, 1785 (1971).

65. *Federal Register*, **44**, 69532–52 (Dec. 3, 1979).

66. Feser, K., and W. Kögler, *J. Chro. Sci.*, **17**, 57 (1979).

67. Field, F. H., *Accts. Chem. Res.*, **1**, 42 (1968).

68. Field, F. H., M. S. B. Munson, and D. A. Becker, *Adv. Chem. Ser.*, **58**, 167 (1960).

68a. Field, F. H., *Biomed. Mass Spectrom.*, **12**, 626 (1985).

69. Futrell, J. H., K. R. Ryan, and L. W. Sieck, *J. Chem. Phys.*, **43**, 1832 (1965).

69a. Games, D. E., *Biomed. Mass Spectrom.*, **8**, 454 (1981).

70. Games, D. E., A. H. Jackson, D. S. Millington, and M. Rossiter, *Biomed. Mass Spectrom.*, **1**, 5 (1974).

70a. Gaskell, S. J., C. Guenat, D. S. Millington, D. A. Maltby, and C. R. Roe, *Anal. Chem.*, **58**, 2801 (1986).

71. Geissmann, U., H. J. Heinen, and F. W. Rollgen, *Org. Mass Spectrom.*, **14**, 177 (1979).

71a. Glish, G. L., and S. A. McLuckey, *Anal. Chem.*, **58**, 1889 (1986).

72. Gohlke, R. S., *Anal. Chem.*, **34**, 1332 (1962).

73. Goldstein, E., *Berlin Monats.* 234 (1876).

74. Gomer, R., *Field Emission and Field Ionization*, Harvard University Press, Cambridge, Mass., 1961.

74a. Gross, M. L., E. K. Chess, P. A. Lyon, F. W. Crow, S. Euans, and H. Tudge, *Int. J. Mass Spectrom. Ion Phys.*, **42**, 243 (1982).

74b. Grotjohn, L., R. Frank, and H. Blocker, *Nucleic Acids Res.*, **10**, 4671 (1982).

75. Haddon, W. F., *Anal. Chem.*, **51**, 983 (1979).

75a. Haddon, W. F., *Org. Mass Spectrom.*, **15**, 539 (1980).

76. Hammond, G., *J. Am. Chem. Soc.*, **77**, 334 (1955).

77. Hansen, G., and B. Munson, *Anal. Chem.*, **50**, 1130 (1978).

78. Harris, L. E., W. L. Budde, and J. W. Eichelberger, *Anal. Chem.*, **46**, 1912 (1974).

78a. Haering, N., J. A. Settlage, S. W. Sanders, and R. Schuberth, *Biomed. Mass Spectrom.*, **12**, 197 (1985).

79. Haskins, N. J., D. E. Games, and K. T. Taylor, *Biomed. Mass Spectrom.*, **1**, 423 (1974).

80. Heller, S. R., and G. W. A. Milne, EPA/NIH Mass Spectral Data Base, NSRDS-NBS 63, U.S. Government Printing Office, Washington D.C., 1978.

81. Heller, S. R., H. M. Fales, R. S. Heller, and A. McCormick, *Biomed. Mass Spectrom.*, **1**, 206 (1974).

81a. Henion, J. D., "Micro LC/MS Coupling", in P. Kucera, Ed., *Microcolumn High Performance Liquid Chromatography*, Elsevier, Amsterdam, 1984.

81b. Heller, S. R., *J. Am. Chem. Soc.*, **110**, 3336 (1988).

82. Hennenberg, D., *Adv. Mass Spectrom.*, **8**, 1511 (1980).

83. Hertz, H. S., R. A. Hites, and K. Biemann, *Anal. Chem.*, **43**, 681 (1971).

84. Hipple, J. A., and E. U. Condon, *Phys. Rev.*, **69**, 347 (1946).

85. Holland, J. F., *Org. Mass Spectrom.*, **14**, 291 (1979).

86. Holland, J. B., B. Soltmann, and C. C. Sweeley, *Biomed. Mass Spectrom.*, **3**, 340 (1976).

87. Hoover, H., and H. W. Washburn, *Calif. Oil World*, **34**, 21 (1941).

88. Howe, I., and D. H. Williams, *Chem. Comm.*, 1968, 220.

89. Howe, I., and F. W. McLafferty, *J. Am. Chem. Soc.*, **93**, 99 (1971).

90. Hunt, D. F., *Adv. Mass Spectrom.*, **6**, 517 (1974).

91.   Hunt, D. F., *Prog. Anal. Chem.*, **6**, 359 (1973).

92.   Hunt, D. F., T. M. Harvey, and J. W. Russell, *J. Chem. Soc. Chem. Commun.*, **5**, 151 (1975).

92a.   Hunt, D. F., J. Shabanowitz, T. M. Harvey, and M. L. Coates, *J. Chromatogr.*, **271**, 93 (1983).

93.   Hunt, D. F., J. Shabanowitz, F. K. Boty, and D. A. Brent, *Anal. Chem.*, **49**, 1160 (1977).

94.   Hunt, D. F., G. C. Stafford, Jr., F. W. Crow, and J. W. Russell, *Anal. Chem.*, **48**, 2098 (1976).

95.   Hunt, D. F., G. C. Stafford, Jr., F. W. Crow, and J. W. Russell, *Anal. Chem.*, **48**, 2098 (1976).

96.   "Index of Mass Spectral Data," American Society for Testing and Materials, Philadelphia, 1969.

97.   Isenhour, T. L., B. R. Kowalski, and P. C. Jurs, *Critical Rev. Anal. Chem.*, **4**, 1 (1974).

98.   Jelus, B. L., B. Munson and C. Fenselau, *Biomed. Mass Spectrom.*, **1**, 96 (1974).

99.   Jennings, K. R., in R. A. W. Johnstone, Ed., *Specialist Periodical Reports, Mass Spectrometry*, Vol. 4, Chemical Society, London, 1977, Chapter 9.

100.   Jennings, K. R., in M. L. Gross, Ed., *High Performance Mass Spectrometry: Chemical Applications*, American Chemical Society, Washington, D.C., 1978.

101.   Jennings, K. R., *J. Chem. Phys.*, **43**, 4176 (1965).

102.   Johnson, E. G., and A. O. Nier, *Phys. Rev.*, **91**, 10 (1953).

102a. Johnson, J. V., and R. A. Yost, *Anal. Chem.*, **57**, 785A (1985).

103.   Karasek, F., *Res. Dev.*, **25**, 30 (1974).

104.   Kemp, D. L., R. G. Cooks, and J. H. Beynon, *Int. J. Mass Spectrom. Ion Phys.*, **21**, 93 (1976).

105.   Kim, M. S., and F. W. McLafferty, *J. Am. Chem. Soc.*, **100**, 3279 (1978).

106.   Kingston, D. G. I., J. T. Bursey, and M. M. Bursey, *Chem. Rev.* **74**, 215 (1974).

107.   Kingston, D. G. I., B. W. Hobrock, M. M. Bursey, and J. T. Bursey, *Chem. Rev.*, **75**, 693 (1975).

108.   Kondrat, W., G. A. McCluskey, and R. G. Cooks, *Anal. Chem.*, **50**, 2017 (1978).

109.   Kondrat, W., and R. G. Cooks, *Anal. Chem.*, **50**, 81A (1978).

110.   Kruger, T. L., J. F. Litton, R. W. Kondrat, and R. G. Cooks, *Anal. Chem.*, **48**, 2113 (1976).

111.   Kwok, K. S., R. Venkataraghavan, and F. W. McLafferty, *J. Am. Chem. Soc.*, **95**, 4185 (1973).

112.   Kwok, K. S., R. Venkataraghavan, and F. W. McLafferty, *J. Am. Chem. Soc.*, **95**, 4185 (1973).

113.   Lacey, M. J., and C. G. Macdonald, *Org. Mass Spectrom.*, **12**, 587 (1977).

113a. Lacey, M. J., and C. G. Macdonald, *Anal. Chem.* **51**, 691 (1979).

113b. Lai, S.-T. F., *Gas Chromatography-Mass Spectrometry Operation*, Realistic Systems, East Longmeadow, Mass., 1988.

114.   Lau, C. A., in A. I. Burlingame, Ed., *Topics in Organic Mass Spectrometry*, 1970, pp. 93–120.

114a. Lee, T. D., W. R. Anderson, Jr., and G. Doyle Daves, Jr., *Anal. Chem.*, **53**, 304 (1981).

114b. Ligon, W. V., and R. J. May, *Anal. Chem.*, **58**, 558 (1986).

114c. Ligon, W. F., and S. B. Dorn, *Int. J. Mass Spectrom. Ion Proc.*, **78**, 99 (1986).

114d. Lyon, P. A., Ed., *Desorption Mass Spectrometry. Are SIMS and FAB the Same?*, Am. Chem. Soc., Washington, D.C., 1985.

115.   Linden, H. B., E. Hilt, and H. D. Beckey, *J. Phys. E: Sci. Instrum.*, **11**, 1033 (1978).

116.   Linden, H. B., E. Hilt, and H. D. Beckey, *J. Phys. E*, **11**, 1033 (1978).

117.   Lohr, L. L., and M. B. Robin, *J. Am. Chem. Soc.*, **92**, 7241 (1970).

118.   Lossing, F. P., A. W. Tickner, and W. A. Bryce, *J. Chem. Phys.*, **19**, 1254 (1951).

118a. Lynn, D. G., J. C. Steffens, V. S. Karmut, D. W. Graden, J. Shabanowitz, and J. L. Riopel, *J. Am. Chem. Soc.*, **103**, 1868 (1981).

118b. Magee, C. W., *Int. J. Mass Spectrom. Ion Phys.*, **49**, 211 (1983).

118c. Markey, S. P., *Biomed. Mass Spectrom.*, **8**, 426 (1981).

118d. Martinsen, D. P., and B-H. Song, *Mass Spectrom. Rev.* **4**, 461 (1985).

119. Mather, R. E., and J. F. J. Todd, *Int. J. Mass Spectrom. Ion Phys.*, **30**, 1 (1979).
120. Matsuo, T., H. Matsuda, and I. Katakuse, *Anal. Chem.*, **51**, 69 (1979).
121. Mattauch, J., and R. F. K. Herzog, *Z. Physik*, **89**, 786 (1934).
122. McClusky, G. A., R. W. Kondrat, and R. G. Cooks, *J. Am. Chem. Soc.*, **100**, 6045 (1978).
123. McFadden, W. H., H. L. Schwartz, and S. Evans, *J. Chromatogr.*, **122**, 389 (1976).
124. McFadden, W. J., *J. Chromatogr. Sci.*, **17**, 2 (1979).
124a. McFadden, W. J., *Techniques of Combined Gas Chromatography/Mass Spectrometry*, Wiley, N.Y. 1973.
125. McFarlane, R. D., in G. R. Waller and O. C. Dermer, Eds., *Biochemical Applications of Mass Spectrometry. First Supplementary Volume* Wiley-Interscience, New York, 1980, Chapter 38, p. 1209.
125a. McFarlane, R. D., *Biomed. Mass Spectrom.*, **8**, 449 (1981).
126. McFarlane, R. D., and D. F. Torgerson, *Science*, **191**, 920 (1976).
127. McKinney, J. D., *Ecol. Bull.*, **27**, 53 (1978).
128. McLachlan, N., *Theory and Applications of Mathieu Functions*, Oxford University Press, London, 1951.
129. McLafferty, F. W., *Phil. Trans. R. Soc. London, A*, **293**, 93 (19).
130. McLafferty, F. W., *Acts. Chem. Res.*, **13**, 33 (1980).
130a. McLafferty, F. W., *Biomed. Mass Spectrom.*, **8**, 446 (1981).
130b. McLafferty, F. W., and D. B. Stauffer, *J. Chem. Inf. Comput. Sci.*, **25**, 245 (1985).
130c. McLafferty, F. W., S. Cheng, K. M. Dully, C. J. Guo, I. K. Mun, D. W. Petersen, S. O. Russo, D. A. Solvucci, J. W. Serum, W. Staedeli, and D. B. Stauffer, *Int. J. Mass Spectrom. Ion Proc.*, **47**, 317 (1983).
130d. McLafferty, F. W., and D. B. Stauffer, *Int. J. Mass Spectrom. Ion Proc.*, **58**, 139 (1984).
130e. McLafferty, F. W., and D. B. Stauffer, *J. Chem. Inf. Comput. Sci.*, **25**, 245 (1985).
131. McLafferty, F. W., A. Hirota, M. P. Barbalas, and R. F. Pegues, *Int. J. Mass Spectrom. Ion Phys.*, **35**, 299 (1980).
132. McLafferty, F. W., B. L. Atwater, K. S. Haraki, K. Hosakawa, I. K. Mun, and R. Venkataraghavon, *Adv. Mass Spectrom.*, **8**, 1564 (1980).
133. McLafferty, F. W., and F. M. Bockhoff, *Anal. Chem.*, **50**, 69 (1978).
134. McLafferty, F. W., P. F. Bente III, R. A. Kornfeld, S.-C. Tsai, and I. Howe, *J. Am. Chem. Soc.*, **95**, 2120 (1973).
135. McLafferty, F. W., P. J. Todd, D. C. McGilvery, and M. A. Baldwin, *J. Am. Chem. Soc.*, **102**, 3360 (1980).
136. McLafferty, F. W., and R. Venkataraghavon, in M. L. Gross, Ed., *High Performance Mass Spectrometry*, A. C. S. Symp. Series, Vol. 70, 1978, p. 311.
137. McLafferty, F. W., T. Wachs, C. Lifshitz, G. Innorta, and P. Irving, *J. Am. Chem. Soc.*, **92**, 6867 (1970).
137a. McNeal, C. J., and R. D. Macfarlane, *J. Am. Chem. Soc.*, **103**, 1609 (1981).
138. Meisels, G. G., and R. H. Emmel, *Int. J. Mass Spectrom. Ion Phys.*, **11**, 455 (1973).
139. Meisels, G. G., C. T. Chen, B. G. Geissner, and R. H. Emmel, *J. Chem. Phys.*, **56**, 793 (1972).
140. Melton, C. R., *Principles of Mass Spectrometry and Negative Ions*, Marcel Dekker, New York, 1970.
141. Millard, B. J., Ed., Proceedings of the Third International Symposium on Quantitative Mass Spectrometry in Life Sciences, Gent, Belgium, 1980; *Biomed. Mass Spectrom.* **7** (11, 12) (1980).
141a. Millington, D. S., C. R. Roe, and D. A. Maltby, *Biomed. Mass Spectrom.*, **11**, 236 (1984).
142. Milne, G. W. A., and M. J. Lacey, *Crit. Rev. Anal. Chem.*, **4**, 145 (1974).

142a. Milne, G. W. A., W. L. Budde, S. R. Heller, D. P. Martinsen, and R. G. Oldham, *Org. Mass Spectrom.*, **17**, 547 (1982).

143. Milberg, R. M., and J. Carter Cook, *J. Chro. Sci.*, **17**, 17 (1979).

144. Mitchum, R. K., G. F. Moler, W. A. Kormacher, and W. W. Holtman, *Anal. Chem.*, **52**, 2278 (1980).

144a. Morgan, R. P., C. J. Potter, and J. H. Beynon, *Org. Mass Spectrom.*, **12**, 735 (1977).

145. Mumma, R. O., and F. J. Vastola, *Org. Mass Spectrom.*, **6**, 1373 (1972).

146. Munson, B., *Anal. Chem.*, **43**(13), 28A (1971).

147. Munson, M. S. B., and F. H. Field, *J. Am. Chem. Soc.*, **88**, 2621 (1966).

148. Munson, M. S. B., and F. H. Field, *J. Am. Chem. Soc.*, **88**, 4337 (1966).

149. Nier, A., E. Ney, and M. Ingraham, *Rev. Sci. Instrum.*, **18**, 294 (1947).

150. Ohashi, M., K. Tsujimoto, and A. Yasuda, *Chem. Lett. (Japan).*, **1976**, 439.

151. Okayama, F., and H. D. Beckey, *J. Chem. Phys.*, **69**, 2110 (1978).

152. Paul, W., and H. Sternwedel, *Z. Naturforsch.*, **8A**, 448 (1953).

152a. Pelli, B., P. Traldi, F. Tagliaro, G. Lubli, and M. Marigo, *Biomed. Env. Mass Spectrom.*, **14**, 63 (1987).

153. Pesyna, G. M., and F. W. McLafferty, in F. C. Nachod, J. J. Zuckerman, and E. W. Randall, Eds., *Determination of Organic Structures by Physical Methods*, Vol. 6, 1976.

153a. Roach, J. A. G., A. J. Malatesta, J. A. Sphon, W. C. Brumley, D. Andzejewski, and P. A. Deifuss, *Int. J. Mass Spec. Ion Phys.*, **39**, 151 (1981).

154. Ronn, W. H., *J. Sci. Instrum.*, **16**, 241 (1939).

155. Reed, R. I., and W. K. Reid, *J. Chem. Soc.*, **1963**, 5933.

156. Robinson, P. J., and K. A. Holbrook, in E. Kendrick, Ed., *Advances in Mass Spectrometry*, Vol. 4, Institute of Petroleum, London, 1968, pp. 523–545.

157. Rosenstock, H. M., and C. E. Melton, *J. Chem. Phys.*, **26**, 314 (1957).

158. Rosenstock, H. M., M. B. Wallenstein, A. L. Wahrhaftig, and H. Eyring, *Proc. Natl. Acad. Sci. U.S.A.*, **38**, 667 (1952).

159. Rosenstock, H. M., and M. Krauss, in F. W. McLafferty, Ed., *Mass Spectrometry of Organic Ions*, Academic Press, New York, 1963, p. 2.

160. Roy, T. A., F. H. Field, Y. Y. Lin, and L. L. Smith, *Anal. Chem.*, **51**, 272 (1979).

161. Russell, D. H., D. H. Smith, R. J. Wahmack, and L. K. Bertram, *Int. J. Mass Spectrom. Ion Phys.*, **35**, 381 (1980).

162. Schoen, A. E., R. G. Cooks, and J. L. Wiebers, *Science*, **203**, 1249 (1979).

163. Schulten, H. R., *Int. J. Mass Spectrom. Ion Phys.*, **32**, 97 (1979).

164. Schulten, H. R., and U. Schurath, *J. Phys. Chem.*, **79**, 51 (1975).

165. Schulten, H. R., W. D. Lehmann, and D. Haaks, *Org. Mass Spectrom.*, **13**, 361 (1978).

166. Semrau, G., and J. Heitbaum, *Anal. Chem.*, **51**, 1998 (1979).

167. Simons, D. S., B. N. Colby, and C. A. Evans, Jr., *Int. J. Mass Spectrom. Ion Phys.*, **15**, 291 (1974).

167a. Small, G. W., *Anal. Chem.*, **59**, 535A (1987).

168. Smit, A. L. C., and F. H. Field, *J. Am. Chem. Soc.*, **99**, 6471 (1977).

169. Smith, D. H., *Computer Assisted Structure Elucidation*, American Chemical Society, Washington, D.C., 1977.

170. Smith, D. H., and R. E. Carhard, in M. L. Gross, Ed., *High Performance Mass Spectrometry*, Vol. 70, A. C. S. Symp. Series, 1978, p. 325.

170a. Smith, D. H., N. A. B. Gray, J. G. Nourse, and C. W. Crandall, *Anal. Chem. Acta*, **133**, 471–497 (1980).

171. Smith, R. M., P. W. O'Keefe, K. M. Aldous, and J. E. O'Brien, *Environ. Sci. Technol.* **17**, 6 (1983).

172. Soltmann, B., C. C. Sweeley, and J. F. Holland, *Anal. Chem.*, **49**, 1164 (1977).

173. Stenhagen, E., S. Abrahamson, and F. W. McLafferty, *Registry of Mass Spectral Data*, Wiley-Interscience, New York, 1974.

174. Stephens, W. E., *Phys. Rev.*, **45**, 513 (1934).

175. Stevenson, D. P., *Disc. Faraday Sco.*, **10**, 35 (1951).

176. Stimpson, B. P., D. S. Simon, and C. A. Evans, Jr., *J. Phys. Chem.*, **82**, 660 (1978).

177. Stimpson, B. P., and C. A. Evans, Jr., *Biomed. Mass Spectrom.*, **5**, 52 (1978).

177a. Stoll, R., and F. W. Rollgen, *J. Chem. Soc. Chem. Commun.*, 789 (1980).

177b. Sundquist, B., and R. D. Macfarlane, *Mass Spectrom. Rev.*, **4**, 421 (1985).

177c. Sundquist, B., P. Roepstorff, J. Fohlman, A. Hedin, P. Hakansson, I. Kamensky, M. Lindberg, M. Solehpour, and G. Save, *Science*, **226**, 696 (1984).

178. Tannenbaum, H. P., J. D. Roberts, and R. C. Dougherty, *Anal. Chem.*, **47**, 49 (1975).

178a. Terwilliger, D. T., A. L. Behbehani, J. C. Ireland, and W. L. Budde, *Biomed. Environ. Mass. Spectrom.*, **14**, 263 (1987).

179. Thompson, J. J., *Rays of Positive Electricity and Their Application to Chemical Analyses*, Longmans Green, London, 1913.

180. Thompson, J. J., *Philos. Mag. VI*, **13**, 561 (1907).

180a. USEPA, "Test Methods for Evaluating Solid Waste. Vol. 1B Laboratory Manual. Physical/Chemical Methods," November, 1986.

180b. USEPA, "Interim Evaluation of Health Risks Associated with Emissions of Tetrachlorinated Dioxins from Municipal Waste Resource Recovery Facilities," November 1981.

180c. Tondeur, Y., W. N. Niederhut, J. E. Campana, and S. R. Missler, *Biomed. Environ. Mass Spectrom.*, **14**, 449 (1987).

180d. Voyskner, R. D., J. R. Hass, G. W. Sovocool, and M. M. Bursey, *Anal. Chem.*, **55**, 744 (1983).

180e. Verbueken, A. H., F. J. Bruynseels, and R. E. Van Grieken, *Biomed. Mass Spectrom.*, **12**, 438, (1985).

180f. Vestal, M. L., *Science*, **226**, 275 (1984).

180g. Tomer, K. B., N. J. Jensen, and M. L. Gross, *Anal. Chem.*, **58**, 2429 (1986).

181. V G Micromass, Winsford, England, Varian Associates, Palo Alto, California.

182. Wachs, T., C. C. Van de Sande, P. F. Bente III, P. P. Dymerski, and F. W. McLafferty, *Int. J. Mass Spectrom. Ion Phys.*, **23**, 21 (1977).

183. Wachs, T., P. F. Bente III, and F. W. McLafferty, *Int. J. Mass Spectrom. Ion Phys.*, **9**, 333 (1972).

184. Warren, J. W., *Nature*, **165**, 810 (1950).

185. Washburn, H. W., H. F. Wiley, and S. M. Rock, *Ind. Eng. Chem. Anal. Ed.*, **15**, 541 (1943).

186. Washburn, H. W., H. F. Wiley, S. M. Rock, and C. E. Berry, *Ind. Eng. Chem. Anal. Ed.*, **17**, 74 (1945).

187. Weston, A. F., K. R. Jennings, S. Evans, and R. M. Elliott, *Int. J. Mass Spectrom. Ion Phys.*, **20**, 317 (1976).

188. Wiley, W. C., and I. H. McLaren, *Rev. Sci. Instr.*, **26**, 1150 (1955).

188a. Willoughby, R. C., and R. F. Browner, *Anal. Chem.*, **56**, 2626 (1984).

189. Wilson, M. S., I. Dzidic, and J. A. McCloskey, *Biochem. Biophys. Acta*, **240**, 632 (1971).

190. Wilson, M. S., and J. A. McCloskey, *J. Am. Chem. Soc.*, **97**, 3436 (1975).

191. Winkler, H. U., and H. D. Beckey, *Org. Mass Spectrom.*, **6**, 655 (1972).

192. Yeo, A. N. H., and D. H. Williams, *J. Am. Chem. Soc.*, **92**, 3797 (1970).

193.　Yost, R. A., and C. G. Enke, *Anal. Chem.*, **51**, 1251A (1979).

194.　Yost, R. A., and C. G. Enke, *J. Am. Chem. Soc.*, **100**, 2274 (1978).

195.　Yost, R. A., C. G. Enke, D. C. McGilvery, D. Smith, and J. D. Morrison, *Int. J. Mass Spectrom. Ion Phys.*, **30**, 127 (1979).

196.　Youssefi, M., R. G. Cooks, and J. L. McLaughlin, *J. Am. Chem. Soc.*, **101**, 3400 (1979).

196a.　Zerbi, G., and J. H. Beynon, *Org. Mass Spectrom.*, **12**, 3 (1977).

196b.　Zakett, D., A. E. Schoen, R. W. Kondrat, and R. G. Cooks, *J. Am. Chem. Soc.*, **101**, 6781 (1979).

Chapter 2

RECENT DEVELOPMENTS IN EXPERIMENTAL FOURIER TRANSFORM–ION CYCLOTRON RESONANCE

By C. D. HANSON, E. L. KERLEY, and D. H. RUSSELL

Department of Chemistry, Texas A&M University, College Station, Texas

Contents

I. INTRODUCTION

The field of instrument development has major impact in all areas of science, and much of the progress in chemistry during the 1970s and the early part of the 1980s can be traced to the development of new instrument methods or instrument concepts. For example, sensitive chromatographic methods and spectroscopic probes have greatly advanced the fields of biochemistry and more recently molecular biology. Similarly, a large number of new experimental methods have been introduced in the field of surface chemistry and catalysis, and our understanding in these areas is advancing at a rapid rate.

Since the early 1970s technological advances have led to the development of a number of instruments for mass spectrometric analysis. The commercial availability of low-cost, reliable mass spectrometers fostered rapid development of gas chromatography–mass spectrometry (GC–MS) (140), and that trend is now continuing with the development of liquid chromatography–mass spectrometry (LC–MS). Concurrently, there has been rapid development of sophisticated high-resolution instruments and multisector instruments for mass spectrometry–mass spectrometry (MS–MS). Advancements in ionization methods suitable for nonvolatile, thermally labile, and polar compounds (viz. fast-atom bombardment (FAB) ionization) have greatly advanced the development of the new generation of sector mass spectrometers.

It is against this backdrop, a period of rapid evolution of long established instrument concepts, that Fourier transform–ion cyclotron resonance (FT–ICR) emerged. Although ICR originated in the field of ion–molecule reactions, the concept was immediately recognized as potentially useful analytically (160). The analytical utility of FT–ICR can be attributed to the performance characteristics of the instrument [e.g., sensitivity (76), mass resolution (4), mass range (24), multichannel advantage (108), the versatility of the instrument for various ionization methods (96)] and the ability to perform tandem mass spectrometry experiments (117). Since Comisarow and Marshall demonstrated the concept of FT–ICR (38), the performance characteristics predicted for the technique have for the most part been demonstrated (98, 154). In a review published in 1984, Gross and Remple discussed in detail the issues of mass resolution, high mass limit, dynamic range, and accurate mass measurements (64). Although ultra-high resolution and simultaneous ion detection over a wide mass range are frequently cited strengths, another potential advantage of FT–ICR as an analytical method is related to its intrinsic versatility. For example, the ability to form ions by a variety of ionization methods [e.g., electron impact ionization, ion–molecule reactions (159), multiphoton ionization (112), laser desorption (146, 162), Cs^+ ion SIMS (6, 22, 23), and ^{252}Cf fission fragment ionization (101)], store the ions for extended periods of time, and probe the ions using a variety of structural characterization methods [e.g., ion–molecule reaction (14), collision-induced dissociation (CID) (31), and photodissociation (44)] has many advantages for identification and characterization of complex samples. Following the adaptation of suitable ionization methods for large involatile, thermally labile biomolecules, a large thrust in the research surrounding FT–ICR development has emphasized detection at high mass and high resolution at high mass (24). The motivation for high-mass applications can be attributed to the method of ion detection (i.e., the detector does not discriminate against high mass), the extended mass range of the instrument, and the promise of high resolution at high mass (160). Ion detection by ICR methods is mass independent (24). The mass range of FT–ICR is directly proportional to the magnetic field strength and inversely dependent on the minimum detectable frequency [see Eq. (1)].

$$\left(\frac{m}{z}\right)_{max} = 1.536 \times 10^7 \frac{B}{f_{(min)}} \tag{1}$$

It is the interference by low-frequency (< 10 kHz) noise that ultimately limits the upper mass range of the instrument. With suitable precautions to shield the instrument from low-frequency noise it is possible to operate in the 3–4 kHz range (22), corresponding to an upper mass limit of ca. 35,000 ($B = 3$ T). Inherent flexibility and wide mass range combined with the ultra-high-mass resolution makes FT–ICR attractive as an analytical method. In contrast to sector instruments, in which mass resolution is dependent on the dimensions of mechanical slits, the FT–ICR mass resolution is dependent on the magnetic field, the observation time of the time-domain signal (t), and inversely dependent on the mass of the ion [Eq. (2)].

$$\frac{M}{\Delta M} = 1.773 \times 10^4 \frac{Bt}{M} \tag{2}$$

The time-domain signal is damped by ion-neutral collisions, therefore the collisional damping rate scales linearly with pressure (typical values for the collisional damping are 1×10^{-2} cm^2 s^{-1} molecule^{-1} torr^{-1}). For example, at pressures of 1×10^{-7} torr and 1×10^{-8} torr, the average time between ion-neutral collisions is approximately 300 ms and 3 s, respectively. The mass resolution that can be obtained at these pressures differ by an order of magnitude. The spectrum shown in Fig. 1 illustrates the sensitivity and mass resolution that can be obtained by FT–ICR (76). The spectrum was acquired from a 100-pmol sample of neurotensin ($m/z = 1674$ amu) dissolved in a glycerol–thioglycerol matrix. In the high-resolution narrow-band spectrum (inset of Fig. 1), a mass resolution of ca. 90,000 is obtained with a signal-to-noise ratio of greater than 500:1.

It is also important to keep in mind that FT–ICR is equally suited for more conventional analytical applications. For example, the typical FT–ICR analyzers are maintained at ultra-high vacuum (e.g., 10^{-9}–10^{-11} torr), thus the systems are relatively free of background except for residual gases. The ultra-high vacuum also facilitates the analysis of low-vapor-pressure samples. As an example, Marshall has demonstrated the advantages of ultra-high-vacuum systems for the analysis of low-volatility organometallics (109). The amount of heating required to generate 10^{-6} torr vapor pressure of many organics and

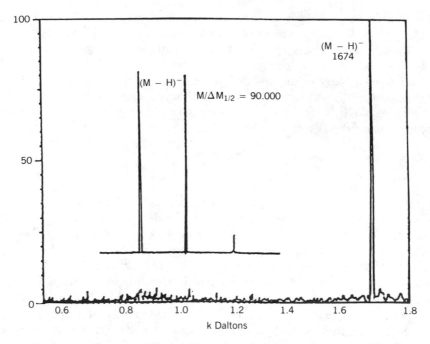

Fig. 1.   A spectrum of neurotensin (m/z 1674 amu) obtained from a 100-pmol sample. The high-resolution narrow-band spectrum (inset) is obtained at a mass resolution of ca. 90,000 with a signal-to-noise ratio of greater than 500:1. [Reprinted by permission (76)]

organometallics in conventional mass spectrometer ion sources causes thermal degradation; conversely, it is possible to sublime many of these samples at 10^{-9}–10^{-10} torr. Similar arguments can be developed for other uses of FT–ICR. These advantages have been discussed in detail in the open literature and in recent reviews (64, 96, 109, 137). We will not attempt to duplicate all the information, instead our approach will be to emphasize two very recent modifications of the FT–ICR instrument, viz. two-section ion cells (61) and external ion sources (3, 92, 94). These devices play an integral part in the future development of the method, both in terms of general analytical applications and the analysis of high-mass molecules. In the subsequent section, recent advances in chemical studies by FT–ICR will be presented with emphasis on (a) experimental techniques, (b) cluster chemistry, and (c) collision-induced dissociation and photodissociation studies.

II.   INSTRUMENTAL CONSIDERATIONS

A.   PRINCIPLES

The application of Fourier transform methods to ion cyclotron resonance has greatly changed the analytical utility of the method (35, 38). The major limitation of the early ICR was slow data acquisition rates, due to the fact that the spectrum was acquired by sweeping the magnetic field and the resonance signal was detected by a marginal oscillator. The method of ion detection limited the mass resolution and the upper mass range (ca. m/z 200). Transform methods provide advantages in both data collection and data manipulation (108). For example, the signal-to-noise ratio can be enhanced by filtering the digitized data, and resolution can be varied by signal observation time or by increasing the number of data points per time-domain transient. In addition, simultaneous acquisition of an entire spectrum yields an increase in the signal-to-noise ratio of up to $N^{1/2}$ (i.e., Fellgett's advantage), where N is the number of data points in the spectrum.

To illustrate the factors affecting the FT–ICR experiment, it is helpful to examine the physics of ion motion in an ICR cell. A charged particle in a magnetic field (B) moves in a circular orbit of angular frequency (ω) in a plane perpendicular to the magnetic field lines. The forces acting on an ion in a static magnetic field are described by Eq. (3).

$$\mathbf{F} = qv_{xy}\mathbf{B} \tag{3}$$

$\mathbf{F}$ is the Lorentz force generated by the magnetic field (B) acting on an ion of charge q, and v_{xy} is the velocity of the ion perpendicular to the magnetic field lines (i.e., the component of the velocity vector in the $X–Y$ plane). The radius of the cyclotron orbit is obtained by equating the Lorentz force to the centrifugal force [Eq. (4)],

$$qv_{xy}\mathbf{B} = \frac{mv_{xy}^2}{r} \tag{4}$$

thus,

$$r = \frac{mv_{xy}}{q\mathbf{B}}$$

where r is the radius of the orbit and m is the mass of the ion. The motion of the ion parallel to the magnetic field lines (i.e., the component of the velocity vector along the Z-axis) is superimposed on the radial motion [Eq. (4)] causing the ion to have a helical motion (see Fig. 2) (128). The pitch of the helix (h) is a function of the velocity of the ion parallel to the magnetic field (v_z). Because the frequency of the cyclotron motion is inversely proportional to m [Eq. (5)], the m/z value of an ion can be determined by measuring the frequency of the cyclotron motion in a static magnetic field.

$$\omega_c = \frac{v_{xy}}{r} = \frac{q\mathbf{B}}{m} \tag{5}$$

An essential feature of this relationship is that the cyclotron frequency is related to the mass-to-charge ratio and the magnetic field strength and independent of the translational energy. It is also important to note that for a given mass-to-charge ratio and magnetic field strength the radius of the cyclotron orbit is related only to the momentum in the $X–Y$ plane. Signal detection was

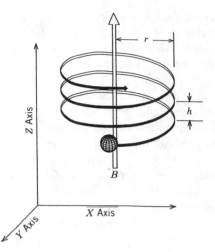

Fig. 2.   The motion of an ion in a magnetic field (B) is a helix of pitch h and radius r.

accomplished on early ICR instruments by measuring the power absorption of an ion using a marginal oscillator detector. The FT–ICR signal is obtained by measuring the image current induced in a radio-frequency (rf) receiver. The detected image current is generated by a coherently cycloiding packet of ions. Thus, the key component of the FT–ICR experiment is excitation of an initially random ensemble of ions into a motion that is coherent.

Ions having a particular mass-to-charge ratio are accelerated by applying an oscillating electric field resonant with the cyclotron frequency. As ions gain translational energy in the X–Y plane, their cyclotron frequency remains the same so that they are accelerated to a larger cyclotron orbit (14, 18). The translational energy (T_{x-y}) of the ions as a result of irradiation by an rf field such that ω_f (the frequency of the rf field) $= \omega_c$ is given by Eq. (6).

$$T_{x-y} = \frac{q^2 E^2 t^2}{8m} + \frac{qEv_0 t}{4}(\cos \gamma) \qquad (6)$$

E is the amplitude of the oscillating electric field, t is the time of irradiation, v_0 is the initial magnitude of the velocity, and γ is the phase relationship of the cyclotron frequency with respect to the phase of the rf excitation (see Fig. 3). Because the ions are typically formed over a period of several milliseconds (many cyclotron orbits), the ions have a random distribution of phase angles (γ) between 0 and 2π. The significance of the second term in Eq. (6), is that rf excitation ($\omega_f = \omega_c$) produces translationally excited ions having a range of energies (i.e., varying radii) that is dependent on the initial energy (v_0) and phase angle (γ). Because γ is random between 0 and 2π, the average contribution of the second term of Eq. (6) will be 0. Equation (6) can therefore be solved to obtain the average translational energy of the ion population by ignoring the second term.

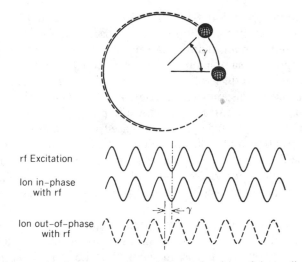

Fig. 3.   The phase angle γ between two ions relative to the phase of the applied rf excitation.

The energy spread, and therefore radial distribution, induced by γ increases as the initial energy (v_0) increases. The radial distribution caused by excitation is superimposed on the initial spatial distribution of the ion population prior to excitation. For example, the ion population formed by electron impact has a small initial spatial distribution relative to the ion cell and is produced with near thermal translational energies (v_0 is small). These initial conditions corresponds to a narrow energy spread and therefore small radial distribution of ions after excitation. A small spatial distribution relative to the radius of the cyclotron orbit corresponds to a coherent packet of ions. It is the motion of a coherent packet of ions that is essential to the capacitance bridge detection utilized by FT–ICR.

B.   PRESSURE LIMITATIONS

1.   Pressure and Resolution

As noted in Eq. (2), the mass resolution that can be achieved by FT–ICR is determined by the duration of the time-domain transient signal. Because the duration of the time-domain signal is a function of the pressure of the analyzer, ultra-high-vacuum conditions must be used so that the coherent motion of the ion population is not lost through ion-neutral collisions. Therefore, a principal instrumental limitation of ion detection is the high-vacuum requirements of the analyzer. Because of the inverse pressure detection relationship, ion sources are by definition incompatible with FT–ICR detection.

The most logical solution for eliminating the pressure limitations of FT–ICR is to separate sample introduction and mass analysis. Separation of the sample introduction and mass analysis can be achieved by separating the two steps in time or by spatial separation. Sack and Gross were the first to propose time separation of sample introduction by use of pulsed valves (139). The approach used was to admit a small quantity of sample by the pulsed valve, ionize the sample at the optimum sample pressure level, then delay mass analysis until the analyzer pressure had returned to normal (e.g., 10^{-8} torr or less). An additional advantage realized by this approach is control of the dynamic range afforded by performing ionization at various points in the pressure profile. Frieser further demonstrated the advantage of the pulsed-valve method for studying ion–molecule reaction and collision-induced dissociation (21). In each of the approaches the desired result is coupling a high-pressure (10^{-7} torr or greater) experimental step (e.g., ionization, ion–molecule reaction, or collision-induced dissociation) with ion detection under optimum, low-pressure (10^{-8} torr) conditions.

Additional steps may be taken to separate sample introduction from mass analysis. However, it is desirable to retain as many of the basic features of the ICR analyzer as possible. That is, in separating the ion source and analyzer regions it is imperative that ions are efficiently transported to the analyzer. For instance, the low transmission efficiencies of ion sources for sector instruments can be circumvented by operating the ion source under conditions that form large numbers of ions. However, such options are less useful for FT–ICR because of the sensitivity of the method to space charge effects. Although it is possible to operate the ICR ion cell under conditions where large ion densities are produced, prior to detection the ion population must be reduced by either ion ejection or lowering the electrostatic trapping potential to allow some fraction of the ions to escape along the Z axis.

2. Two-Section Cells

A second approach for linking a high-pressure source region with an ultra-high-vacuum analyzer region is the two-section ion cell developed by Ghaderi and Littlejohn (61). The two-section cell is comprised of two contiguous cells separated by a vacuum restriction. Normally, the two vacuum regions are differentially pumped (see Fig. 4). A pressure differential between the two ion cells of approximately 3 orders of magnitude can be maintained for a 1-mm radius

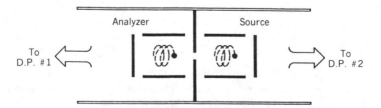

Fig. 4.   A schematic of a two-section ion cell comprised of two contiguous cells that share a common center trap plate. An orifice in the center trap plate acts as a conductance limit and therefore permits differential pumping of the two regions.

orifice in the conductance limit. Ions are produced in the high-pressure source region and partitioned between the two regions by drifting through the aperture of the conductance limit. The partitioning is accomplished by grounding the center trap plate, thus permitting the ions to be accelerated by the trapping field in the Z direction. Since the motion of the ions in the Z direction can be approximated by a harmonic oscillator, this type of partitioning of ions creates an oscillation of the observed signal in the analyzer region as they pass back and forth between the cells (152).

Although the principles of a two-section cell are straightforward, the experiment does introduce complications, for example, increased Z-axis translational energy of the ions and increased longitudinal interaction of the ions with magnetic field lines. Thus, the alignment of the ion cell with the magnetic field lines is critical. During the partitioning event, the electrostatic field lines defining the ion trap are altered to allow the ions to pass through the orifice of the conductance limit. The field lines that exist during partitioning form a gradient that accelerates the ions in the Z direction (see Fig. 5). This electrostatic acceleration corresponds to an increase in the translational energy of the trapped ions. The extent of the electrostatic acceleration is a function of the trapping voltage (V_{TR}) and the location of the ion on the electrostatic gradient. Therefore, ions partitioned prior to detection will have a translational energy spread between near-thermal translational energy (e.g., translational energy of ions formed in a single-cell experiment) and V_{TR}.

The force imposed on an ion in a magnetic field is defined by Eq. (7):

$$\mathbf{F} = qv\mathbf{B} \sin \phi \tag{7}$$

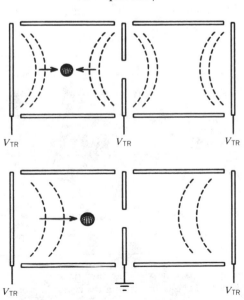

Fig. 5. Electric field gradients caused by the trapping voltage accelerate ions upon partitioning.

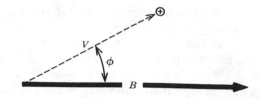

Fig. 6.   The force imposed on a ion in a magnetic field is defined by $\mathbf{F} = q v \mathbf{B} \sin \phi$ where ϕ is the angle between the ion's velocity vector and the direction of the magnetic field.

where ϕ is the angle between the direction of the ion motion and the direction of the magnetic field (see Fig. 6). Ions moving parallel with the magnetic field lines (e.g., $\phi = 0$) experience no force due to the magnetic field. Because the Lorentz force of the magnetic field increases with ϕ, ions tend to follow the magnetic field lines. Ions produced along a magnetic field line that does not transverse the orifice of the conductance limit are lost by collisions with the cell wall (see Fig. 7). The ability to partition ions between cells is a function of the relative dimensions of the initial ion population, the dimension of the orifice of the conductance limit, and the alignment of the conductance limit with respect to the magnetic field lines (θ). This effect can be modeled by the motion of an ensemble of ions produced along the $Z = 0$ axis of the ion cell and then partitioned through conductance limit. At small θ all ions are partitioned through the conductance limit orifice (Fig. 8a). At larger angles of θ (e.g., misalignment of the ion cell) ion losses occur because of collisions with the cell wall (Fig. 8b).

In the case of the single cell, ions are produced with near-thermal kinetic energies in the confines of an electrostatic trap. Because of the strong dependence of electrostatic and magnetic fields with respect to distance (e.g., $1/r^2$ and $1/r^3$,

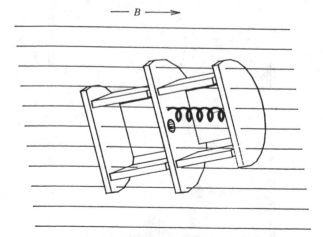

Fig. 7.   Ions formed along a magnetic field line that does not transverse the orifice of the conductance limit will be lost due to collisions with the center trap plate. Because of the strong dependence of ion motion on the direction of magnetic field lines, alignment of a two-section cell is important.

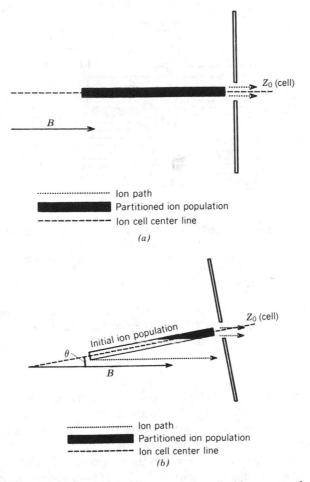

Fig. 8.   The effect of the alignment angle of the two-section cell with respect to the magnetic field lines on partitioning efficiency in (a) a perfectly aligned two-section cell and (b) a relative alignment angle of θ.

respectively), field imperfections generated by the cell walls will have negligible effect on the ions in the trap. The impact of field imperfections becomes increasingly important in terms of ion loss as the ions pass through the orifice of the conductance limit. The effect of magnetic inhomogeneities as ions are partitioned through an orifice composed of magnetically susceptible material will limit the efficient transportation of ions from a source to analyzer region (20). This effect is modeled by the redirection of the magnetic field lines by magnetically susceptible material (see Fig. 9). Owing to the magnetic permeability of 316 stainless steel, magnetic field lines preferentially permeate the cell material. Because ions follow the magnetic field lines, ions produced along these redirected field lines are lost by collisions with the cell wall resulting in a

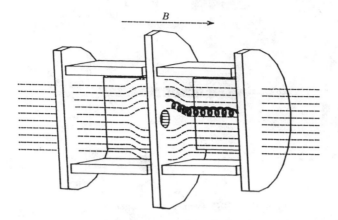

Fig. 9. Ions formed along a magnetic field line that does not transverse the orifice of the conductance limit will be lost due to collisions with the center trap plate. Because of the strong dependence of ion motion on the magnetic field lines, the magnetic susceptibility of the material comprising the two-section cell is crucial.

reduction of the effective dimensions of the conductance limit orifice. Reducing the effective dimension of the orifice of the conductance limit will increase the critical nature of the cell alignment.

3.   External Ion Sources

The third approach to coupling high-pressure source regions with ultra-high-vacuum analyzers suitable for FT–ICR is the differentially pumped external ion source. This approach is similar to the two-section ion cell technique in that the source and detection regions are spatially separated and differentially pumped. Ions are produced in a source external to the magnetic field and guided into the ion cell by means of either a quadrupole (76, 115) or electrostatic ion guides (3, 92, 94). Because the ions are produced external to the magnetic field, secondary electrons produced during the ionization/desorption are not trapped by the magnetic field.

The introduction of the tandem quadrupole Fourier transform mass spectrometer (115) (Q–FTMS) opened new possibilities for analytical applications in GC–MS and the analysis of large involatile, thermally labile biomolecules. To date there have been no examples published in the open literature on GC or LC external ion sources. However, this is an area where major contribution can be realized. As the capabilities for LC–MS interfaces are extended to polar biomolecules, the strengths of FT–ICR (viz, mass range and mass resolution) become of even greater importance. The first real successes of the external ion source systems with biomolecules used a liquid SIMS ionization on a tandem quadrupole FT–ICR instrument (76) (see Fig. 10). Although the transmission and trapping efficiency of sample ions in the cell has not been fully characterized, it has been demonstrated to be comparable to other methods of mass spectro-

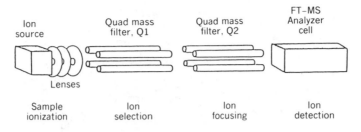

Fig. 10.   A schematic of the tandem-quadrupole Fourier transform mass spectrometer. [Reprinted by permission (76)]

metry. An additional advantage of the Q–FTMS system is the photodissociation of large peptides. The spectra shown in Fig. 11 demonstrate the impressive mass resolution for ions of 1500–2000 amu (76). Although the potential utility of FT–ICR for high-mass detection is indicated by the separation of bovine (m/z 5733) and porcine (m/z 5777) insulins (see Fig. 11a) and the detection of horse cytochrome C (m/z 12,384; see Fig. 11b), the loss of mass resolution at high mass has not yet been solved. At $m/z > 2,500$ amu, the duration of the time-domain signal is insufficient for acceptable mass resolution to be obtained. More recently, Wilkins has reported data for LD/FT–ICR for high-mass polymers with impressive mass resolution (77).

Following the introduction of ions into the ICR cell by a tandem quadrupole injection systems, Wanczek and coworkers demonstrated that ions could be efficiently introduced by means of electrostatic ion guides (94). By utilizing the fringing magnetic field lines of the solenoid magnet and low translational energies, ions could be focused into the magnetic funnel by means of an Einzel lens and then follow the magnetic field lines into the ion cell (see Fig. 12). By reducing the translational energies required to move the ions from the source to the analyzer region to near thermal, an enhancement in the mass resolution has been observed.

The deceleration and translational cooling of ions has recently been investigated by Smalley and coworkers (3), who have introduced bare-metal cluster ions into an FT–ICR cell from an external ion source (see Fig. 13). The metal ion clusters are generated in a supersonic beam by laser vaporization in a pulsed nozzle. Injection of the ions is accomplished with a series of Einzel lenses that direct the ions along the magnetic field lines leading into the magnet. Ions are given approximately 400 eV of translational energy to transport them to the ion cell. Prior to introduction and trapping of the ions to the ion cell, the ions are decelerated to a few electron volts of energy. A trapping efficiency of 60% has been reported for this system.

The deceleration of the ions in Smalley's system is accomplished by the use of a pulsed decelerator as shown in Fig. 14. This type of decelerator provides mass selective deceleration of the ions but is limited in terms of analytical utility. As (Fig. 14a) ions leave the source region, they acquire appreciable translational energy. The ions spatially separate according to mass as they drift through the

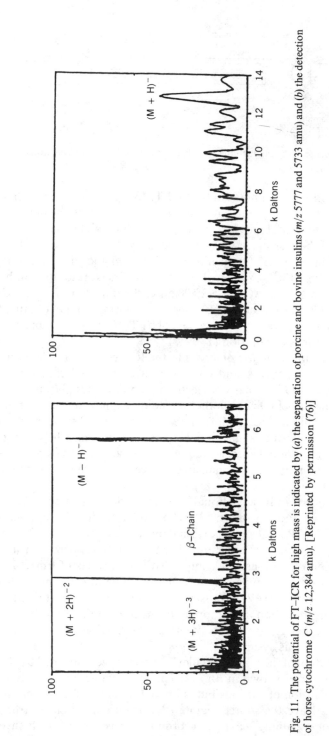

Fig. 11. The potential of FT–ICR for high mass is indicated by (*a*) the separation of porcine and bovine insulins (*m/z* 5777 and 5733 amu) and (*b*) the detection of horse cytochrome C (*m/z* 12,384 amu). [Reprinted by permission (76)]

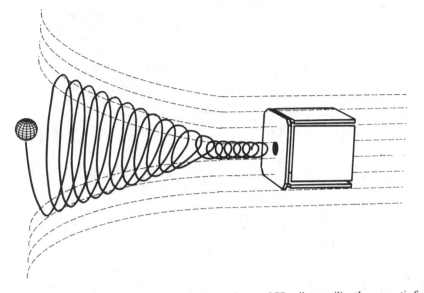

Fig. 12.   Introduction of low translational energy ions into an ICR cell can utilize the magnetic field lines.

field-free region between the source and deceleration regions (Fig. 14b). The spatially separated ions are temporally decelerated by electrostatic lines of force generated by the potential difference of the decelerator with respect to the surrounding grids (Fig. 14c). Ions that have passed through the decelerator are reaccelerated by the force lines encountered in the second deceleration region and exit the ion cell. As the ions of interest enter the pulsed decelerator, the voltage is pulsed such that the lines of force are reversed. Since the interior of the decelerator is electrostatically discrete, no field lines exist. Thus, ions in the interior of the pulsed decelerator when the pulse occurs will not be affected by the changing voltage. Ions that have not entered the decelerator prior to the pulse will not undergo deceleration in the primary region and will also exit the ion cell. Ions that are in the cell will again encounter decelerating force lines upon exiting and undergo secondary deceleration to a few electron volts of translational energy (Fig. 14d). These ions will then drift into the ion cell where they can be trapped by simple electrostatic methods (Fig. 14e). Further experiments by Smalley, et al. demonstrate the ability to fill the ion cell with multiple ion injections and "put it on ice" by collisionally cooling the ions excess translational energies prior to detection by standard FT–ICR methods (95). Before detection by FT–ICR methods can be considered to be routine, fundamental information concerning the role of initial ion translational energy and location must be determined.

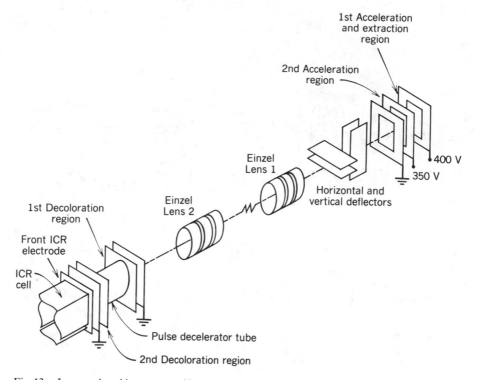

Fig. 13.   Ions produced in an external ion source are injected into the ion cell using electrostatic ion guides and a pulsed decelerator. [Reprinted by permission (3)]

C.   ION MOTION

The potential analytical utility of ICR has been widely anticipated following the development of an excitation scheme compatible with Fourier transform data analysis methods (38). The potential of FT–ICR is based on the equations governing ion motion, which have been discussed elsewhere (14, 145). The understanding of ion detection is based on theory and experimental data for ions existing under conditions corresponding to "ideal." The fundamental motion of an ion and deviations from theoretical ion motion affect ion detection. Ion motion in an ICR cell is a function of the initial conditions, for example, spatial location of the ions in the ICR cell, translational energy, and the forces acting on ions in $\mathbf{E} \times \mathbf{B}$ fields (145). Production of a coherent packet of ions that has a detectable cyclotron orbit is therefore based on the initial conditions of the ions prior to excitation. The problems incurred in FT–ICR will be assessed here by (i) describing ions existing under theoretically ideal conditions, (ii) the nature of the ion trap, and (iii) the effect of deviations from the ideal ion conditions on ion detection.

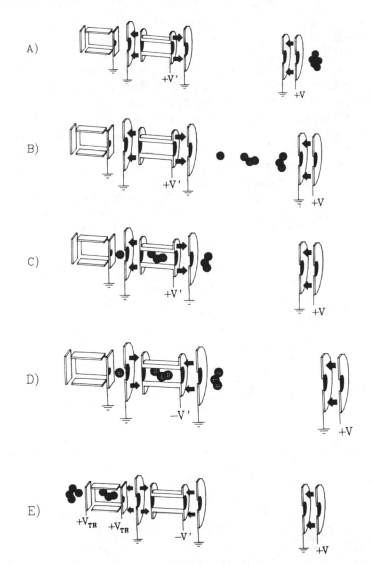

Fig. 14.   The sequence of selective pulse deceleration: (*a*) Ions are produced in the external source; (*b*) ions are spatially separated as they travel down the drift region; (*c*) selected ions enter the pulsed decelerator; (*d*) the pulsed decelerator is switched to a different potential to further decelerate the selected ions; (*e*) the selected ions are trapped in the ICR cell.

1. Ion Formation

The original mode of sample ionization was electron impact. Ions produced in this manner have near-thermal translational energies. The initial dimensions of the ion population are defined in the $X–Y$ plane by the volume of the electron beam, and in the Z direction by location of the electrostatic trap plates (see Fig. 15). Diffusion into the $X-Y$ plane is a function of the ion density and the magnetic field strength. During the course of ionization, the ion population is maintained such that space charge effects are negligible. Consequently, the diffusion of the ion population into the $X-Y$ plane will also be negligible.

The motion of the ions along the Z axis is confined by the electrostatic potential well created by the two trap plates positioned perpendicular to the $X-Y$ plane (see Fig. 16a). Ions are given kinetic energy in the Z direction by the potential gradient of the imposed trapping field. Under collisionless conditions, the ions oscillate along the Z axis. The amplitude of the Z axis oscillation is determined by the initial condition of the neutral and its location in the potential gradient upon ionization. By operating under conditions of a shallow trapping well (e.g., for low trapping potentials relative to the distance between the trapping plates, see Fig. 16b), ions existing within the boundaries of the trap gain negligible energies in the Z direction.

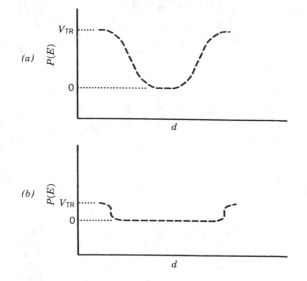

Fig. 15.   The initial dimensions of the ion population is defined in the $X-Y$ plane by the volume of the electron beam and in the Z direction by the location of the trap plates.

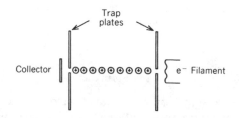

Fig. 16.   Potential wells formed by the electric field gradients of the trap plates.

Optimized conditions therefore produce an ion population of near-thermal translational energies. The ion population resides along the Z axis and in the geometric center of the $X-Y$ plane. This situation is consistent with the theoretical model described by the original equations. Owing to the strong correlation between observed and predicted results, the effects of nonideal ion conditions can be ignored. It is the consequence of an ion population whose initial condition is nonideal, which presently limits the FT–ICR experiment. Based on this premise, it is necessary to characterize the complex motion of ions and the nature of an ion trap.

2. Ion Traps

The fundamental concept for ion detection in an ion trap is the ability to trap ions in an ordered manner. The ability to trap and manipulate ions allows the opportunity to study a wide variety of phenomena. The applications of ion trapping in FT–ICR ranges from ion–molecule reactions to MS^n scans (5). Of principal consideration when detecting ions in ion traps is the fundamental difference between ions that are trapped and ions that are merely stored. In detection by FT–ICR methods ions must be arranged in a controlled manner (i.e., conditions corresponding to ideal) prior to excitation and detection. Once ions are trapped in a controlled manner, a variety of manipulations may be performed. The most common manipulation in FT–ICR is the ability to make ion selections. In the absence of an ordered trapping of ions, such manipulations or even satisfactory ion detection is not possible.

Ions trapped in the quadrupole ion storage trap (QUISTOR) are sensitive to the same kind of requirements for satisfactory ion detection (163). Although charged particulates have been stored indefinitely in these ion traps (149), the limiting factor in detection is loss of resolution due to unstable or incoherent initial ion conditions. This technique is not capable of initially producing a thermal assembly of ions in a discrete manner as easily as its ICR counterpart. Owing to the constant rf of the active trapping process, it is difficult to project a well-controlled electron beam through the cell. To alleviate this obstacle, a relatively high pressure of helium buffer gas is used to cool the ion motion (41). The helium damps the ion motion, increasing the concentration of ions in the center of the trap. By using collisional relaxation methods, ions that have initial spatial locations near the center of the QUISTOR are detected by a capacitance detection technique. This type of detection can be compatible with Fourier transform data analysis methods (147).

A QUISTOR operated without a collision gas is analogous to the situation for ions trapped in an ICR cell in a random manner with a wide distribution of translational energies. This condition can arise from ion injection from an external source, ion emission from a surface near the ion cell (i.e., laser desorption or Cs^+ ion SIMS), or spatial and energy randomization caused by field inhomogeneities or complex coupling of ion motions. Cooling the excess translational energy of ions trapped in an ICR cell should result in an increase in the performance of the technique due to increased coherence of the detected ion packet. The use of collisional quenching of excess ion energy and ion relocation to the $z=0$ axis has been reported (135). Additionally, a recent communication reports the ability to trap multiple ionization/injection events in an ICR cell by using bursts of a buffer gas to collisionally relax the injected ions (95).

3. Deviations from Theoretical Ideality

Under situations where ion motion is no longer synonymous with the theoretical ideal, loss of mass resolution or frequency shifts can occur. It has been shown that the motion of ions formed by electron impact ionization and trapped in an electrostatic trapping well can be viewed as a combination of three fundamental motions (145, 154). Ions created in this manner have motion that is a function of their natural cyclotron motion, ω_c, magnetron or drift motion, ω_d, and the trapping oscillation, ω_t. By raising the electric field placed on the trap plates, the effect of the harmonic trapping oscillation, ω_t, increases according to Eq. (8):

$$\omega_t \propto (V_t - V_0)^{1/2} \tag{8}$$

This will affect the observed cyclotron frequency, ω_{eff}, according to Eq. (9):

$$\omega_{eff} = (\omega_c^2 - \omega_t^2)^{1/2} \tag{9}$$

Thus, by increasing electric field and thereby the trapping oscillation, a shift in the cyclotron frequency is observed. For ions created under ideal conditions (i.e., electron impact ionization, near-thermal energies, and low trapping voltages) the observed shift in the frequency is negligible ($>.15\%$) (145). The magnitron motion is the tendency of the center of the cyclotron motion to precess in the $X-Y$ plane due to redirection by electrostatic field lines. The consequence of this motion on the ICR spectrum manifests itself as sidebands flanking the principle resonance peak. These effects are well understood and the theoretical description is in good agreement with experimental results (43).

Additionally, ion motion in an ICR cell is affected by ion density. In order to produce a signal of adequate signal-to-noise ratio throughout the mass spectrum, 10^2–10^6 ions are required. Consequently, space charge effects lead to complex ion motion. The effects of Coulombic repulsion on mass calibration was discussed by Jeffries et al. in terms of an ellipsoidal ion cloud (88). The influence of the harmonic interaction of the ions causes a shift in the observed cyclotron frequency. A further relationship between ion mass and effective cyclotron frequency was derived by Gross et al. in which the errors due to Coulombic repulsion decrease with the square of the magnetic field strength (99). That is, the relative force caused by Coulombic repulsion of thermal ions will be small in comparison to the Lorentz force. However, ion–ion Coulomb repulsion caused by Z-axis oscillation has been shown to cause spectral line broadening, for example, loss of resolution (155). Figure 17 depicts the effects of Lorentz and

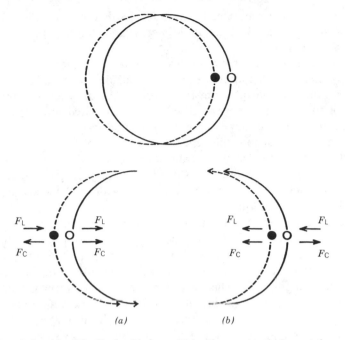

Fig. 17.   The relationship of the Coulombic forces (F_C) with respect to the Lorentz forces (F_L) acting on an ion pair causes spectral line broadening.

Coulombic forces on the cyclotron motion of two ions. As shown, the Lorentz force always points to the center of the ion's orbit. Therefore the Lorentz force remains independent of the relative location of the two ions. Conversely, the Coulomb repulsion is dependent on the relative location of the other ion (e.g., the direction of the Coulombic force will oppose the location of the other ion). In Fig. 17a, both the Coulombic force (F_c) and the Lorentz force (F_L) are acting on ion 1 (denoted by a solid orbit) in the same direction. Conversely, the forces acting on ion 2 (denoted by a dashed orbit) are opposing each other. During this half of the cyclotron orbit, ion 1 is accelerated and ion 2 is decelerated. During the second half of the cyclotron orbit (Fig. 17b), ion 2 is accelerated and ion 1 is decelerated. Since the acceleration and deceleration do not exactly cancel, the result is slightly different cyclotron frequencies for ions 1 and 2, thus causing broadening of the detected frequency. Since mass detection is defined by the cyclotron frequency, frequency broadening results in loss of mass resolution.

The induced motion of the ions (viz. motion due to rf excitation or harmonic oscillations due to the trapping field) is a function of the electrostatic potential lines created by the cell geometry. Equations for calibrating mass frequency must contain constants that are a function of the cell geometry (32). Recently the dynamics of ion motion in an elongated cell and the affect on signal shape was discussed by Nikolaev and Gorshov (125). Based on a careful study of electrostatic interaction on ion motion, a correlation between kinetic energy and spectral peak shape was obtained. These results demonstrate the effect of ion conditions (viz. ion location and kinetic energy) and electrostatic field geometry on signal detection.

The most dramatic example of electrostatic field interaction is the complex coupling of the ion excitation in the $X-Y$ plane to the Z-axis harmonic oscillation leading to unstable ion trajectories (73, 93). The mechanism for Z excitation is momentum transfer caused by synchronized interaction between the ion's motion and the shape of the equipotential lines of the excitation field. The angle at which an ion interacts with the lines of force generated by an electrostatic field can create a net average force in the Z direction to increase the harmonic oscillation along the Z axis (see Fig. 18). Ions can gain sufficient energy in this manner such that their translational energy along the Z axis exceeds the trapping field, leading to ion loss by either ejection or by loss of coherence of the ion packet. Signal loss can occur when the ion packet interacts with the inhomogeneous electrostatic fields generated by the trapping field and the temporal excitation field. Huang et al. propose that Z excitation can be minimized by eliminating either the synchronized motion or the high amplitude of the rf excitation pulse (73). Therefore, if ions residing on or near the $Z=0$ axis of the ion cell underwent excitation in a series of low-amplitude pulses, Z excitation would be minimized. Alternatively, if the cell were redesigned to diminish the electrostatic inhomogeneities causing the Z-axis component of the electrostatic force vectors, the effect of Z excitation would be diminished.

It is important to keep in mind the effects that electrostatic and magnetic inhomogeneities have on the FT–ICR experiment. As already indicated, electrostatic field effects influence the ability to maintain a detectable density of ions in the ion cell. Field effects contribute to signal broadening and loss of mass resolution. The contribution of field inhomogeneities (both magnetic and electrostatic) has been examined both theoretically and experimentally (20, 98). The net effect of magnetic inhomogeneity on ion detection is negligible (98). This

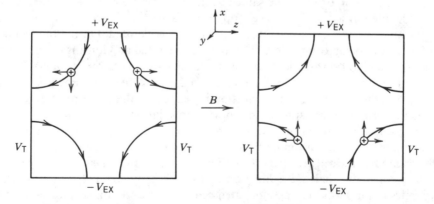

Fig. 18.   The synchronization between the rf excitation and the ion cyclotron motion can lead to Z-axis excitation. [Reprinted by permission (76)]

is not unreasonable considering that ions are detected following excitation by the applied rf. At the point at which ions encounter the magnetic inhomogeneities (i.e., near the cell walls), they have gained appreciable translational energy, and the effect of the magnetic perturbations is insignificant. In the case of ions that have not been excited or ions having low translational energies prior to encountering the magnetic perturbations, the effect of field inhomogeneities can no longer be neglected. This is the case for ion partitioning in a two-section cell or for ions of low translational energy being injected into the ion cell from an external ion source. In this case, the ions must pass through a region of field inhomogeneity and be redirected into the $X–Y$ plane. In terms of ion trapping and ion motion, this is an additional complication.

In the simplest case, which is the partitioning of ions between two sections of the same cell, the effect of magnetic field inhomogeneity has been studied both theoretically and experimentally (20). In Fig. 19, the calculated magnetic field lines for a 316 stainless steel two-section cell indicate a redirection of the magnetic field lines into the cell material. Owing to the significant magnetic permeability of the stainless steel, the magnetic field flux will be higher in the cell material. Consequently, the ions are redirected upon passing through the magnetic bottleneck. This effect is observed by comparing ion partitioning in a stainless steel ICR cell to the ion partitioning in an oxygen-free copper (OFC) ICR cell. Owing to the lower magnetic permeability of the copper material, the

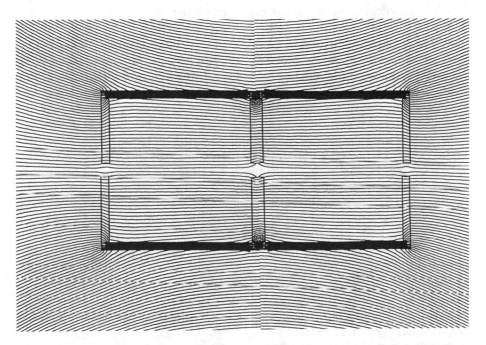

Fig. 19. The magnetic field lines calculated for a 316 stainless steel two-section cell indicate redirection of the magnetic lines into the cell material.

degree of magnetic perturbation is minimized. In these experiments, ion partitioning in the ICR cell made of stainless steel resulted in no detectable signal from the transported ions. By replacing stainless steel trap plates with OFC plates (thus virtually removing magnetic inhomogeneities from the cell), ions produced in the source region were routinely partitioned, trapped, and detected in the analyzer region of the ion cell.

Considering the complex combination of factors influencing ion motion and consequently ion detection, the problems encountered in trying to expand the flexibility of the technique may not be adequately modeled by studies of ions formed by electron impact. In order to reach the anticipated potential of the method, the fundamental dynamics and physics describing the mechanism of ion detection must be defined such that the method may evolve to new levels.

D.   EXTENDING THE DYNAMIC RANGE OF FT–ICR

One of the principal limiting factors of FT–ICR is the relatively small dynamic range compared to other types of mass spectrometers. The cause of this limitation is twofold. First, the limits of detection by the capacitance methods employed requires approximately 100 ions to be observed. On the other extreme, as the ion population exceeds 100,000, space charge effects cause peak broadening. This yields a useful dynamic range of 1000. Second, the dynamic range of an analog-to-digital converter (ADC) limits the digital dynamic range to about 12 bits. The net effect of the digital dynamic range is that in the presence of a strong ICR signal, smaller signals are lost.

The most practical method for extending the dynamic range is the successive ejection of unwanted peaks (24). This procedure can also cause undesirable effects. In order to achieve suitable resolution during the ejection sequence a long irradiation time is necessary. This large amount of time can cause loss of the desired signal due to ion–molecule reactions or dissociative processes. If the ejection sequence is performed in a rapid sweep, frequency overlap occurs.

By tailoring the excitation, Marshall and coworkers have demonstrated the ability to use a inverse Fourier transform to increase the resolution of the excitation pulse (156). In this manner the ejection sequence can be performed rapidly while maintaining high selectivity. The advantages of stored waveform inverse Fourier transform (SWIFT) excitation are illustrated in Fig. 20. Two major advantages indicated in Fig. 20 are (a) the resolution of the excitation pulse is increased, and (b) the power curve of the excitation is uniform throughout the ejected mass range (iii) Furthermore, since the SWIFT time-domain waveform is turned on only once, as opposed to N times for a multiple ejection sweep sequence, the SWIFT ejection will be N times more effective for the same period of excitation.

Further enhancement of the FT–ICR mass resolution can be achieved by simply phase correcting the spectrum (37, 39). By implementing this procedure, the resolution can be increased by a factor of 2 at a given magnetic field strength, mass, and pressure (37). The phase distortions arise from the solution of the

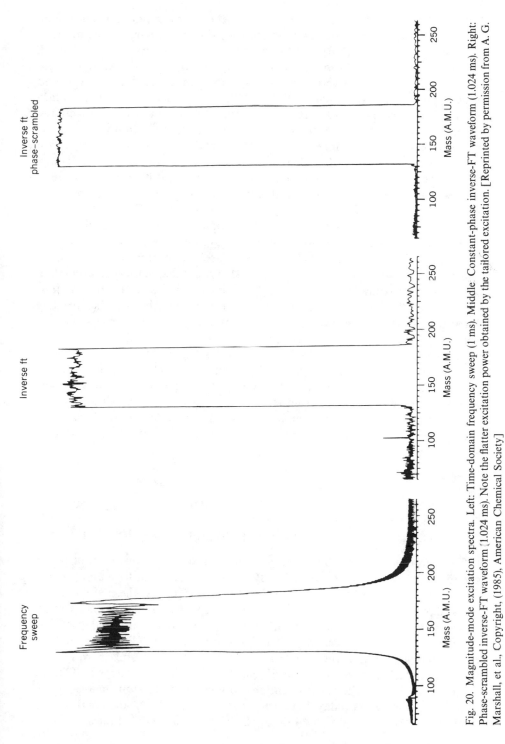

Fig. 20. Magnitude-mode excitation spectra. Left: Time-domain frequency sweep (1 ms). Middle Constant-phase inverse-FT waveform (1.024 ms). Right: Phase-scrambled inverse-FT waveform (1.024 ms). Note the flatter excitation power obtained by the tailored excitation. [Reprinted by permission from A. G. Marshall, et al., Copyright, (1985), American Chemical Society]

Fourier transform of the continuous form of the digitized time-domain signal. The Fourier transform results in two frequency-domain spectra, the continuous absorption spectrum $[A(\omega)]$ and the continuous dispersion spectrum $[D(\omega)]$. The absolute-value spectrum or magnitude spectrum is an admixture of these two separate spectra. It is this magnitude spectrum that is commonly measured in acquiring a mass spectrum by FT–ICR. Owing to the increased resolution without loss of signal-to-noise ratio, it is preferable to measure exclusively the absorption spectrum. The distortion of the absorption spectrum is due to the phase correlation of the admixture with the dispersion spectrum. The dispersion versus absorption method (DISPA) corrects for the phase addition of the distinct parts of the magnitude spectrum. This phase correction has been achieved over a narrow mass range by determining the variation of phase with frequency (37). The DISPA method also simplifies quantitation in an FT–ICR spectrum since the number of ions is proportional to absorption mode FT–ICR peak area.

One major limiting factor in the mass range obtainable by FT–ICR is the electronic limitations of the possible sampling rates. For example, the lower mass limit for a maximum sampling rate of 5.2 MHz and a magnetic field strength of ca. 3 T is approximately m/z 18 amu. This lower limit is set by the Nyquist theorem, which states that the highest signal frequency that can be correctly determined by discrete Fourier transformation of the time-domain signal is half that of the maximum sampling rate. If the time-domain signal is sampled at a lower frequency than that required, then the peaks obtained will appear at higher frequencies. It has been demonstrated by Verdun et al. (153) that if two offset time-domain signals are interleaved together, then the time-domain waveform that is produced is equivalent to a waveform that was produced at a sampling rate that is twice as fast as the sampling rate used. In this manner the lower mass limit can be extended to below m/z 12 amu with a sampling rate of only 4 MHz in a 3-T magnet (153).

III. CHEMICAL STUDIES

A. INTRODUCTION

The purpose of this section is to illustrate the tone and direction of recent chemical studies performed using FT–ICR spectrometry and to summarize the advances in experimental hardware that have precipitated these studies. The work discussed is comprised of that which appeared in the open chemical literature from 1985 to the present as well as selected drafts and preprints supplied by the cited researcher. The basic concepts of FT–ICR as well as representative chemical studies are discussed in reviews by Comisarow, Freiser, Gross, Marshall, and others (34, 59, 60, 64, 107). Since the early days of ICR, the principal driving force for instrument development has been ion chemistry; this same trend is apparent today. However, as already discussed, much of this same development work has potential for application in analytical studies.

The high rate of data acquisition, the ability to selectively isolate and store ions for seconds, the good to excellent mass resolution, and multichannel detection have established FT–ICR as an important technique for studies of ion–molecule reactions. Ion–molecule reaction studies by FT–ICR differ from other methods. Specifically, the high-vacuum conditions used in FT–ICR give rise to the ability to control the ion–molecule reaction time (ranging from periods of milliseconds to seconds), which, in turn, allows for the control of the degree of collisional stabilization of the reactant ion. The extent to which ions are collisionally stabilized prior to ion–molecule reactions and/or irradiation with a high-intensity photon source to induce photodissociation provides information about relaxation phenomena as well as the effects of internal energy on reactivity. Questions regarding the internal energies of ions are important for ion chemistry studies (47,48), and such questions are also important considerations for structural characterization by collision-induced dissociation and photodissociation. This discussion will be concerned primarily with studies of ion–molecule reactions of monatomic ions, ligated mononuclear ionic complexes, and ionic clusters. A second section concerned with the photodissociation and collision-induced dissociation studies performed in the FT–ICR will follow. Whenever appropriate, recent reviews are cited in lieu of extensive discussion.

B. ION–MOLECULE CHEMISTRY

1. Method Development

In the past several years, a number of innovations of significance to ion–molecule reaction studies have been introduced to FT–ICR, for example, state-selective ionization methods (78), neutral reactant introduction methods (21), and ion ejection methods (15, 36, 110, 127). The basic ion chemistry study consists of at least five fundamental steps: (a) ion formation and introduction, (b) reactant ion isolation, (c) reactant neutral introduction, (d) reactant ion reaction, and (e) product ion detection. These steps can be overlapped, rearranged, or iterated. Each step is an integral part of the experiment and adds flexibility and scope to the ion chemistry experiment. Owing to the introduction of novel software procedures and hardware, it is now feasible to perform extensive studies of complex systems controlling both the reactant ion and neutral as well as the reaction energy.

a. ION PRODUCTION

Advances in methods of ion production and introduction to the FT–ICR cell were already considered in detail. Some of these same methods possess significant utility for ion–molecule reaction studies. Control over ionization is important because different methods of ion production lead to structurally and/or energetically different ions. Ideally, one would like to produce a given ion in a specific state and with specific energy for ion–molecule chemistry studies. However, there are practical limits on the existing experimental capabilities. This problem is being addressed in the area of cluster ion reaction studies where it is necessary to generate both ligated and bare clusters ranging from trimers to clusters with nuclearities in the tens of atoms. Both positive and negatively charged clusters of silicon, carbon, tungsten, tungsten–carbon, and antimony–bismuth have been formed in situ by direct vaporization of stationary targets mounted in or near the FT–ICR analyzer cell using a Nd:YAG laser (40,102–104,114,105,133). It must be understood that species generated in this fashion generally possess ill-defined internal energies (126). In one configuration, a silicon target is located on an end trap plate of a differentially pumped two-section cell (32,150). The laser beam travels parallel to the magnetic field and traverses both regions of the cell before striking the target (Fig. 21) (102–104,131). Ions are partitioned to the analyzer region by Coulombic repulsion between the ions generated by laser ablation rather than by the conventional method of grounding the conductance limit. In addition, Smalley has successfully coupled a supersonic beam cluster ion source with an FT–ICR for studies of bare clusters of nuclearities ranging from 2 to 100 atoms. These clusters are generated by laser ablation of a rotating source mounted in a supersonic nozzle. Internally cool (estimated to be < 50 K) cluster ions of virtually all elements can be generated in this apparatus (3).

Collision-induced dissociation (CID) has been employed to form unique, nonligated transition metal cluster fragment (83) and monatomic metal ions (142). This method consists of coupling conventional ionization techniques (e.g., electron bombardment or laser ablation) with CID. In the procedure for forming

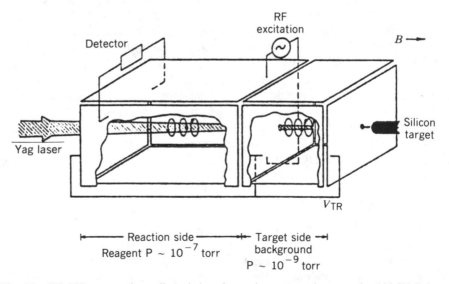

Fig. 21. FT–ICR two-section cell depicting the stationary sample target for Nd:YAG laser evaporation. The target resides in the low-pressure side of the cell to prevent contamination of the surface by the neutral reagent gas that is maintained at static pressure in the reaction region of the cell. [Reprinted by permission (104)]

cluster fragments, precursor atomic metal ions are formed and allowed to react with transition metal carbonyls and nitrosyls to yield metal carbonyl–nitrosyl clusters. These products are then subjected to CID, which causes the ligands to be successively removed. Using this method, the chemistry of $FeCo^+$ (83, 85, 86), $FeCo_2^+$ (84), VFe^+ (66), Fe_2^+, Co_2^+, $FeCo_2^+$, Co_3^+ (81, 83, 85, 86), $CuFe^+$ (148), $RhFe^+$, $RhCo^+$, and $LaFe^+$ (71) with various reactants has been probed, and values for bond energies of these fragments as well as with specific ligands have been bracketed. In these studies, the collision gas is typically argon, and the pressure in the cell is maintained at mid 10^{-6} torr pressures. Under these conditions, the fragments of interest undergo thermalizing collisions with neutrals prior to reaction. It is important to note that the internal energy of species formed by this method is unknown. It is quite possible that the relatively few thermalizing collisions that occur do not completely thermalize the ions.

b.   Reactant Ion Isolation

The most dramatic evolution of techniques useful for ion–molecule reaction studies performed by FT–ICR is in the area of reactant ion isolation. To be useful for such studies, the isolation technique must be highly selective both in terms of mass *and* energy resolution. Early in the development of ICR, reactant ions of interest were isolated by using rf sweeps over a range of frequencies corresponding to unwanted ions (15,62). Whereas the method is effective and has been widely used, it suffers from three weaknesses: (a) overall slow speed, (b) potential frequency overlap of the cyclotron frequencies of reactant ions causing undesired translational excitation, and (c) poor ejection resolution as a result of (b). The problems of frequency overlap and poor ejection resolution can be overcome by decreasing the speed of the ejection sweep while using lower excitation amplitudes. Unfortunately, reducing the ejection sweep rate diminishes the efficacy of the technique for ion–molecule reaction studies because reactions of interest then occur on the ejection timescale. Recently, four new techniques have been developed to overcome the inherent weaknesses of ion ejection. The first of these techniques, selective on-resonant excitation, follows directly from the principles governing translational excitation of ions trapped in an $\mathbf{E} \times \mathbf{B}$ field. This method is useful only for ejection of an ion having a single m/z value close to that of the ion of interest. Energy absorption for an off-resonant ion goes as a function $\sin^2(kt)$; and therefore, at properly selected times, these off-resonance ions receive no net translational excitation. This null excitation time window is given by Eq. (10) where f_1 and f_2 are the frequencies of the resonant and nonresonant ions:

$$t = \frac{1}{|f_1 - f_2|} \tag{10}$$

respectively. Any single-frequency excitation performed at the resonant frequency for multiples of the period given by Eq. (10) yields net unexcited nonresonant ions. Using this method, Cody and Goodman selectively isolated

$C_2F_4^+$ from a doublet with $C_5H_8O_2^+$ (nominal mass 100 amu) (33). This represents a mass resolution of about 1700. By itself, this technique is of limited utility for ion–molecule studies because of the ability to eject ions of only a single mass.

Three novel procedures for isolating a specific reactant ion from a population of ions have been developed: (a) front-end resolution enhancement using tailored sweeps (FERETS) (51), (b) mass and energy selective ion partitioning, and (c) stored waveform inverse Fourier transform (SWIFT) (111). FERETS is an extension of the conventional ion ejection technique already described. FERETS requires a high-power rf excitation amplifier (400 Vpp) and a wide range of software-selectable attenuators (61 dB). In addition, a large number of excitation sweeps must be allowed by the software driving the experiment. The tailored sweeps are made up of "hard (high-power) sweeps" and "soft (low-power) sweeps." A standard hard sweep is attenuated -10 dB from 400 Vpp and consists of 16-μs duration rf pulses spaced at 500-Hz intervals over the desired mass range. As an example, operating under these parameters, it requires 25 ms to cover the mass range from 150 to 3000 amu. A soft sweep typically is attenuated -35 dB from 400 Vpp, and consists of 2-ms pulses separated by 200 Hz across the mass range of interest. The overall ejection experiment requires 350 ms—the problem of speed is not directly addressed by FERETS, however, mass resolution for selection of the reactant ion (normalized to mass 100 amu) is about 10,500. Using FERETS in a very large FT–ICR cell, $(MS)^5$ was demonstrated for various reactions of $Re_2(CO)_{10}$ (Fig. 22) (51).

Figure 22a consists of a normal electron impact (70 eV) spectrum of $Re_2(CO)_{10}$. From this ion population, $Re_2(CO)_7^+$ (all isotopes) are isolated by

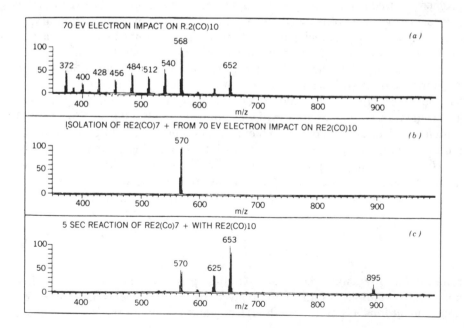

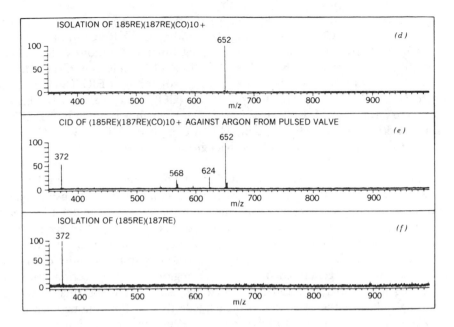

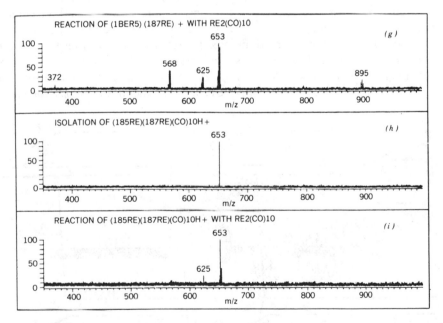

Fig. 22. MS^5 sequence. (a) Positive ion electron impact spectrum of Re$_2$(CO)$_{10}$ at 70 eV; (b) isolation Re$_2$(CO)$_7^+$ cluster group; (c) 5-s reaction after isolation of Re$_2$(CO)$_7^+$; (d) FERETS isolation of ^{185}Re ^{187}Re$_2$(CO)$_{10}^+$ formed in (c); (e) addition of argon using a pulsed-valve inlet and the collisional decomposition of $m/z = 652$; (f) FERETS isolation of CID product $m/z = 372$; (g) 5-s reaction of $m/z = 372$ with background Re$_2$(CO)$_{10}$; (h) FERETS isolation of $m/z = 653$ product formed in (g); (i) 5-s reaction of m/z isolated in (h). [Reprinted by permission (167)]

using FERETS as shown in Fig. 22b. $Re_2(CO)_7^+$ is allowed to react with the parent neutral, $Re_2(CO)_{10}$, for 5 s (Fig. 22c), and $^{185}Re^{187}Re(CO)_{10}^+$ is isolated by using the FERETS technique (Fig. 22d). At this point, argon is allowed to enter the cell via a pulsed valve and $^{185}Re^{187}Re(CO)_{10}^+$ undergoes CID to form $^{185}Re^{187}Re^+$ (Fig. 22e), which is subsequently isolated by FERETS (Fig. 22f). $^{185}Re^{187}Re^+$ then reacts with the parent $Re_2(CO)_{10}$ and water in the background to form $^{185}Re^{187}Re(CO)_{10}H^+$ (Fig. 22g), which is isolated by FERETS as depicted in Fig. 22h. At this point, MS^5 is complete. $^{185}Re^{187}Re(CO)_{10}H^+$ then reacts with $Re_2(CO)_{10}$ and the background to give the ion distribution found in Fig. 22i.

Mass- and energy-selective ion partitioning is a novel technique that is limited to two-section cell instruments. Ion selection is accomplished by translationally exciting an ion to an energy such that the radius of the cyclotron orbit is larger than the aperture in the conductance limit. For instance, for a two-section cell with a conductance limit aperture of 1 mm radius, an ion of m/z 100 must be excited to greater than 4.4 eV in order for partitioning to be disallowed. In this technique, ions are generated in the source region of a two-section cell and undesired ions are translationally excited by using rf sweeps such that the resulting radius of the cyclotron orbit exceeds that of the conductance limit aperture. It is also possible to further enhance the mass resolution of the ion selection process by use of the selective on-resonant ejection method defined by Eq. (10). Following radio frequency excitation, the conductance limit is grounded for the optimum partitioning period (152) causing the ions to partition to the analyzer region of the cell. Ions that have enlarged cyclotron orbits cannot partition and are quenched by collision with the conductance limit (see Fig. 23). A secondary advantage of this technique is that an upper limit can be assigned to

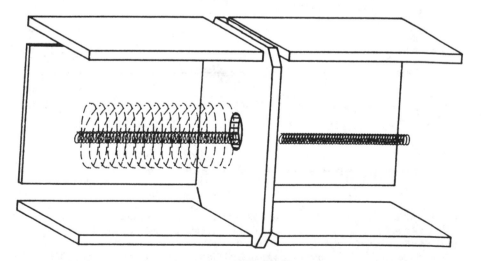

Fig. 23. Mass and energy selective ion partitioning. Translationally excited ions (large orbit) cannot partition through the conductance limit aperture (89).

the energy of the cyclotron motion of those ions that are partitioned. A further advantage of this method is found in the fact that an ion, once partitioned, can be removed from the reactive environment containing its parent neutral. Using this routine, the reaction of $^{54}Fe^+$ [formed by electron impact of $Fe(CO)_5$] with $Co(CO)_3NO$ forming $FeCo(CO)(NO)^+$ was monitored (see Fig. 24) (89).

Finally, SWIFT (see Section II.F) promises to be a standard ion ejection technique on second-generation FT–ICR instruments (25). This method of generating excitation waveforms offers the means for and selectively ejecting ions at any number of m/z values while leaving the ion of interest virtually unperturbed (123). As an example, using SWIFT $Os_3(CO)_5^+$ was separated from all $H_xOs_3(CO)_y^+$ ions for subsequent study. Figure 25 depicts a comparison of waveforms for such selective ion isolation. It is clear from Fig. 25a that using "chirp" excitation $Os_3(CO)_5^+$ ions receive significant excitation due to frequency overlap. The SWIFT excitation waveforms of Fig. 25b have virtually zero amplitude at the m/z ratio corresponding to $Os_3(CO)_5^+$. The benefit of such efficient and selective reactant ion isolation for ion–molecule reaction probes is obvious.

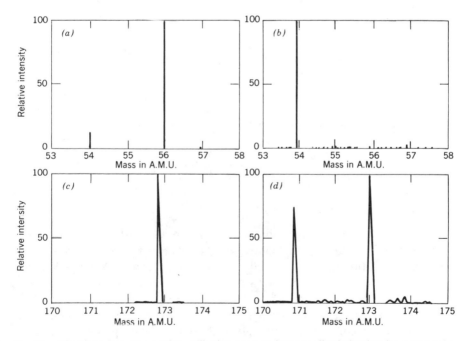

Fig. 24. Chemistry in a two-section cell using mass and energy discrimination for reactant ion isolation. (a) $^{56}Fe^+$ isolated in the analyzer region ($^{54}Fe^+$ is not partitioned); (b) reaction of $^{56}Fe^+$ isolated in (a) with $Co(CO)_3(NO)$ to form $^{56}FeCo(CO)(NO)^+$ ($m/z = 173$); (c) isolation of $^{54}Fe^+$ in the analyzer region; (d) reaction of $^{54}Fe^+$ with $Co(CO)_3(NO)$ to form $^{54}FeCo(CO)(NO)^+$ ($m/z = 171$), the peak at $m/z = 173$ corresponds to $Co(CO)_3(NO)^+$ formed via a minor charge exchange pathway.

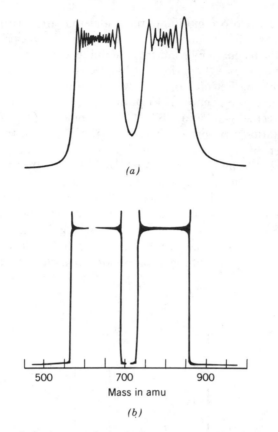

(a)

(b)

Fig. 25. Comparison of selective mass ejection for frequency sweep (a) and SWIFT (b) excitation, for equal excitation periods (8.4 ms). Both excitation waveforms were designed to eject all $H_xOs_3(CO)_y^+$ species except for $Os_3(CO)_5^+$. Note the much higher mass selectivity of the SWIFT method. [Reprinted by permission from S. L. Mullen, et al., Copyright (1988), American Chemical Society]

c.   REACTANT NEUTRAL INTRODUCTION

Gaseous neutral reactants are admitted to the ion–molecule reaction region in FT–ICR by three means: (a) controlled leak valve, (b) pulsed valve, and (c) dynamic partitioning. A controlled leak valve consists of a hard, optically flat surface (usually sapphire) in contact with a malleable seal (usually copper). The pressure of the seal against the flat surface is variable, and, therefore a variable orifice can be achieved allowing reagents to be admitted to the vacuum system at a constant rate. A constant leak rate gives rise to a constant pressure of neutral reactant allowing rate constants to be measured. Unfortunately, the neutral bled into the vacuum system using a leak valve is present throughout the entire FT–ICR experiment sequence, which often causes complications due to masking reactions with the neutral. In CID studies, a relatively high pressure (10^{-6}–10^{-5} torr) of collision gas is usually required. Unfortunately, pressures in this range give short signal transients and, hence, poor resolution. In order to circumvent the complication of masking reactions and high pressure during detection, pulsed-valve techniques were introduced for use with FT–ICR (21,139). Typically, the valve is opened for a few milliseconds during which time the pressure within the FT–ICR vacuum system increases dramatically. While the neutral pressure remains high, reaction studies or CID can be performed. In this way, experimental steps that are not compatible can be separated temporally. While pulsed-valve studies are of great utility in both CID and reaction studies, the rapidly changing pressures associated with this method forbids the determination of rate constants. Dynamic partitioning combines the advantages of both pulsed valves and controlled-leak valves.

Dynamic partitioning is a additional benefit derived from the nature of the two-section cell. The differentially pumped two-section cell has demonstrated analytical utility for coupling high-pressure sources to a high-vacuum FT–ICR cell (30), but the two-section cell configuration also has an advantage over that of single cells in that it allows spatial separation of neutrals used in ion–molecule reaction studies. The spatial separation of neutrals is of great benefit to FT–ICR studies because both masking reactions and high pressure during detection can be avoided. For example, studies using laser ablation as a source of ions suffer from interfering species that arise from reaction of neutrals at the surface of the target (104). Using a two-section cell allows neutral reactants to be admitted to a differentially pumped reaction/analyzer cell at a static pressure while ions of interest are formed in a separate source cell and then isolated and partitioned as described previously. By using this approach, the need for pulsed-valve sample introduction can be eliminated as well as the sample pressure ambiguities associated with the pulsed-valve method. These ambiguities are due to the rapidly changing neutral reagent pressure (and, therefore, ion-neutral collision frequency) inside the cell for 100–500 ms after the valve pulses. Using a differentially pumped configuration has the advantage of maintaining constant pressure for the neutral, thus rate constant measurements can be made. Operating in this mode, it was possible to study ligand exchange reactions of $Cr_x(CO)_n^+$ ($x = 2$–4, $n = 1$–6) with a number of ligands and extract kinetic data for the ligand exchange reactions (Fig. 26) (144).

d. REACTANT ION REACTION

The reactant ion reaction step of an ion–molecule reaction study is no more than a time window, and thus cannot be fundamentally altered. One interesting approach to this time window was employed by Wronka et al. (166). Normally, a quench pulse is used to purge all charged species in the cell at the beginning of each reaction pulse sequence. In this study, however, the quench step was removed. Product ions were allowed to build up as more and more reactant ions were formed during successive reaction cycles. In this mode, the reaction time is as long as the entire signal averaging experiment. By coupling this "quench-off" technique with selective FERETS ion ejections, the reaction channels of

$Re_2(CO)_{10}$ were elucidated (98). Note that the energy of reactant ions cannot be assumed to be low nor can kinetics be measured using this method.

e.   PRODUCT ION DETECTION

Ion–molecule studies place several requirements on detection in FT–ICR. First, the limited dynamic range of FT–ICR often causes the products of less

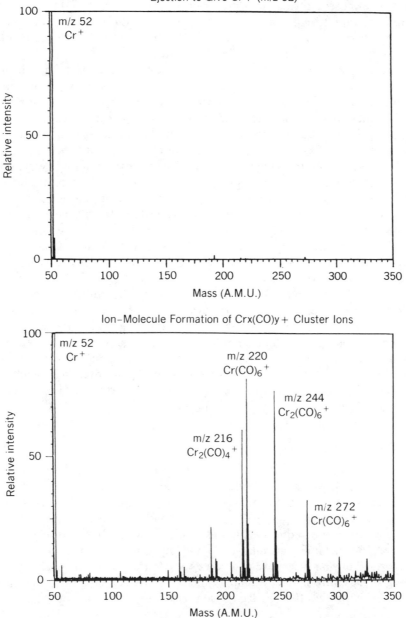

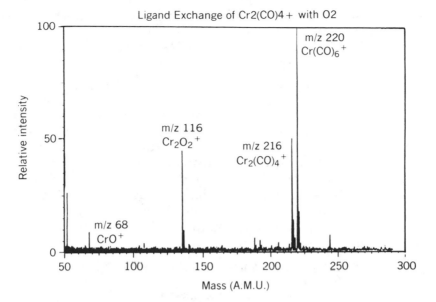

Fig. 26. Dynamic partitioning for ligand exchange studies. (a) Cr^+ formed by electron impact ionization of $Cr(CO)_6$ is isolated in the source region of a two-section cell; (b) Cr^+ isolated in (a) is allowed to cluster with $Cr(CO)_6$ in the source region of the cell; (c) $Cr_2(CO)_4^+$ is partitioned to the analyzer region of the cell that contains a static pressure of O_2 and allowed to ligand exchange yielding $Cr_2O_2^+$ (144).

favored reaction channels to be obscured. This handicap can usually be overcome by ejection of dominant ions (23). Chen and Marshall have proposed a SWIFT simultaneous excitation–ejection detection sequence that appears promising for such dynamic range enhancement (26). Second, mass resolution must be sufficient to resolve ambiguous product compositions [e.g., $Fe_2CO_3^+$ versus $Fe(CO)_5^+$]. Finally, kinetic analysis of ion–molecule reactions requires that relative ion abundances be accurately inferred from the spectra. These requirements are routinely met by available excitation–detection schemes.

Two new detection schemes using chirp or single-frequency excitation have recently been developed—Hadamard transform and two-dimensional FT–ICR. The novelty of these techniques is that excitation and/or detection are modulated in such a way that mother–daughter or reactant–product relationships can be unambiguously determined. The Hadamard transform technique (119) takes advantage of the multichannel detection permitted by FT–ICR. A complete parent ion spectrum for n parent ions giving rise to a given daughter can be derived by first measuring the parent ion intensities for different combinations (the Hadamard "mask") of $0.5n$ of the parent ions. The resulting n simultaneous equations are solved by Hadamard transformation to give the parent ion spectrum for a given daughter ion. This can lead to an increase of the signal-to-noise ration of up to $0.5n^{0.5}$. In addition to parent ion spectra, daughter ion and neutral loss spectra can be obtained by this method in greatly reduced time.

Two-dimensional FT–ICR has been applied to the study of ion–molecule reactions (129). The pulse sequence used for this experiment is fundamentally similar to the NMR "NOESY" experiment (87, 121). The pulse sequence for 2-dimensional FT–ICR is given in Eq. (11):

$$P_1 - t_1 - P_2 - t_m - P_3 - t_2 \qquad (11)$$

where P refers to an rf excitation–deexcitation pulse and t denotes a time delay. P_1 excites all potential reactant ions to a detectable cyclotron orbit, while P_2 deexcites a specific reactant ion to a nondetectable orbit. The P_3 pulse corresponds to an excite-to-detect pulse. The first delay, t_1, defines the time between P_1 and P_2 while t_m is the ion–molecule reaction time. Finally, t_2 corresponds to the detection period. A time-domain two-dimensional signal $s(t_1, t_2)$ is generated by scanning t_1. The frequency (mass) domain 2-dimensional spectrum $S(\omega_1, \omega_2)$ is generated by Fourier transformation. Reaction [1] was shown to occur by using this method.

[1] $\qquad CH_3CO^+ + CH_3COCH_3 \longrightarrow CH_3C^+(OH)CH_3$

Figure 27 contains the two-dimensional spectrum for Reaction [1]. In this experiment, P_1 and P_2 were 79 Hz removed from the cyclotron frequency of CH_3CO^+ while P_3 differed from the cyclotron frequency of $CH_3C^+(OH)CH_3$ by 100 Hz. The set of peaks at $\omega_2 = 100$ Hz spaced at 79 Hz along the ω_1 axis indicate that Reaction [1] occurs for this system.

2. Examples of Ion–Molecule Studies: Clusters and Cluster Fragments

One of the most active areas of ion–molecule research falls in the category of metal ions and clusters. Although there are certainly exceptions, these studies are

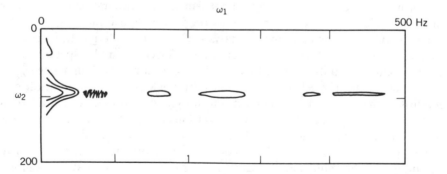

Fig. 27.   Two-dimensional FT–ICR spectrum obtained with the dual-frequency scheme of Eq. (11) to investigate the conversion of CH_3CO^+ to $CH_3C^+(OH)CH_3$ according to Reaction [1]. In the conventional ω_2 domain (ordinate), a single resonance appears at 100 Hz, corresponding to the offset of the cyclotron frequency of $CH_3C^+(OH)CH_3$ with respect to the rf carrier of the excite-to-detect pulse (P_3). In the ω_1 domain (abscissa), a family of sidebands appears at multiples of 79 Hz, corresponding to the offset of the frequency of CH_3CO^+ with respect to the rf carrier of the first two pulses (P_1 and P_2). [Reprinted by permission (129)]

directed primarily toward three goals: (a) understanding of cluster formation, (b) comprehension of the factors that control reactions of these species with a focus on contributing to the understanding of chemically important processes such as homo- and heterogeneous catalysis reaction mechanisms, and (c) physical descriptions of these species (e.g., modeling of catalytically important processes such as C–H and C–C bond insertion reactions). In this section, recent studies contributing to these three subareas will be reviewed with special emphasis given to those investigations that attempt to marry observations made by using FT–ICR to chemical theory. Although it is considered to be non-rigorous, dimers will be considered to be clusters for the purposes of this discussion. The natural progression from mononuclear species to clusters is dictated by mathematics, and no *direct* correlation between these two classifications is implied unless explicitly stated. It has been extensively suggested, however, that studies of homologous series of clusters will ultimately yield a greatly enhanced understanding of clusters, surfaces, and the reactions thereon (2, 158).

a. SILICON

A particularly exciting island in the sea of ionic cluster studies is the area of FT–ICR probes of the cluster chemistry of silicon. An understanding of this silicon chemistry holds great significance for the semiconductor industry especially the processes of laser chemical vapor deposition (LCVD) and surface etching. These processes have recently been followed on a molecular level in an FT–ICR spectrometer, and in some cases direct correlations to theory have been made. As is common to all ion–molecule studies, the overall procedure for monitoring the chemistry of these silicon species is based on kinetic analysis. It was already mentioned that the fourth step in the archetypal ion–molecule reaction experiment is "reactant ion reaction." Interestingly enough, this time window step is at the core of ion–molecule studies performed using FT–ICR. By varying this window and monitoring reactant and product relative ion populations, kinetic schemes for the reactions of any isolatable ion with neutral reactants can be generated. These kinetic schemes in turn suggest values for physical parameters (such as structure or electron deficiency) of the given reactant ion.

A severe lack of information regarding destructive aggregate formation during silane chemical vapor deposition prompted a detailed study of silicon ion–molecule reactions. In this study, the clustering reactions of $^{29}Si^+$ with SiD_4 were monitored at 0.4–4×10^{-6} torr (105). By utilizing a totally labeled reaction scheme, both forward and reverse rate constants were assigned. Figure 28 displays a temporal kinetics plot for the initial reaction of $^{29}Si^+$ with SiD_4. The observed reactions and associated rate constants are tabulated in Table 1. Clustering of $^{29}Si^+$ with SiD_4 proceeds in a specific fashion involving addition of individual SiD_2 units to the reactant ion (with consecutive loss of deuterium) until $Si_4D_6^+$ is formed. Further cluster growth occurs by a slow, bimolecular

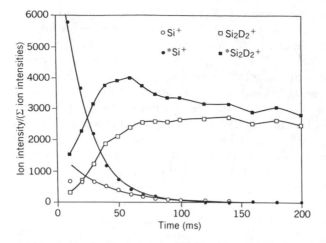

Fig. 28.   Temporal kinetic plot for the reaction of $^{29}Si^+$ with 1.85×10^{-6} torr of SiD_4 yielding $Si_2D_2^+$ and $^{29}Si_2D_2^+$. [Reprinted by permission (105)]

TABLE 1. Sequential clustering reactions of $^{29}Si^+$ with SiD_4. [Reprinted by permission (105)]

Reaction[a]	Product Fraction (%)	Rate Constant[b] ($\times 10^{10}$) (cm^3 molecule^{-1} s^{-1})	Reaction Probability[c]
$^*Si^+ + SiD_4 \rightarrow {}^*Si_2D_2^+ + D_2^{d,e}$	85 ± 4	8.1 ± 0.4^f	0.66 ± 0.04
$Si^+ + {}^*SiD_4^e$	15 ± 4		
$^*Si_2D_2^+ + {}^*SiD_4 \rightarrow {}^*Si_3D_4^+ + D_2$	88 ± 5	0.36 ± 0.04	0.035 ± 0.004
$^*Si_2D_2^+ + SiD_4$	12 ± 5		
$^*Si_3D_4^+ + {}^*SiD_4 \rightarrow {}^*Si_4D_6^+ + D_2$	87 ± 5	2.0 ± 0.3	0.21 ± 0.03
$Si_3D_4^+ + SiD_4$	13 ± 5		
$Si_4D_6^+ + SiD_4 \rightarrow Si_5D_{10}^{+g}$	100	0.0010 ± 0.0003	0.00011 ± 0.00003

[a] Only exothermic reactions are given here. Neutral products are given based on thermodynamical calculations of possible products. Those species that contain one silicon-29 isotope are designated as $(Si_xD_y)^{(+)}$. For example, $^*Si_5D_{10}^+ \equiv {}^{29}Si^{28}Si_4D_{10}^+$.

[b] The rate constant is the total measured reaction rate for the silicon-29 labeled species and includes isotope exchange where observed.

[c] The probability of reaction is calculated as the ratio of the measured reaction rate to the Langevin ion-molecule collision rate.

[d] This exothermic reaction product has been previously reported in ion cyclotron resonance and low-energy ion beam studies of the reactions of Si^+ with SiD_4 and SiH_4.

[e] This isotope exchange process has been observed in low-energy ion beam studies of the reactions of Si^+ with SiD_1 and SiH_1.

[f] This rate constant has also been previously determined (23) and the afore determined values are consistent with our measurements when consistent reference pressure calibrations are used and the isotope exchange process is properly accounted for.

[g] Only unlabeled $Si_4D_6^+$ reactions have been examined.

attachment of SiD_4. Care was taken to assure that the reaction data measured for $^{29}Si^+$ corresponded to the ground electronic state. This was established by two means. First, products arising from reaction of excited ions often gives rise to nonexponential kinetic behavior of the species in question causing slight deviations at early times in the temporal kinetic plot. Correction for this formation of excited-state products removes any deviation from the pseudo-first-order decay expected for the reacting ion. Second, oftentimes reactions of excited ions can be delineated by examining the sensitivity of the reaction to laser ablation power, the amplitude of the radio frequency field used for excitation in double resonance or ion detection, and reagent neutral pressure. Specific details of probing excited-state species in this manner are covered in a separate paper (105). An ab initio study of this aggregation process yielded two-dimensional potential energy surfaces such as the one displayed in Fig. 29 for the initial

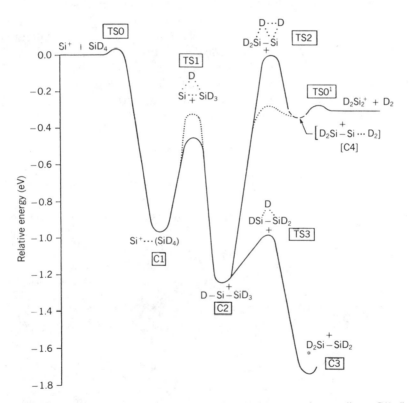

Fig. 29.   Schematic potential energy surface depicting the important intermediates, C(1–4), and transition states, TS(0–3), in the reaction of Si$^+$ with SiD_4. The symmetric structure, C3, responsible for scrambling of the ^{29}Si isotope is highlighted by a star. All of the energies for the products, reactants, C(i), and TS(i) obtained from ab initio electronic structure calculations (130) are indicated by solid lines. The dashed line at C4 and the brackets around the depicted C4 structure indicate that this structure and energy are estimated. Dotted lines at TS1 and TS2 show the energies calculated for these transition states using the phase-space theory that are consistent with experiment. [Reprinted by permission (105)]

reaction of $^{29}Si^+$ with SiD_4 (130). Using experimental values obtained in this study, phase-space theory (27) was applied to these clustering reactions giving reasonable agreement with the energies obtained for the proposed transition state by ab initio methods (see Fig. 29). The numerical results for both the ab initio and the phase-space theory results are given in Table 2. Based on these studies, the sequential clustering reactions of SiD_4 are proposed to encounter a bottleneck, which in turn prevents efficient formation of an aggregation nucleus. Further, it is suggested that this bottleneck arises from the branched nature of $Si_4D_6^+$ (Fig. 30a), which is apparently formed with precedence over its linear counterpart (Fig. 30b) due to an endothermic transition state encountered in the formation of the latter. It was noted that electron counting schemes do not correlate well with reactivity for this system, and, therefore, the structural reorganization due to electronic unsaturation proposed for metal clusters apparently does not occur for cationic hydrogenated silicon clusters (53, 57, 120).

The prototypical etching and deposition reactions of S_n^+ ($n = 2-9$) and Si_n^- ($n = 2-6$) have been modeled in the FT–ICR instrument. Xenon difluoride is observed to both fluorinate and degrade the positive silicon clusters sequentially (104). No reaction is observed for the anions. The scheme for the reactions of the positive silicon clusters and Xenon difluoride is shown in Fig. 31. In a companion study, nitrous oxide was found to sequentially degrade both the positive and negative silicon clusters according to Reaction [2] (102).

$$[2] \qquad Si_n^\pm \longrightarrow NO_2 \rightarrow Si_{n-1}^\pm + SiO + NO$$

TABLE 2. Comparison of PST and *ab initio* calculations of transition state (TSI-2) or intermediate complex (C2) energies for each sequential clustering reaction of $^{29}Si^+$ with SiD_4.[a] [Reprinted by permission (130)]

	Energy (eV)					
	PST Calculations			Ab initio Calculations		
Reaction	TS1	TS2	C2	TS1	TS2	C2
$*Si^+ + SiD_4 \rightarrow \begin{cases} *Si_2D_2^+ + D_2 \\ Si^+ + SiD_1 \end{cases}$	-0.34 ± 0.05	-0.29 ± 0.04	—	-0.46	-0.01	—
$*Si_2D_2^+ + SiD_4 \rightarrow \begin{cases} *Si_3D_4^+ + D_2 \\ Si_2D_2^+ + *SiD_4 \end{cases}$	-0.16 ± 0.02	-0.14 ± 0.02	—	-0.21	-0.05	—
$*Si_3D_4^+ + SiD_4 \rightarrow \begin{cases} *Si_4D_6^+ + D_2 \\ Si_3D_4^+ + *SiD_4 \end{cases}$	-0.27 ± 0.02	-0.19 ± 0.02	—	-0.21	-0.06	—
$Si_4D_4^+ + SiD_4 \rightarrow Si_5D_{10}^{+\,b}$	—	—	-0.80 ± 0.02^c	—	—	-0.98

[a] Zero-point energies are included in the energies listed.
[b] Only unlabeled $Si_4D_6^+$ reactions were examined.
[c] This value is for a PST calculation that includes free internal rotation and sets the collision efficiency factor, β, equal to unity.

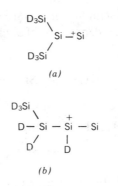

(a)

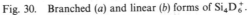

(b)

Fig. 30.   Branched *(a)* and linear *(b)* forms of $Si_4D_6^+$.

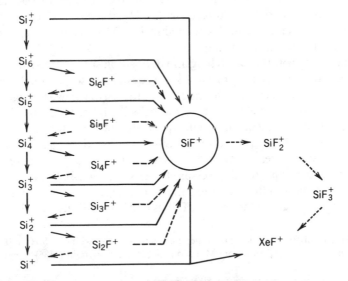

Fig. 31.   Summary of the initial and product ions for the reactions of bare and fluoridated silicon cluster ions with XeF_2. The product channels of the initial reactant clusters, Si_{1-7}^+, are designated by solid lines. The reaction pathways of the fluorinated cluster ion products, $Si_{2-6}F^+$, are shown by dashed lines. The reaction sequence for SiF^+, which leads to the ultimate product, XeF^+, is given by the dashed lines. Each of the lines/arrows represents a single bimolecular reaction with XeF_2. None of the neutral products are shown. [Reprinted by permission (104)]

This cluster etching reaction is proposed to occur via coupling of the unpaired electron on the cluster with an unpaired electron on an oxygen atom of nitrous oxide. It is interesting to note that etching of silicon surfaces is dramatically enhanced by ion bombardment and photochemical activation (50), which apparently increases the number of dangling bonds (radical or coordinatively unsaturated sites) on the surface. This indicates that these prototypical etching reactions of small ionic silicon clusters are good models for etching of the silicon-extended surface.

The reactions of CH_3SiH_3 with Si_n^+ ($n = 1-5$) proceed according to Reactions [3–6] resulting in prototypical deposition of a silicon atom on an ionic silicon cluster (103).

[3] $Si_n^+ + CH_3SiH_3 \longrightarrow Si_{n+1}CH_4^+ + H_2$

[4] $\longrightarrow Si_{n+1}CH_2^+ + 2H_2$

[5] $Si_5^+ + CH_3SiH_3 \longrightarrow Si_6CH_5^+ + H$

[6] $\longrightarrow Si_6H_3^+ + CH_3$

The product distributions for these reactions are given in Table 3. Notice the abrupt switch from molecular to radical products at Si_5^+. Also note that Si_6^+, Si_7^+, and none of the anionic silicon clusters are observed to react with CH_3SiH_3. The cationic silicon clusters that do react with CH_3SiH_3 are proposed to do so in accordance with the mechanism depicted in Fig. 32. This mechanism requires a divalent silicon site on the silicon cluster.

The studies of prototypical etching and deposition summarized above indicate the presence of two different reactive site types on these low nuclearity silicon clusters. Etching by xenon difluoride and nitrous oxide was suggested to be initiated by an unpaired electron on a radical silicon site (102, 104). In contrast, two electrons on a divalent silicon site appear to couple with the electrons in an Si–H bond of CH_3SiH_3 to give an Si–Si–H insertion product (103). The relationship between reactivity and site type is dramatically portrayed by a plot of observed collision probability versus cluster size (Fig. 33). Reactivity requiring a radical site (xenon difluoride and nitrous oxide) remains essentially constant throughout the range of cluster sizes. This is expected because a radical site must

TABLE 3. Product distributions for the exothermic reactions of Si_2^+ with CH_3SiH_3 at 0.6-eV trapping voltage[a]. [Reprinted by permission from M. L. Mandich et al., Copyright (1986) American Chemical Society]

Reactant Ion Si_a^+	$Si_{a+1}CH_5^+$ + H	$Si_{a+1}CH_4^+$ + H_2	$Si_{a+1}CH_3^+$ + $2H_2$	$Si_{a+1}H_2^+$ + CH_3	$Si_aCH_3^+$ + SiH_3	$Si_aCH_2^+$ + SiH_4	Si_aCH^+ + $SiH_3 + H_3$
Si^+	—	0.53	—	—	0.47	—	—
Si_2^+	—	0.28	0.40	—	0.13	0.16	0.03
Si_3^+	—	0.76	0.24	—	—	—	—
Si_4^+	—	0.70	0.30	—	—	—	—
Si_5^+	0.81	—	—	0.19	—	—	—

[a] Distributions are listed as fractions of the total primary product ions and have been corrected for relative isotope abundances. No reaction products were observed for Si_6^+ and Si_7^+ therefore they have been omitted from the table.

[b] Ionic products are listed according to their stoichiometries; formulas are not intended to imply structures.

[c] Calculated from ΔH_f^0 of possible neutral products; for example formation of SiH_3 is favored over $SiH + H_2$ by 44 kcal/mole.

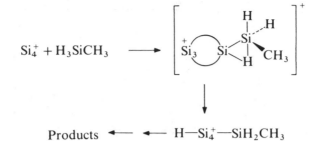

Fig. 32. The first step of the reaction of Si_n^+ with CH_3SiH_3 is proposed to be Si–H insertion as depicted here for Si_4^+ [Reprinted by permission (104)]

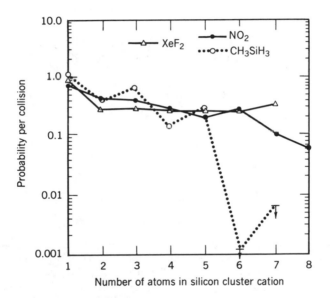

Fig. 33. Probability of reaction per collision for the reaction of Si_{1-8}^+ with various substrates. The cluster size dependence for clustering with CH_3SiH_3 suggests requirement of divalent silicon sites on the bare cluster. [Reprinted by permission (103, 104)]

be present in every positive cluster ion. On the other hand, reactivity of these clusters with CH_3SiH_3 falls off dramatically for $n > 5$. Bare silicon clusters larger than four silicon atoms contain primarily trivalent or tetravalent silicon sites, therefore reactivity requiring divalent sites should be diminished as cluster size increases as is observed.

In a separate work, the reactions of D_2, H_2O, CH_3OH, C_2H_2, NH_3, O_2, and N_2O with Si_n^+ ($n = 1$–6) were monitored (40). A significant attenuation of reactivity was found on going from $n = 3$ to $n = 4$ (Fig. 34). This attenuation was attributed to a switch from linear to cyclic structures for the clusters (40).

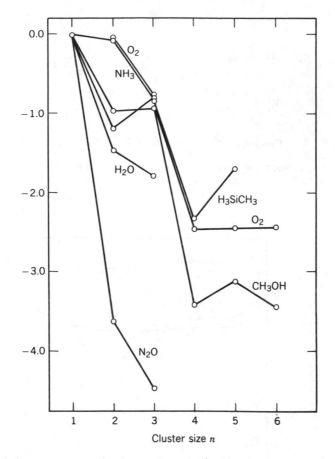

Fig. 34.   Relative rate constants for the reactions of Si_n^+ with the neutral molecules shown as a function of cluster size. The rate constants are normalized to the rate of Si^+ except for the reaction with O_2, which is normalized to the rate for Si_2^+. Rate data for H_3CSiH_3 were taken from Ref. 104. [Reprinted by permission (40)]

Smalley et al. probed the reactivity of Si_n^+ ($n = 7\text{--}65$) using supersonic beam generation of clusters as described above (46). It was found that certain clusters ($n < 14, 20, 25, 33, 39,$ and 45) were particularly inert to reaction with NH_3 and CH_3OH. For example, in the reaction of NH_3, Si_{39}^+ was measured to be two orders of magnitude less reactive than Si_{43}^+. In contrast, reactions of these clusters with NO, O_2, and H_2O showed no such selectivity. It seems that these clusters have "well-defined structures," which supports the work of Mandich et al. (105).

b.   CARBON

Carbon cluster chemistry has recently been the subject of both experimental and theoretical studies (136, 168). A great deal of the interest surrounding these clusters is related to the structural diversity of this class of compounds. In addition, these species are also of relevance to combustion, astrophysics, and organic chemistry in general (113). As in the silicon chemistry presented above, FT–ICR spectrometry has allowed novel and important probes into the reactions of carbon clusters. The internal excitation of C_2^+ was studied by charge transfer–chemical bracketing techniques (111). In these experiments, C_2^+ was formed by electron impact of various parent compounds and allowed to charge exchange with xenon or krypton. By using this experimental approach, it is possible to bracket the excess internal energies of C_2^+ generated from different sources relative to the ionization potentials of the noble gases. The bracketing method used is directly analogous to proton affinities obtained by observing proton transfer between a compound of known acidity and one of unknown acidity (12). C_2^+ formed from C_2N_2 was found to undergo charge transfer with xenon but not krypton. On the other hand, C_2^+ formed from C_2H_2 consisted of two populations: 70% of the ions would not charge exchange with Xe while 30% would. Charge exchange between C_2^+ and Kr was not observed (Fig. 35). The difference in reactivity was attributed to a long-lived excited state of C_2^+ (lifetime > 2 s) formed in 30% yield by electron impact ionization of C_2H_2. On the basis of additional studies the long-lived excited state was assigned as a $^2\Pi_u$ state approximately 1 eV above the $^4\Sigma_g^-$ ground state of C_2^+.

An FT–ICR study of the reactivity of molecular oxygen and deuterium with C_n^+ (n = 3–19) formed by laser ablation of diamond was coupled with a MNDO

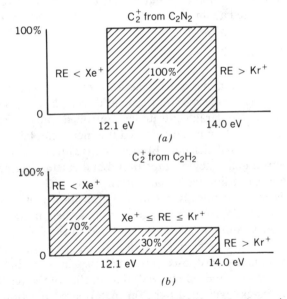

Fig. 35. Schematic mapping of the energy content of (a) C_2^+ formed by electron impact of C_2N_2, and (b) C_2^+ formed by electron impact of C_2H_2. The hatched portions represent the fraction of C_2^+ in each case that have recombination energies below 12.1 eV, between 12.1 and 14.0 eV, or above 14.0 eV, as determined by charge transfer–energy bracketing reactions with Xe and Kr. [Reprinted by permission (126)]

theoretical study yielding structural information for these clusters (113). The major product of the primary reaction of the carbon cation clusters with deuterium is the singly deuterated cluster ion (Reaction [7]):

[7] $$C_n^+ + D_2 \longrightarrow C_nD^+ + D$$
[8] $$C_n^+ + O_2 \longrightarrow C_nO^+ + O$$

Similarly, the major reaction channel for the primary reaction of these carbon clusters with oxygen leads to the monooxygenated carbon cluster (Reaction [8]). Two trends are observed for both Reaction [7] and [8]. First, reactivity of these clusters falls off slowly with increasing cluster size for $n = 3$–9. The reactions of these clusters is proposed to occur by the addition of either deuterium or molecular oxygen to the terminal carbene atoms of the linear carbon clusters with subsequent loss of deuterium or oxygen, respectively. This decrease in reactivity was attributed to a decrease in the carbene character of the terminal carbons of the linear carbon clusters and is supported by MNDO calculations. Second, a sudden drop in reactivity was observed to occur between C_9^+ and C_{10}^+. No clusters larger than $n = 9$ were observed to react with deuterium or molecular oxygen. This drop in reactivity was interpreted to indicate a change from linear to monocyclic structure. The lack of reactivity for the cyclic cluster ions would then be attributed to the lack of terminal carbene sites. A separate theoretical study indicated that this shift in structure occurs between $n = 10$ and $n = 11$ (91). C_7^+ exhibited a bimodal reactivity indicating that both a linear and a monocyclic isomer were present.

c. METALS

FT–ICR studies of ion–molecule reactions of transition metal ions are numerous. The popularity of such studies stems directly from the ability to study unsaturated metal-centered fragments under controlled collision/reaction conditions. Unsaturated transition metal-centered fragments can be characterized by several means: (a) electron deficiency, (b) coordinative unsaturation, and (c) steric availability. Electron deficiency of transition metal cluster fragments relates to electron counting. Coordinative unsaturation deals with the inability of the available ligands to electronically satisfy the transition metal cluster core to which they are bound. Coordinative unsaturation is, then, intimately connected to electron deficiency. Steric availability is related to the fraction of the ligand sphere surrounding the cluster nucleus that is unoccupied. These fragments frequently correspond to proposed intermediates in solution studies, and, therefore, it is hoped (and indeed realized) that probes of the reactivity of these metallic building blocks will shed light on metal-based reaction mechanisms. Because of the space limitations of this review, even a cursory discussion of all the work published in this area in the last three years is not possible. As a partial list, conventional ion–molecule studies of Y^+, La^+ (74); V^+, VO^+ (79); Cu^+, Ag^+, Au^+ (29, 161); Al^+, Ga^+, In^+ (28); $FeOH^+$, $CoOH^+$ (19); $Cr(CO)_3$, $Cr(CO)_4^-$

(63); $CuFe^+$ (148); and $Re_n(CO)_m^+$ (166) reacting with various organic reagents have recently entered the literature. Several unique experiments concerning this topic are presented below.

The state in which an ion is produced is fundamental to the study of ion–molecule reactions. Recent studies of the Cr^+ ion indicate that 70-eV electron impact ionization of $Cr(CO)_6$ forms Cr^+ in two electronic states (134). A kinetic treatment of the data obtained for the reaction of Cr^+ with $Cr(CO)_6$ suggests that 74% of the Cr^+ ions are formed in an excited electronic state having a radiative lifetime in excess of 2 s. This excited state reacts at 4.4 times the rate of the ground state Cr^+ ions. The reactions of Cr^+ formed by electron impact with CH_4 were also followed kinetically by FT–ICR (Reactions [9] and [10]):

[9] $$Cr^{*+} + CH_4 \longrightarrow CrCH_2^+ + H_2$$

[10] $$\longrightarrow Cr^{\circ+} + CH_4^*$$

Cr^{*+} forms $CrCH_2^+$ on reaction with CH_4 [a reaction which is endothermic for ground state Cr^+ (9)]. Least squares analysis of the kinetic data for Reactions [9] and [10] required that the rate constant for the excited-state quenching reaction (Reaction [10]) be $6.2 \pm 2.7 \times 10^{-10}$ $cm^3 s^{-1}$. It has been demonstrated, however, that a homogeneous ground-state population of Cr^+ ions can be generated by multiphoton ionization–multiphoton dissociation of $Cr(CO)_6$ at both 266 and 353 nm (72).

The effects of metal-to-metal and ligand-to-metal bonding on the reactivity of transition metal clusters with alkanes bears heavily on the elucidation of organometallic reaction mechanisms. The presence of a single ligand or an additional metal center can dramatically alter the chemistry of a transition metal species. A clear example of such alteration is found in a recent study of the reactivity of Co_2^+ and Co_2CO^+ with alkanes (55). Co^+ activates both C–C and C–H bonds in the gas phase, but Co_2^+ is unreactive toward alkanes (54). In contrast, Co_2CO^+ efficiently activates C–H bonds of alkanes. For example, Co_2CO^+ dehydrogenates n-butane according to Reactions [11] and [12]:

[11] $$Co_2CO^+ + n\text{-}C_4H_{10} \longrightarrow Co_2COC_4H_6^+ + 2H_2$$

[12] $$\longrightarrow Co_2COC_4H_8^+ + H_2$$

The smallest rate constant for Co_2CO^+ reacting with n-butane is at least 2 orders of magnitude greater than the greatest rate constant for the reaction of Co_2^+ with n-butane. The reaction enhancement for Co_2CO^+ relative to that for Co_2^+ was suggested to be due to Co–Co bond polarization by the carbonyl ligand giving a Co^+-like site on the Co_2CO^+ cluster fragment.

The mode of chemisorption for nitrous oxide on small cobalt clusters (2–4 atoms) was recently probed by following the reactions of the nitrosylated clusters with $^{18}O_2$ (90) (Reactions [13–17]):

[13]　　　　$Co_2NO^+ + {}^{18}O_2 \longrightarrow Co_2{}^{18}O_2 + N^{16}O$

[14]　　　　$Co_3NO^+ + {}^{18}O_2 \longrightarrow Co_3{}^{18}O_2^+ + N^{16}O$

[15]　　　　　　　　　　$\longrightarrow Co_3{}^{16}O^{18}O^+ + N^{18}O$

[16]　　　　$Co_4NO^+ + {}^{18}O_2 \longrightarrow Co_4{}^{18}O_2^+ + N^{16}O$

[17]　　　　　　　　　　$\longrightarrow Co_4{}^{16}O^{18}O^+ + N^{18}O$

The observance of only $N^{16}O$ for Reaction [13] indicates that nitrous oxide is molecularly absorbed to Co_2^+. Isotope scrambling was observed for reaction of Co_3NO^+ and Co_4NO^+ with ${}^{18}O_2$ indicating that nitric oxide is dissociatively chemisorbed on both Co_3^+ and Co_4^+. If the scrambling reactions [14–17] proceed statistically, then a $2:1$ ratio should be observed for Reactions [15] and [17] over Reactions [14] and [16], respectively. While this holds for Co_4NO^+, Co_3NO^+ yields a monolabeled/dilabeled product ratio of $3.2:1$ indicating that Reaction [15] is favored over Reaction [14]. Considering this result, the structure of Co_3NO^+ was proposed consist of three cobalt atoms at the vertices of a triangle with a nitrogen atom occupying one face and the oxygen atom occupying the opposite face.

In most cases, reactions observed by FT–ICR are exothermic. In a recent study, however, endothermic thresholds for the reactions Co^+ and Fe_2^+ with cyclopropane, ethane, and ethene were determined (52). In this experiment, the ion of interest is excited to a known translational energy by rf excitation, and the endothermic reaction thresholds are determined by plotting product ion abundance versus the center-of-mass energy for the reaction. By using this method, the corrected threshold for Reaction [18] was determined to be 0.53 ± 0.3 eV (Fig. 36).

[18]　　　　$Co^+ + c\text{-}C_3H_6 \longrightarrow CoCH_2^+ + C_2H_4$

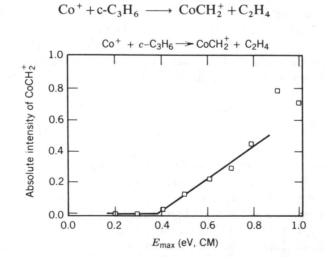

Fig. 36.　Intensity of $CoCH_2^+$ vs. kinetic energy of Co^+ for Reaction [18]. The threshold for this reaction is approximately 0.38 eV (CM) without correction. [Reprinted by permission (52)]

Using this threshold, $D^0(Co^+-CH_2)$ was calculated to be 81 ± 7 kcalmol^{-1} in good agreement with the ion beam result of 85 ± 7 kcalmol^{-1} (8) and the photodissociation result of 84 ± 5 kcalmol^{-1} (67). This technique for determining endothermic thresholds suffers from systematic overestimation and large error limits with respect to measurements made by ion beam instruments. The insensitivity of FT–ICR with respect to beam experiments is the likely source of the overestimation inherent in this technique. One source of error for these measurements as made by FT–ICR is related to the uncertainty of the ion's translational energy upon collision. Equation (6) describes the translational energy of the ion after rf excitation. The second term gives the energy spread for the ions, which is centered around the average value given by the first term. This energy spread can be significant (e.g., for an ion at 30 eV the energy spread is $\pm 5\%$).

The exothermic reactions of $ClCr^+$, $ClMn^+$, and $ClFe^+$ with small alkanes differ significantly from that of the bare-metal ions (106). While Fe^+ is reactive with alkanes larger than propane (65), $ClFe^+$ obtained by electron impact of $CpFe(CO)_2Cl$ is "totally unreactive" as determined by FT–ICR. In contrast, Cr^+ displays no reactivity with alkanes (106), while $ClCr^+$ obtained by electron impact induced fragmentation of $CpCr(NO)_2Cl$ is reactive with alkanes larger than propane. This is the first instance of a ligand causing an unreactive atomic ion to become reactive. It is important to note that this is not the first instance of reactivity enhancement due to ligation of a cluster species as was discussed above for Co_2^+ (55). Neither Mn^+ (65) or $MnCl^+$ appear to activate C–C or C–H bonds. Mandich et al. performed Hartree–Fock and generalized valence bond (GVB) calculations on $ClCr^+$, $ClMn^+$, and $ClFe^+$ to determine the ground state and low-lying excited states for these species. $FeCl^+$ and $MnCl^+$ were found to form a σ bond between a metal s orbital and a chlorine p orbital. $CrCl^+$, on the other hand, was found to have two triple-bonded low-lying bonding states: (a) a covalent σ bond analogous to that of $FeCl^+$ and $MnCl^+$ with a $p\pi$–$d\pi$ dative bonding component, and (b) a covalent π bond formed between a metal d orbital and a singly occupied chlorine p orbital. In this state, doubly occupied p and s orbitals on chlorine still form dative bonds with the respective metal orbitals. The increased reactivity of $CrCl^+$ over that of Cr^+ with alkanes is attributed to the π bonding state. A C–H bond is proposed to add across the π Cr Cl covalent bond in the initial reaction step for $CrCl^+$ reacting with alkanes.

A major goal for ion–molecule reaction studies is the elucidation of chemical structure and processes associated with the surface of transition metal clusters. Several issues are at the heart of these studies: (a) ligand binding modes, (b) ligand fluxionality, (c) metal–metal bonding, (d) metal–metal bonding fluxionality, and (e) cluster unsaturation. These areas can all be probed by FT–ICR ion–molecule reaction studies.

Hydrogen migrations on Fe_2^+ were monitored by following the reactions of Fe_2H^+ and Fe_2D^+ with both labeled and unlabeled ethene and propene (82). The efficiency of the symmetric exchange (Reaction [19]) proceeds with only 8% efficiency (based on the Langevin collision frequency) indicating that both reversible insertion–β-elimination and reversible vinylic C–H bond insertion have significant reaction barriers for $HFe_2(ethene)^+$.

[19] $$Fe_2D^+ + C_2H_4 \longrightarrow Fe_2H^+ + C_2DH_3$$

In contrast, labeling studies of $HFe_2(propene)^+$ indicate that Fe_2^+ insertion into allylic C–H bonds is both reversible and facile resulting in formation of $Fe_2(allyl)^+$ and H_2.

   An FT–ICR study of niobium clusters, Nb_n ($n = 3$–25), strongly suggests that geometrical structure of clusters is very important in determining reactivity (49). These clusters were formed and studied in the tandem supersonic beam/FT–ICR device previously discussed. The clusters were exposed to a constant pressure of H_2, and two types of reactivity were monitored. First, the kinetics of the chemisorption of the first H_2 for each cluster was measured. Clusters containing 8, 10, 12, and 16 atoms were found to have the lowest reaction rates (Fig. 37). Second, the number of hydrogen atoms required to saturate the cluster was counted. Clusters containing 8, 9, 10, 12, 16, and 19 became saturated (passivated) at H–Nb ratios significantly lower than the theoretical value of 1.3 (Fig. 38).

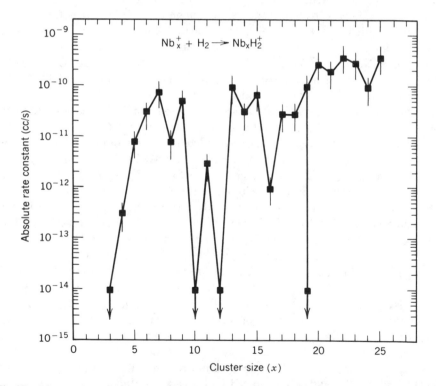

Fig. 37.   Measured reaction rates of Nb_x^+ clusters in the FT–ICR cell as a function of cluster size for the chemisorption of the first H_2 molecule. Two rates are plotted in the case of Nb_{19}^+, corresponding to two distinct structural forms of these clusters. [Reprinted by permission (49)]

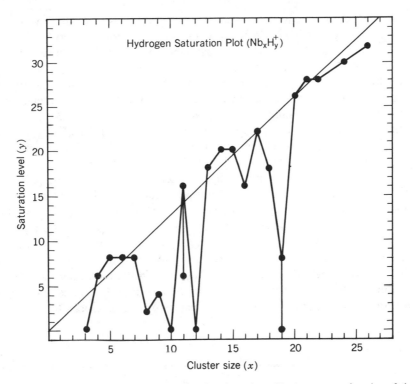

Fig. 38.   Plot of saturation values for niobium chemisorption of hydrogen as a function of cluster size. Note that two forms of clusters were found for Nb_{11}^+ and Nb_{19}^+, both with different values of saturation coverage, [Reprinted by permission (49)]

Nb_{11}^+ and Nb_{19}^+ were found to passivate at two different saturation values suggesting the presence of two geometric forms for these clusters. The geometric forms of Nb_{11}^+ and Nb_{19}^+ having the lower saturation value also have the lowest rate for chemisorption of the initial H_2. These data were interpreted to indicate that geometric factors influence transition metal cluster reactivity much more than previously thought.

The mechanism of cluster formation is of seminal importance to the understanding of clusters and their chemistry. The sequential clustering reactions of several transition metal systems have recently been studied using FT–ICR. Wronka and Ridge studied the clustering reactions of iron carbonyl anions with $Fe(CO)_5$ using conventional ICR (167). On the basis of these studies, a direct relationship between electron deficiency and reactivity of the cluster fragment ions was proposed. In this study, electron deficiency was calculated by subtracting the number of valence electrons (i.e., metal electrons and donated ligand electrons) for the cluster fragment from $18n$ ($n =$ number of iron atoms) and then dividing by n. This deficiency corresponds to the average electron deficiency per metal center with respect to the 18-electron rule. The electrons in metal–metal bonds are counted twice in this scheme. Wronka and Ridge (167) found a smooth monotonic rise for the log of the reaction rate as a function of electron deficiency as long as $Fe_2(CO)_x^+$ $(x = 5-7)$ are considered to contain double metal–metal bonds. The plot of log of the reaction rate versus electron deficiency for the iron carbonyl anions leveled off at an electron deficiency of 2, indicating that the reaction rate for these species reaches a maximum when each metal center possesses a $2\,e^-$ donor site.

A study of the clustering reactions of the $Fe_n(CO)_x^+/Fe(CO)_5$ and the $Cr_m(CO)_y^+/Cr(CO)_6$ systems directly analogous to the anion clustering study described above has been reported (56). Electron deficiencies calculated for product cluster fragments by assuming (a) structures based on boron polyhedra, (b) single metal–metal bonds, and (c) conventional ligand binding modes (e.g., carbon monoxide as a $2\,e^-$ donor) do not always correlate in a monotonic manner with the log of the reaction rate. Figure 39 is a plot of the log of the reaction rate versus the electron deficiency for the sequential clustering reactions of Fe^+ with $Fe(CO)_5$ based on these assumptions. $Fe_4(CO)_8^+$, $Fe_3(CO)_6^+$, and $Fe_2(CO)_3^+$ have reactivities far lower than predicted. By assuming the existence of multiple metal–metal bonds or unusual ligand binding modes (e.g., carbon monoxide as a $4\,e^-$ donor) for this same clustering system, a monotonic function

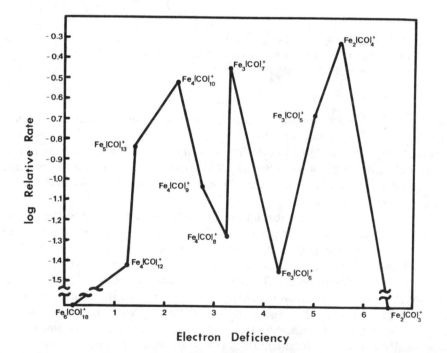

Fig. 39. Plot of log relative rate vs. calculated electron deficiencies for the $Fe^+/Fe(CO)_5$ system. Calculated electron deficiencies are based on simple polyhedra models of boron hydrides. [Reprinted by permission from D. A. Fredeen et al., Copyright (1985), American Chemical Society]

of electron deficiency was obtained showing the same "leveling off" at an electron deficiency of 2 as for the iron carbonyl anions (Fig. 40). Similar experiments were reported for the $Ni_x(CO)_c^+$–$Ni(CO)_4$ and the $Co_z(CO)_d(NO)_e^+$–$Co(CO)_3(NO)$ systems (57).

The electron deficiency model as just discussed is useful for correlating electron saturation and clustering reactivity for ionic transition metal clusters. Unfortunately, the clustering mechanism cannot be derived from this approach. The CVMO theory proposed by Lauher (97) is helpful for determining the geometry of transition metal clusters that, in turn, suggests possible mechanisms for cluster formation. According to the CVMO model, the valence orbitals of the metals belong to two classes: (a) high lying antibonding orbitals (HLAO) and cluster valence molecular orbitals (CVMO). The CVMO are available for either metal–metal or ligand–metal bonding. HLAO are unoccupied due to their strongly antibonding nature. The number of CVMO and HLAO for a given cluster is a function of the geometry alone—the identity of both ligands and metals are not important. Analogous to the electron deficiency model, clusters that do not have two cluster valence electrons (CVE) in each CVMO will be unsaturated electronically, and, therefore, these clusters will be reactive. One basic difference between the electron deficiency and the CVMO models is that

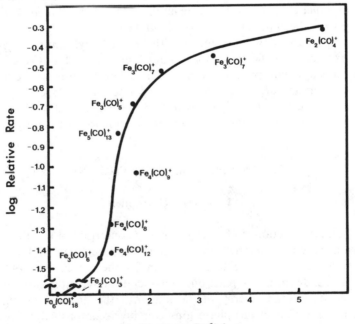

Fig. 40. Plot of the relative rate vs. electron deficiency for the $Fe^+/Fe(CO)_5$ system. In this plot, the electron deficiencies were calculated by assuming that some of the cluster fragments contain multiple metal–metal bonds and/or carbonyl ligands acting as $4\,e^-$ donors. [Reprinted by permission from D. A. Fredeen et al., Copyright (1985), American Chemical Society]

the electron deficiency model relates the *average* electron deficiency per metal atom to reactivity, whereas the CVMO model relates overall cluster valence electron deficiency (CVED $\equiv 2 \cdot$ CVMO–CVE) to reactivity. Using the CVMO approach, the log relative rates were related to the CVED for the clustering reactions of the $Co(CO)_3(NO)$–$Ni(CO)_4$, $Fe(CO)_5$–$Ni(CO)_4$, and $Fe(CO)_5$–$Co(CO)_3(NO)$ systems (58).

For unsaturated cluster fragments, the trend for clustering reactions is toward electronic saturation. For the Ni, Fe, Co, and Cr carbonyl–nitrosyl cationic clustering reactions, this drive for saturation proceeds in steps of 14 e$^-$ additions (56–58). For example, in the Cr^+–$Cr(CO)_6$ system, $Cr_2(CO)_4^+$ (19 CVE) reacts with $Cr(CO)_6$ to give $Cr_3(CO)_8^+$ (33 CVE) by addition of $Cr(CO)_4$ (14 potential CVE) (56). This, according to Mingos, corresponds to addition of a bridging cluster fragment to the growing cluster in a local raft structure (122). CVMO theory also satisfactorily explains the cessation of clustering in these systems. For example, in the $Fe_2(CO)_4^+$ clustering reactions, clustering ceases at $Fe_6(CO)_{18}^+$. $Fe_6(CO)_{18}^+$ has 83 CVE, which corresponds to electronic saturation of a cationic bicapped tetrahedron (97).

The clustering reactions of $H_2Os_3(CO)_{10}$ were studied at very low pressures ($< 10^{-8}$ torr) by FT–ICR (123). In these reactions, the Os_3 central core remained intact such that oligomers of this central fragment were formed up to pentamers. The structures of the dimer and trimer cluster cations were proposed using kinetic arguments. The log of the reaction rate for the dimers varies monotonically with the number of carbonyls in $H_xCo_6(CO)_y^+$ suggesting that the electron deficiency changes smoothly with the carbonyl number. This smooth variation of electron deficiency implies that the geometry of the Os_6 core changes from a capped square pyramid to a bicapped tetrahedron on the loss of two hydrogens. Similar arguments lead to a proposed tricapped ($\square^3$) trigonal prismatic structure for the trimers with two hydrogens and ≥ 20 carbonyls. A tricapped octahedron is suggested for trimers with 15–19 carbonyls.

C.   CID AND PHOTODISSOCIATION STUDIES

1.   General Description

Ion–molecule reaction studies are normally comprised of kinetic mapping of exothermic reaction channels in an effort to categorize and quantify the reactivity of an ion. Structural information for the ion can be obtained from studies of its dissociation reactions. Dissociation reactions are induced by increasing the internal energy of the ion until the dissociation threshold is exceeded. Excitation of gas-phase ions can be accomplished by two means: (a) collisional activation (CA) and (b) photon absorption. CA occurs when translationally excited ions are allowed to collide inelastically with a target gas; conversion of translational energy into internal energy of the ion yields a vibrationally excited reactant ion. Ions are translationally excited in ICR by rf irradiation at the characteristic cyclotron frequency for a given value of m/z [Eq. (6)] (14). The upper limit on translational excitation in FT–ICR is imposed by the radius of the cell and the flux density of the magnetic field. Typically, ions can be accelerated to energies between 10 eV and a few hundred eV; however, it is possible to extend the upper limit to several keV (17). Alternately, lasers or arc lamps can be used as photon sources (IR, Visual, UV) for performing photo-excitation of the ion. For both collisional activation and photodissociation, structural information is obtained by comparing the relative abundance of the collision-induced dissociation product ions or the photofragment ions.

2. Collisional Activation and Collision-Induced Dissociation

The collision between a translationally excited ion and a neutral is a complicated reaction resulting in three possible processes: (a) collisional activation of the incident ion, (b) formation of an ion–molecule product ion via an endothermic reaction channel, or (c) charge transfer between the incident ion and the neutral target. We have ignored other processes that are not observable in this experiment, for example, collisional activation without ionization of the target or production of excited states that decay by radiative or nonradiative decay channels other than dissociation. The relative cross sections for these processes depend on the kinetic energy of the incident ion and the nature of the target neutral. For example, at low kinetic energies (1–10 eV) and using complex polyatomic targets, endothermic ion–molecule reactions are frequently observed. At higher kinetic energies (10–100 eV), the cross sections for formation of ion–molecule products decrease, and collisional activation becomes an important reaction channel, especially for rare-gas target neutrals (e.g., He, Ne, Ar, etc.). In the energy range between a few hundred eV and several keV, the cross section for CA is appreciable; however, recent studies have demonstrated that charge transfer reactions are also important at these energies. For example, in a recent study on the $C_6H_6^{+\cdot}$ (benzene) ions colliding with argon at 500 eV to 3.2 keV (laboratory frame of reference), charge transfer to argon was observed. This result shows that large amounts of energy can be transferred to the neutral target during CA (e.g., Reaction [20] is endothermic by approximately 6 eV).

[20] $$[C_6H_6^+]^* + Ar \longrightarrow C_6H_6 + Ar^+$$

[21] $$[C_6D_6^+]^* + C_6H_6 \longrightarrow C_6D_6 + C_6H_6^+$$

In addition, energetic collisions between perdeuterobenzene $C_6D_6^{+\cdot}$ and benzene neutral (Reaction [21]) demonstrate that for translational energies in excess of 500 eV, dissociative charge transfer of benzene is favored over CID of $C_6D_6^{+\cdot}$ by a factor of 10 (138).

Ion–molecule reactions of translationally excited ions by FT–ICR are used (a) to study dissociation reactions of ion–molecule product ions, (b) to determine structures of ionic species formed by direct ionization, and (c) to study energy requirements of ion–molecule reactions in order to obtain bond strengths or

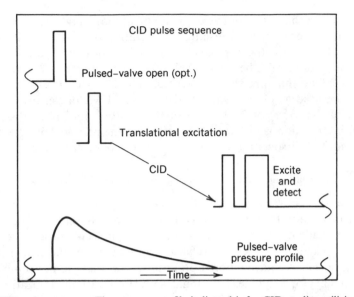

Fig. 41.   CID pulse sequence. The pressure profile indicated is for CID studies utilizing a pulsed valve.

reaction threshold energies. The generic CID pulse sequence depicted in Fig. 41 can be used at any point in the overall FT–ICR pulse sequence.

As an example of such studies, the mechanism of dissociation for various collisionally activated N-alkylpyridinium cations has been investigated (157). The primary objective of this study was to determine the most facile route of dissociation for both CID and infrared multiphoton dissociation. Two mechanisms for dissociation were observed. If an alkyl hydrogen β to the nitrogen on the pyridine ring is available, then dissociation yields pyr-H$^+$ ion by loss of an olefin. If a β alkyl hydrogen is not present, then simple cleavage of the N-alkyl bond yields an alkyl carbocation and loss of neutral pyridine. The results of CID are analogous to infrared multiphoton absorption dissociation for this system.

The effectiveness of ion–molecule reaction studies hinges directly on resolving structures of reaction products. Such elucidation is the greatest utility for CID. The ions formed by ion–molecule reactions in the FT–ICR are almost never isolable species formed in minute abundance; and, therefore, little structural information exists for these ions. Freiser et al. introduced CID by FT–ICR as a method for studying the structures of ion–molecule reaction products (31). Since that time, this technique has been employed for such studies, especially for the study of the structures of organometallic product ions. For example, the mode of oxygen binding in $Cr(CO)_3O_2^-$, $Cr(CO)_3O^-$, and CrO_3^- formed by reaction of $Cr(CO)_5^-$ with O_2 has been probed using CID (17a). $Cr(CO)_3O_2^-$ was observed to lose O_2^- during CID, indicating that molecular oxygen is bound to $Cr(CO)_3$. $Cr(CO)_3O^-$ undergoes consecutive loss of carbonyl ligands with no loss of O observed. The CID data indicates that atomic oxygen binds to the chromium anion more strongly than does carbon monoxide. CrO_3^- loses all three oxygen atoms upon CA suggesting that the oxygens of CrO_3^- are not bound as O_2.

CID was used to study two isomers of $LaC_4H_6^+$ formed from the reaction of La^+ with n-butane and iso-butane (74). Figures 42a and 42b show the CID spectra for each isomer as a function of collision energy on argon. Although similar fragments are observed for both isomers, the relative abundances of the fragment ions are characteristic for each isomer. Labeling studies as well as reaction of La^+ with 1,3-butadiene indicate that the product of La^+ reacting with n-C_4H_{10} has a 1,3-butadiene structure (Fig. 43a) while the product of La^+ reacting with iso-C_4H_{10} has a trimethylene-methane structure (Fig. 43b).

CID techniques can also be used to bracket bond energies within ionic species. As an example, observation of Reaction [22]

[22] $CuFe(C_4H_6)^+ + N \longrightarrow [CuFe(C_4H_6)^+]^* \longrightarrow CuFe^+ + C_4H_6$

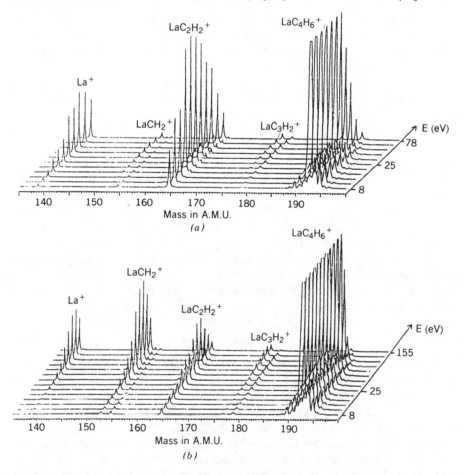

Fig. 42.   CID of $LaC_4H_6^+$; (a) formed from La^+ and n-butane, and (b) La^+ and isobutane. [Reprinted by permission from Y. Huang, et al., Copyright (1987), American Chemical Society]

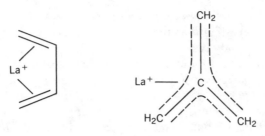

Fig. 43.   Two structures for $LaC_4H_6^+$; (a) La^+ is proposed to give the 1,3-butadiene structure on reaction with n-C_4H_{10}, while (b) La^+ is proposed to give the trimethylene-methane structure on reaction with iso-C_4H_{10}.

by CA suggests that $D^0(Fe^+\text{–}Cu) > D^0(Fe^+\text{–}C_4H_6)$ (148)$=48\pm5$ kcalmol^{-1} (80). This approach assumes that the collision energy is randomized throughout the ion-neutral collision complex such that the weakest bond breaks (165). It is obvious that the bond strengths against which these measurements are weighted must be determined by other methods.

CID suffers from three fundamental problems related to conditions within the FT–ICR cell. First, unlike sector instruments, the translational energy of the incident ion cannot be determined accurately. This was discussed previously with reference to endothermic threshold determinations by CA. Second, CID efficiency falls off sharply with mass; this is due to the increasing degrees-of-freedom with increased size of the molecule. The low CID efficiency limits the utility of CID for large molecules (e.g., biomolecules). Finally, efficient CID is a multiple collision process, but the pressure required for multiple collision conditions ($>1 \times 10^{-6}$ torr) result in poor resolution for FT–ICR. The limitation on resolution imposed by pressure can be circumvented by use of pulsed valves or a two-section cell where CID is performed in a low-vacuum source region and the fragment ions are partitioned to the high-vacuum analyzer region for detection (164). Isobaric daughter ions (m/z 105) formed by CID of a mixture of acetophenone and 1,3,5-trimethylbenzene were resolved ($m/\Delta m = 211,000$) using a two-section cell (164).

3. Photodissociation

Photodissociation is ideally suited for ICR dissociation studies. Ions are trapped in the ICR for seconds, allowing photodissociation of novel ionic species to be studied. The three fundamental limitations concerning studies performed by CA/CID are circumvented by using photon absorption as a means of ion excitation. Namely, (a) ion excitation by photon absorption is not limited to low mass ions; (b) photodissociation does not require a bath gas, hence such studies can be performed under optimal detection conditions for FT–ICR; and (c) photon absorption corresponds to discrete energy deposition such that the internal excitation of photodissociating ions can be specified.

FT–ICR photodissociation studies are broad in scope and have been recently reviewed (42, 44). The discussion here represents a selective sample of such studies. Photodissociation studies can be loosely assigned to three different regimes depending on the information desired: (a) branching ratios among reaction channels, (b) the photon energy dependence for dissociation, and (c) studies concerned with time dependence for both radiative and nonradiative decay of excited ions.

The first category of photodissociation studies corresponds to analytical studies by FT–ICR (e.g., peptide sequence elucidation). In such studies, the energy and number of the incident photons is of secondary importance to structural information. This approach has been demonstrated by consecutive isolation–photodissociation of Gly–Phe–Ala–CH_3/CD_3 and its daughter ions (16). In this study, protonated parent ions of the tripeptide (m/z 308 and 311) were formed by low-pressure chemical ionization of the tripeptide neutral. The parent ions were then irradiated by 193-nm photons, which yields four daughter ions at m/z 104/107 (H_3N–$CH(CH_3)$–$C(O)$–CH_3/CD_3^+), 120 (H_2N=$CHCH_2(C_6H_5)^+$), 91 ($C_7H_7^+$), and 30 (H_2N=CH_2^+). The sequence of dissociation processes was confirmed by isolation of the m/z 120 fragment using rf sweeps and allowing it to absorb a second 195-nm photon yielding grand-daughter ions at both m/z 77 ($C_6H_6^+$) and 91($C_7H_7^+$), indicating the presence of phenylalanine. In a similar experiment, Hunt, et al. were able to obtain sequence information for a 15-residue peptide (1771 amu) using only 20 pmol of sample (75). These experiments were performed on a tandem quadrupole–FT–ICR instrument. The ions were trapped in the cell and photodissociated by 193-nm photons from an argon fluoride laser. By studying both the peptide and its methyl ester, the sequence Phe–Gln–Glu–Thr–Phe–Glu–Asp–Val–Phe–Ser–Ala–Ser–Pro–Lxx–Arg was determined. The first three N-terminal residues were confirmed by Edman degradations.

The second category of photodissociation studies is concerned with the relationship of photodissociation efficiency to wavelength of incident light. In these experiments, the number of photons absorbed must be known. By monitoring the extent of photodissociation with respect to wavelength for an ion absorbing single photons, one generates the photodissociation spectrum for the ion. The onset (longest wavelength) of photodissociation is believed to be an accurate estimate of the dissociating bond strength as long as an electronic state lies close in energy to the dissociation threshold. If no excited electronic states lie near the dissociation threshold for the ion, then the photodissociation onset is spectroscopically limited and only an upper limit for the dissociating bond energy may be determined (67). Hettich and Freiser have employed the photodissociation technique to estimate bond energies for ML^+ (67, 69), ML_2^+ (70), and MFe^+ (68) (M is a transition metal, L is any ligand). For most transition metal containing species, the density of electronic states is high enough to yield thermodynamically limited values for bond strengths. For example, Fig. 44 depicts the photodissociation spectrum for loss of CH_3 from $CoCH_3^+$ (70). The dissociation threshold for $CoCH_3^+$ is taken to be 500 nm corresponding to a

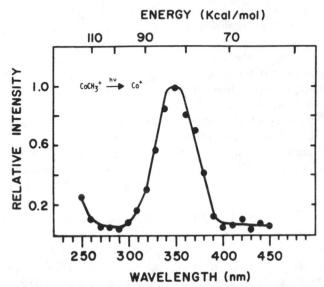

Fig. 44. The photodissociation spectrum of $CoCH_3^+$ obtained by measuring the intensity of the signal for Co^+ as a function of wavelength. [Reprinted by permission from R. L. Hettich et al., Copyright (1986), American Chemical Society]

Co^+–CH_3 bond strength of 57 ± 7 kcalmol^{-1}. This is in excellent agreement with the value of 61 ± 4 kcalmol^{-1} determined previously (7).

The collisional relaxation of photoexcited bromobenzene ions was studied using pressure-dependent two-photon photodissociation (1, 100). The wavelength of the incident photons (514.5 nm) was selected such that two photons were needed to reach the dissociation threshold for bromobenzene ion. The pressure of the system was maintained sufficiently high (up to 10^{-5} torr) such that collisions occurred on the timescale of the laser irradiation pulse, viz. 1 s. Therefore, collisional relaxation of photoexcited bromobenzene ions by the collision gas and on neutral bromobenzene molecules gives rise to a relaxation process that competes with photodissociation. The collisional quenching rates for a given collision gas can be determined from kinetics plots at several pressures. Using this approach, Ahmed and Dunbar measured the quenching efficiency for 35 neutral partners (1). As an example, He quenches excited bromobenzene ions with 0.38 times the efficiency of N_2, whereas CH_3F quenches bromobenzene ions at 2.4 times the rate of N_2.

Finally, time-resolved photodissociation studies of gas-phase ions yield information concerning internal energy content or ion structure as well as the mechanism of both dissociation and relaxation processes. Three types of such studies are represented by the typical experimental pulse sequences depicted in Figs. 45a–45c. The key difference between these experiments consists in the number of times the ions are irradiated during the experiment.

The time-resolved single-pulse experiment (Fig. 45a) is performed in two modes. The first mode consists of monitoring photodissociation as a function of

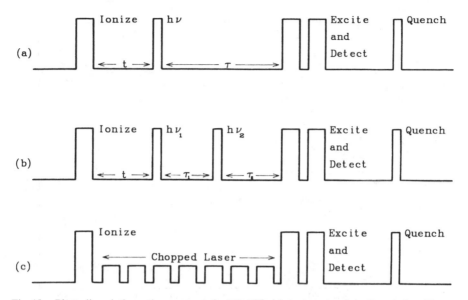

Fig. 45.   Photodissociation pulse sequences for FT–ICR: (a) single-pulse photodissociation, (b) two-pulse photodissociation, and (c) chopped photodissociation.

the time between the ionization event and photoexcitation of the ion (t). This mode allows one to monitor the net excitation of the ion due to the ionization event itself. Using this method, the relaxation of excited thiophenol ions formed by electron impact (17 eV) was probed under collisionless and collisional conditions (10). In this experiment, 2.4-eV photons (515 nm) were used for photoexcitation. The dissociation threshold of thiophenol ions is 3.2 eV, and, therefore, single-photon photodissociation occurs if the internal energy of the ions exceeds 0.8 eV. If the energy of an ion drops below the 0.8-eV threshold, then two photons are required for photodissociation. By varying t for different pressures, one can monitor the relaxation of thiophenol ions to an energy below 0.8 eV due to both radiative and collisional relaxation. At 0.6×10^{-8} torr sample pressure, relaxation of thiophenol ions is found to be primarily radiative at a rate of 3 s^{-1}. At 2.1×10^{-8} torr, the relaxation rate includes a collisional component causing this rate to increase to 4.5 s^{-1}. In both cases 40% of the thiophenol ions formed by electron impact (17 eV) have energies in excess of 0.8 eV above thermal.

The single-pulse photodissociation experiment can be performed in a second mode where the extent of photodissociation is monitored with respect to the time between irradiation and detection, τ. Using a rapid FT–ICR detection sequence, Dunbar was able to measure the time-resolved threshold photodissociation of chlorobenzene ion with a time resolution of about 10 μs (42a). Figure 46 contains one such time-resolved plot for the production of m/z 77 ions from $C_6H_5Cl^+$ weighted against the dissociation parameter at $\tau = 65\ \mu$s. The dissociation

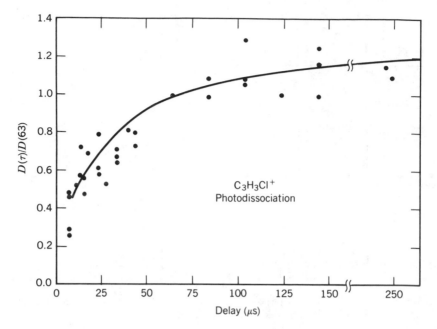

Fig. 46. Extent of production of product ion $m/z = 77$ as a function of delay time τ. The ordinate plots the dissociation parameter $D = -\ln[(\text{light-on signal})/(\text{light-off signal})]$, normalized to the D value at $\tau = 63\,22\ \mu\text{s}$. [Reprinted by permission from R. Dunbar, Copyright (1987), American Chemical Society]

parameter, D, is defined as $-\ln[(\text{light-on signal})/(\text{light-off signal})]$. The data in Fig. 46 was obtained for chlorobenzene ions formed by electron impact (12 eV) followed by a 2-s delay (t) to allow for thermalization. The neutral pressure was maintained at 2.5×10^{-7} torr. The solid curve in the figure represents a calculation of the photodissociation rate assuming that the ion contains 0.03 eV above the cell ambient temperature of 350 K. It was found that small amounts of excess energy dramatically affect the threshold photodissociation rate. A residual excess energy of 0.03 eV corresponds to an 25% increase in the photodissociation rate over that calculated for thermal ions.

Experiments corresponding to the two-pulse photodissociation experimental sequence found in Fig. 45b allow one to monitor the relaxation of ions of *known* internal energy. In this mode, ions are allowed to thermalize for time t and then are subjected to photoexcitation. The amount of excitation for the first pulse must be kept below the dissociation threshold for the ion and generally corresponds almost exactly to the energy of a single photon. Photodissociation of the ions is induced by a second light pulse. The photodissociation kinetics are measured as a function of τ_2. The relaxation of photoexcited iodobenzene ions was studied in this fashion (11). Photoexcited ions can relax radiatively by one of two mechanisms; the rate process mechanism and the cascade mechanism. Figure 47 depicts the energy levels pertinent to the iodobenzene system as well as the proposed relaxation mechanisms. The rate process model assumes that the

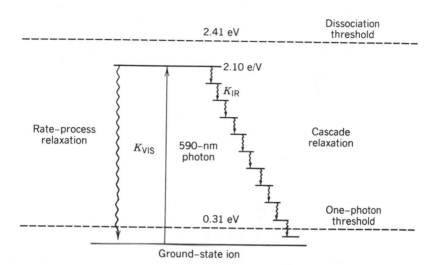

Fig. 47.   Energy levels in the two-photon photodissociation of iodobenzene. Visible photon energy is 2.10 eV (590 nm). The relaxation can follow either rate process relaxation (emission of a single visible photon, left) or cascade relaxation through stepwise emission of IR photons, (right). When iodobenzene has dropped below 0.31 eV, it will require two 590 nm photons to dissociate. [Reprinted by permission from B. Asamoto and R. C. Dunbar, Copyright (1987), American Chemical Society]

photoexcited ions will decay by loss of a single photon. In contrast, the cascade model assumes that ion relaxation occurs in discrete steps each consisting of emission of an IR photon. It is important to note that 2.10-eV photons (510 nm) were used in this experiment for both v_1 and v_2. The time-resolved (τ_1) photodissociation plot for iodobenzene for $t = 0.3$ s under collisionless conditions is found in Fig. 48. Predicted photodissociation curves for both the rate process and the cascade model are inscribed on the data. The presence of an induction period in Fig. 48 suggests that relaxation of photoexcited iodobenzene ions follows a cascade mechanism. The overall relaxation rate for this system is 1.4 ± 0.3 s^{-1}.

The chopped laser photodissociation experiment (Fig. 45c) is fundamentally the same as the two-pulse experiment described above. The time between pulses is a function of the chopping rate, and, therefore, measuring the photodissociation at different chopping rates while keeping the average power level constant allows one to make quantitative relaxation rate determinations. By fitting the chopper-rate dependence of the photodissociation signal to a simulation of two-photon kinetics, the relaxation rate of an ion is found by varying the relaxation rate to give a proper fit. Dunbar used this method to measure the collisionless relaxation rate for photoexcited benzene ions (45). For 488-nm photons, the collisionless relaxation rate was found to be 11 s^{-1}.

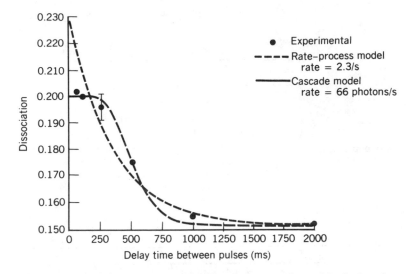

Fig. 48. Comparison of the rate process model and the cascade model of relaxation to the experimentally measured photodissociation. The quantity labeled "dissociation" is as in Fig. 46. [Reprinted by permission from B. Asamoto and R. C. Dunbar, Copyright (1987), American Chemical Society]

IV. CONCLUSIONS

Although FT–ICR is still undergoing rapid changes, especially in terms of instrumentation and experimental concepts, substantial progress has been made in developing a new analytical mass spectrometer. The versatility of FT–ICR is based on the concept of the ion trap. Ion storage and ion manipulation in the ICR cell allows routine chemical studies. Coupling this intrinsic versatility with a detection scheme compatible with Fourier transform data analysis methods makes FT–ICR a versatile mass spectrometer with capabilities for high sensitivity and high resolution and a working mass range comparable to that of time-of-flight instruments.

The utilization of FT–ICR for analytical applications has been hampered by the fact that too much emphasis has been placed on aspects of the method that work and insufficient amounts of time have been dedicated to understanding, at first principles levels, those aspects of the method that do not work. For example, the external ion source systems (e.g., the tandem quadrupole FT–ICR and electrostatic ion guide systems) are attractive for analytical FT–ICR applications; however, the issues concerning injecting ions from an external ion source are not yet resolved. Injected ions enter the ICR cell with an appreciable initial velocity, and it is unclear how the initial velocity of the ion influences the time-domain signal. Preliminary studies from our laboratory using computer simulations of the ion trajectories show clearly that the initial velocity of the ion greatly influences the degree of coherence of the ensemble of ions following excitation for detection. It is important to remember that FT–ICR is a relatively new technique compared to other mass analysis methods. Systematic evaluation of the dynamics of ion motion and its effects on ion trapping and detection will lay the foundation for the future evolution of the technique.

At some point during the pleistocene era, a neanderthal held up a sharp stone and proclaimed that man now had the ultimate tool. For that moment in time, the stone tool was state-of-the-art. As new problems were discovered, the stone tool had to evolve to much the arising needs. FT–ICR has emerged at the brink of its own potential. Just like the first stone tools, FT–ICR must evolve in order to answer the new demands and requirements of present and future research.

REFERENCES*

1.   Ahmed, M. S., and R. C. Dunbar, *J. Am. Chem. Soc.*, **109**, 3215 (1987).

2.   Alford, J. M., F. D. Weiss, R. T. Laaksonen, and R. E. Smalley, *J. Phys. Chem.*, **90**, 4480 (1986).

3.   Alford, J. M., P. E. Williams, D. J. Trevor, and R. E. Smalley, *Int. J. Mass. Spectrom. Ion Proc.*, **72**, 33 (1986).

4.   Allemann, M., Hp. Kellerhals, and K. P. Wanczek, *Int. J. Mass Spectrom. Ion Proc.*, **46**, 139 (1983).

5.   Allison, J., and R. M. Stepnowski, *Anal. Chem.*, **59**, 1072A (1987).

6.   Amster, J., J. A. Loo, J. J. Furlong, and F. W. McLafferty, *Anal. Chem.*, **59**, 313 (1987).

7.   Armentrout, P. B., and J. L. Beauchamp, *J. Am. Chem. Soc.*, **103**, 784 (1981).

8.   Armentrout, P. B., and J. L. Beauchamp, *J. Chem. Phys.*, **74**, 2819 (1981).

9.   Armentrout, P. B., L. F. Halle, and J. L. Beauchamp, *J. Am. Chem. Soc.*, **103**, 6501 (1981).

10.   Asamoto, B., and R. C. Dunbar, *Chem. Phys. Lett.*, **139**, 225 (1987).

11.   Asamoto, B., and R. C. Dunbar, *J. Phys. Chem.*, **91**, 2804 (1987).

12.   Bartmess, J. E., and R. T. McIver, Jr., in M. T. Bowers, Ed., *Gas Phase Ion Chemistry*, Vol. 2, "Classical Ion–Molecule Collision Theory", Academic Press, New York, 1979, p. 87.

13.   Baykut, G., and J. R. Eyler, *Trends Anal. Chem.*, **5**, 44 (1986).

14.   Beauchamp, J. L., *J. Chem. Phys.*, **46**, 1231 (1967).

15.   Beauchamp, J. L., and T. J. Armstrong, *Rev. Sci. Instr.*, **40**, 123 (1969).

16.   Bowers, W. D., S.-S. Delbert, and R. T. McIver, Jr., *Anal. Chem.*, **58**, 972 (1986).

17.   Bricker, D. L., T. A. Adams, Jr., and D. H. Russell, *Anal. Chem.*, **55**, 2417 (1983).

17a.   Bricker, D. A., and D. H. Russell, *J. Am. Chem. Soc.*, **109**, 3910 (1987).

18.   Buttrill, S. E., *J. Chem. Phys.*, **50**, 4125 (1969).

19.   Cassady, C. J., and B. S. Freiser, *J. Am. Chem. Soc.*, **108**, 5690 (1986).

20.   Castro, M. E., E. L. Kerley, C. D. Hanson, and D. H. Russell (unpublished results).

21.   Carlin, T. J., and B. S. Frieser, *Anal. Chem.*, **55**, 574 (1983).

22.   Castro, M. E., and D. H. Russell, *Anal. Chem.*, **56**, 578 (1984).

23.   Castro, M. E., and D. H. Russell, *Anal. Chem.*, **57**, 2290 (1985).

24.   Castro, M. E., D. H. Russell, S. Ghaderi, R. B. Cody, I. J. Amster, and F. W. McLafferty, *Anal. Chem.*, **58**, 483 (1985).

---

*Some references listed are not cited in text.

25. *Chem. Eng. News*, March 7, 1988, p. 81.

26. Chen, L., and A. G. Marshall, *Int. J. Mass Spectrom. Ion. Proc.*, **79**, 115 (1987).

27. Chesnavich, W. J., and M. T. Bowers, in M. T. Bowers, Ed., *Gas Phase Ion Chemistry*, Vol. 1, "Statistical Methods in Reaction Dynamics", Academic Press, New York, 1979, p. 119.

28. Chowdhury, A. K., and C. L. Wilkins, *Int. J. Mass Spectrom. Ion Proc.* (in press).

29. Chowdhury, A. K., and C. L. Wilkins, *J. Am. Chem. Soc.*, **109**, 5336 (1987).

30. Cody, R. B., *ACS Symp. Ser.*, **359**, 59 (1987).

31. Cody, R. B., R. C. Burnier, and B. S. Frieser, *Anal. Chem.*, **54**, 96 (1982).

32. Cody, R. B., J. A. Kinsinger, S. Ghaderi, I. J. Amster, F. W. McLafferty, and C. E. Brown, *Anal. Chim. Acta*, **178**, 43 (1985).

33. Cody, R. B., and S. D. Goodman, private communication.

34. Comisarow, M. B., *Adv. Mass Spect.*, **8**, 1698 (1980).

35. Comisarow, M. B., *Fourier, Hadamard, and Hilbert Transformations in Chemistry*, A. G. Marshall, Ed., Plenum, New York, 1982.

36. Comisarow, M. B., V. Grassi, and G. Parisod, *Chem. Phys. Lett.*, **57**, 413 (1978).

37. Comisarow, M. B., and J. Lee, *Anal. Chem.*, **57**, 464 (1985).

38. Comisarow, M. B., and A. G. Marshall, *Chem. Phys. Lett.*, **25**, 282 (1974).

39. Craig, E. C., I. Santos, and A. G. Marshall, *Rapid Communications in Mass. Spec.*, **2**, 33 (1987).

40. Creasy, W. R., A. O. O'Keefe, and J. R. McDonald, *J. Phys. Chem.*, **91**, 2848 (1987).

41. Dawson, P. H., *Mass Spec. Rev.*, **5**, 1 (1986).

42. Dunbar, R. C., in *Gas Phase Ion Chemistry*, Vol. 3, M. T. Bowers, Ed., Academic Press, New York, 1984, p. 130.

42a. Dunbar, R., *J. Phys. Chem.*, **91**, 2801 (1987).

43. Dunbar, R. C., *Int. J. Mass Spectrom. Ion Proc.*, **56**, 1 (1984).

44. Dunbar, R. C., *Molecular Ions: Spectroscopy, Structure and Chemistry*, T. A. Miller and V. E. Bondybey Ed., "Photofragmentation of Molecular Ions", North-Holland, 1983, p. 231.

45. Dunbar, R. C., *Chem. Phys. Lett.*, **125**, 543 (1986).

46. Elkind, J. L., J. M. Alford, F. D. Weiss, R. T. Laaksonen, and R. E. Smalley, *J. Chem. Phys.*, **87**, 2397 (1987).

47. Elkind, J. L., and P. B. Armentrout, *J. Am. Chem. Soc.*, **108**, 2765 (1986).

48. Elkind, J. L., and P. B. Armentrout, *J. Chem. Phys.*, **84**, 4862 (1986).

49. Elkind, J. L., F. D. Weiss, J. M. Alford, R. T. Laaksonen, and R. E. Smalley, *J. Chem. Phys.*, **88**, 5215 (1988).

50. Flamm, D. L., V. M. Donnelly, and J. A. Mucha, *J. Appl. Phys.*, **52**, 3633 (1981).

51. Forbes, R. A., F. H. Laukien, and J. Wronka, *Int. J. Mass Spectrom. Ion Proc.*, **83**, 23 (1988).

52. Forbes, R. A., L. M. Lech, and B. S. Freiser, *Int. J. Mass Spectrom. Ion Proc.*, **77**, 107 (1987).

53. Foster, M. S., and J. L. Beauchamp, *J. Am. Chem. Soc.*, **97**, 4808 (1975).

54. Freas, R. B., and D. P. Ridge, *J. Am. Chem. Soc.*, **102**, 7129 (1980).

55. Freas, R. B., and D. P. Ridge, *J. Am. Chem. Soc.*, **106**, 825 (1984).

56. Fredeen, D. A., and D. H. Russell, *J. Am. Chem. Soc.*, **107**, 3762 (1985).

57. Fredeen, D. A., and D. H. Russell, *J. Am. Chem. Soc.*, **108**, 1860 (1986).

58. Fredeen, D. A., and D. H. Russell, *J. Am. Chem. Soc.*, **109**, 3903 (1987).

59. Freiser, B. S., *Analytica Chim. Acta*, **178**, 137 (1985).

60. Freiser, B. S., *Talanta*, **32**, 697 (1985).

61. Ghaderi, S., and D. Littlejohn, *Proceedings of the 33rd, Ann. Conf. on Mass Spectrometry and Allied Topics, San Diego*, American Society for Mass Spectrometry, p. 727.

62. Goode, G. C., A. J. Ferrer-Correia, and K. R. Jennings, *Int. J. Mass Spectrom. Ion. Phys.*, **5**, 229 (1970).

63. Gregor, I. K., *Org. Mass Spect.*, **22**, 644 (1987).

64. Gross, M. L., and D. L. Rempel, *Science*, **226**, 261 (1984).

65. Halle, L. F., and J. L. Beauchamp, *Organomet.*, **1**, 963 (1982).

66. Hettich, R. L., and B. S. Freiser, *J. Am. Chem. Soc.*, **107**, 6222 (1985).

67. Hettich, R. L., and B. S. Freiser, *J. Am. Chem. Soc.*, **108**, 2537 (1986).

68. Hettich, R. L., and B. S. Freiser, *J. Am. Chem. Soc.*, **109**, 3537 (1987).

69. Hettich, R. L., and B. S. Freiser, *J. Am. Chem. Soc.*, **109**, 3543 (1987).

70. Hettich, R. L., T. C. Jackson, E. M. Stanko, and B. S. Freiser, *J. Am. Chem. Soc.*, **108**, 5086 (1986).

71. Huang, Y., S. W. Buckner, and B. S. Freiser (unpublished).

72. Huang, S. K., and M. L. Gross, *J. Phys. Chem.* **89**, 4422 (1985).

73. Huang, S. K., D. L. Remple, and M. L. Gross, *Int. J. Mass Spectrom. Ion Proc.*, **72**, 15 (1986).

74. Huang, Y., M. B. Wise, D. B. Jacobson, and B. S. Freiser, *Organomet.*, **6**, 346 (1987).

75. Hunt, D. F., J. Shabanowitz, and J. R. Yates III (unpublished).

76. Hunt, D. F., J. Shabanowitz, J. R. Yates, N. Zhu, D. H. Russell, and M. E. Castro, *Proc. Natl. Acad. Sci.*, **84**, 620 (1987).

77. Ijames, C. F., and C. L. Wilkins, *J. Am. Chem. Soc.*, **110**, 2687 (1988).

78. Irion, M. P., W. D. Bowers, R. L. Hunter, F. S. Rowland, and R. T. McIver, Jr., *Chem. Phys. Lett.*, **93**, 375 (1982).

79. Jackson, T. C., T. J. Carlin, and B. S. Freiser, *J. Am. Chem. Soc.*, **108**, 1120 (1986).

80. Jackson, T. C., R. L. Hettich, E. M. Stanko, and B. S. Freiser, *J. Am. Chem. Soc.*, **108**, 5086 (1986).

81. Jacobson, D. B., *J. Am. Chem. Soc.*, **109**, 6851 (1987).

82. Jacobsen, D. B., *Organomet.*, **7**, 568 (1988).

83. Jacobson, D. B., and B. S. Freiser, *J. Am. Chem. Soc.*, **106**, 4623 (1984).

84. Jacobson, D. B., and B. S. Freiser, *J. Am. Chem. Soc.*, **106**, 5351 (1984).

85. Jacobson, D. B., and B. S. Freiser, *J. Am. Chem. Soc.*, **107**, 1581 (1985).

86. Jacobson, D. B., and B. S. Freiser, *J. Am. Chem. Soc.*, **108**, 27 (1986).

87. Jeener, J., B. H. Meier, P. Bachmann, and R. R. Ernst, *J. Chem. Phys.*, **71**, 4546 (1979).

88. Jeffries, J. B., S. E. Barlow, and G. H. Dunn, *Int. J. Mass Spectrom. Ion Proc.*, **54**, 169 (1983).

89. Kerley, E. L., and D. H. Russell (unpublished results).

90. Klaassen, J. J., and D. B. Jacobsen, *J. Am. Chem. Soc.*, **110**, 974 (1988).

91. Knight, R. D., R. A. Walch, S. C. Foster, T. A. Miller, S. L. Mullen, and A. G. Marshall, *Chem. Phys. Lett.*, **129**, 331 (1986).

92. Kofel, P., M. Allemann, Hp. Kellerhals, and K. P. Wanczek, *Int. J. Mass Spectrom. Ion Proc.*, **65**, 97 (1985).

93. Kofel, P., M. Allemann, Hp. Kellerhals, and K. P. Wanczek, *Int. J. Mass Spectrom. Ion Proc.*, **74**, 1 (1986).

94. Kofel, P., M. Allemann, Hp. Kellerhals, and K. P. Wanczek, *Int. J. Mass Spectrom. Ion Proc.* (unpublished).

95. Laaksonnen, R. T., F. K. Weiss, J. L. Elkind, J. M. Alford, and R. E. Smalley, private communication.

96. Laude, D. A., C. L. Johlman, R. S. Brown, D. A. Weil, and C. L. Wilkins, *Mass Spect. Rev.*, **5**, 107 (1986).

97. Lauher, J. W., *J. Am. Chem. Soc.*, **101**, 2604 (1979).

98.   Laukien, F. H., *Int. J. Mass Spectrom. Ion Proc.*, **73**, 81 (1986).

99.   Ledford, E. B., D. L. Remple, and M. L. Gross, *Anal. Chem.*, **56**, 2744 (1984).

100. Lev, N. B., and R. C. Dunbar, *J. Phys. Chem.*, **87**, 1924 (1983).

101. Loo, J. A., E. R. Willams, I. J. Amster, J. J. Furlong, B. H. Wang, F. W. McLafferty, B. T. Chait, and F. H. Field, *Anal. Chem.*, **59**, 1882 (1987).

102. Mandich, M. L., W. D. Reents, Jr., and V. E. Bondybey, *J. Chem. Phys.*, **86**, 4245 (1987).

103. Mandich, M. L., W. D. Reents, Jr., and V. E. Bondybey, *J. Phys. Chem.* **90**, 2315 (1986).

104. Mandich, M. L., W. D. Reents, Jr., and V. E. Bondybey, *Mat. Res. Soc. Symp. Proc.*, **75**, 467 (1987).

105. Mandich, M. L., W. D. Reents, Jr., and M. F. Jarrold, *J. Chem. Phys.*, **88**, 1703 (1988).

106. Mandich, M. L., M. L. Steigerwald, and W. D. Reents, Jr., *J. Am. Chem. Soc.*, **108**, 6197 (1986).

107. Marshall, A. G., *Accts. Chem. Res.*, **18**, 316 (1985).

108. Marshall, A. G., *Fourier, Hadamard, and Hilbert Transformations in Chemistry*, A. G. Marshall, Ed., Plenum, New York, 1982.

109. Marshall, A. G., *Spectroscopy in the Biomedical Sciences*, CRC Press, 1986, p. 87.

110. Marshall, A. G., M. B. Comisarow, and G. Parisod, *J. Chem. Phys.*, **71**, 4434 (1979).

111. Marshall, A. G., T. C. L. Wang, and T. L. Ricca, *J. Am. Chem. Soc.*, **107**, 7893 (1985).

112. McCrery, D. A., and M. L. Gross, *Analytica Chim. Acta*, **178**, 91 (1985).

113. McElvany, S. W., B. I. Dunlap, and A. O'Keefe, *J. Chem. Phys.*, **86**, 715 (1987).

114. McElvany, S. W., M. M. Ross, and A. P. Baronavski, *Anal. Instr.* (in press).

115. McIver, R. T., R. L. Hunter, and W. D. Bowers, *Int. J. Mass Spectrom. Ion Proc.*, **64**, 67 (1985).

116. McLafferty, F. W., *Anal. Chem.*, **59**, 2212 (1987).

117. McLafferty, F. W., *Int. J. Mass Spectrom. Ion Proc.*, **72**, 85 (1986).

118. McLafferty, F. W., and I. J. Amster, *Int. J. Mass Spectrom. Ion Proc.*, **72**, 85 (1986).

119. McLafferty, F. W., D. B. Stauffer, S. Y. Loh, and E. R. Williams, *Anal. Chem.*, **59**, 2212 (1987).

120. Meckstroth, W. K., D. P. Ridge, and W. D. Reents, Jr., *J. Phys. Chem.*, **89**, 612 (1985).

121. Meier, B. H., and R. R. Ernst, *J. Am. Chem. Soc.*, **101**, 6441 (1979).

122. Mingos, D. M. P., *Acc. Chem. Res.*, **17**, 311 (1984).

123. Mullen, S. L., and A. G. Marshall, *J. Am. Chem. Soc.*, **110**, 1766 (1988).

124. Nibbering, N. M., *Recl. Trav. Chem. Pays-Bas*, **105**, 245 (1986).

125. Nikolaev, E. N., and M. N. Gorshov, *Int. J. Mass Spectrom. Ion Proc.*, **64**, 115 (1985).

126. O'Keefe, A. O., S. W. McElvany, and J. R. McDonald, *Chem. Phys.*, **111**, 327 (1987).

127. Parisod, G., and M. B. Comisarow, *Adv. Mass Spectrom.*, **8**, 212 (1980).

128. Paszkowski, B., *Electron Optics*, translation by R. C. G. Leckey, Iliffe Books, LTD. Chapters 1 & 7, 1968.

129. Pfändler, P., G. Bodenhausen, J. Rapin, R. Houriet, and T. Gäumann, *Chem. Phys. Lett.*, **138**, 195 (1987).

130. Raghavachari, K., *J. Chem. Phys.* (submitted).

131. Reents, Jr., W. D., M. L. Mandich, and M. F. Jarrold, *J. Chem. Phys.*, **88**, 1703 (1988).

132. Reents, Jr., W. D., M. L. Mandich, and V. E. Bondybey, *Chem. Phys. Lett.*, **131**, 1 (1986).

133. Reents, Jr., W. D., A. M. Mujsce, V. E. Bondybey, and M. L. Mandich, *J. Chem. Phys.*, **86**, 5568 (1987).

134. Reents, W. D., F. Strobel, R. B. Freas, J. Wronka, and D. P. Ridge, *J. Phys. Chem.*, **89**, 5666 (1985).

135. Rempel, D. L., S. K. Huang, and M. L. Gross, *Int. J. Mass Spectrom. Ion Proc.*, **70**, 163 (1986).

136. Rohlfing, E. A., D. M. Cox, and A. Kaldor, *J. Chem. Phys.*, **81**, 3322 (1984).

137. Russell, D. H., *Mass Spec. Rev.*, **5**, 167 (1986).

138. Russell, D. H., and D. L. Bricker, *Anal. Chim. Acta.*, **178**, 117 (1985).

139. Sack, T. M., and M. L. Gross, *Proceedings of the 31st Ann. Conf. on Mass Spectrometry and Allied Topics. Boston*, American Society for Mass Spectrometry, p. 396.

140. Sack, T. M., D. A. McCrery, and M. L. Gross, *Anal. Chem.*, **57**, 1290 (1985).

141. Sack, T. M., D. L. Miller, and M. L. Gross, *J. Am. Chem. Soc.*, **107**, 6795 (1985).

142. Sallans, L. Lane, K. R. Lane, R. R. Squires, and B. S. Freiser, *J. Am. Chem. Soc.*, **107**, 4379 (1985).

143. Sawyer, D. T., T. S. Calderwood, C. L. Johlman, and C. L. Wilkins, *J. Org. Chem.*, **50**, 1409 (1985).

144. Sellers-Hann, L., and D. H. Russell (unpublished results).

145. Sharp, T. E., J. R. Eyler, and E. Li, *Int. J. Mass Spec. Ion Phys.*, **9**, 421 (1972).

146. Shomo, R. E., A. Chandrasekaran, A. G. Marshall, R. H. Reuning, and L. W. Robertson, *Biomed. Environ. Mass Spec.*, **14**, 1 (1987).

147. Syka, J. E. and W. J. Fies, *Proceedings of the 35th Ann. Conf. on Mass Spectrometry and Allied Topics, Denver*, American Society for Mass Spectrometry, p. 767.

148. Tews, E. C., and B. S. Freiser, *J. Am. Chem. Soc.*, **109**, 4433 (1987).

149. Todd, J. F. J., *Dynamic Mass Spectrometry*, Vol. 6, Heyden & Son, London, 1981, p. 44.

150. U.S. Pat. Appl. Ser. No. 610502.

151. van der Wel, H., N. M. M. Nibbering, J. C. Sheldon, R. N. Hayes, and J. H. Bowie, *J. Am. Chem. Soc.*, **109**, 5823 (1987).

152. Verdun, F. R., and C. Giancaspro, *Anal. Chem.*, **58**, 2099 (1986).

153. Verdun, F. R., T. L. Ricca, and A. G. Marshall, *Appl. Spect.* (in press).

154. Wanczek, K. P., *Int. J. Mass Spectrom. Ion Proc.* (in press).

155. Wang, T. I., and A. G. Marshall, *Int. J. Mass Spectrom. Ion Proc.*, **68**, 287 (1986).

156. Wang, T. I., T. L. Ricca, and A. G. Marshall, *Anal. Chem.*, **58**, 2935 (1986).

157. Watson, C. H., G. Baykut, Z. Mowafy, A. R. Katritzky, and J. R. Eyler (unpublished).

158. Whyman, R., in Johnson, B. F. G. Ed., *Transition Metal Clusters*, Wiley, New York, 1980.

159. Wilkins, C. L., D. A. Weil, C. L. Yang, and C. F. Ijames, *Anal. Chem.*, **57**, 520 (1985).

160. Wilkins, C. L., and M. L. Gross, *Anal. Chem.*, **53**, 1661A (1981).

161. Wilkins, C. L., and D. A. Weil, *J. Am. Chem. Soc.*, **107**, 7316 (1985).

162. Wilkins, C. L., and C. L. Yang, *Int. J. Mass Spectrom. Ion Proc.*, **72**, 195 (1986).

163. Wineland, D. J., W. M. Itano, and R. S. VanDyck, *Adv. Atomic Mol. Phys.* **19**, 135 (1983).

164. Wise, M. B., *Anal. Chem.*, **59**, 2289 (1987).

165. Wood, K. V., S. A. McLuckey, and R. G. Cooks, *Org. Mass Spect.*, **21**, 11 (1986).

166. Wronka, J., R. A. Forbes, F. H. Laukien, and D. P. Ridge, *J. Phys. Chem.*, **91**, 6450 (1987).

167. Wronka, J., and D. P. Ridge, *J. Am. Chem. Soc.*, **106**, 67 (1984).

168. Zhang, Q. L., S. C. O'Brien, J. R. Heath, Y. Liu, R. F. Curl, H. W. Kroto, and R. E. Smalley, *J. Phys. Chem.*, **90**, 525 (1986).

Chapter 3

SPARK SOURCE MASS SPECTROMETRY

By W. W. HARRISON

Department of Chemistry, University of Florida, Gainesville, Florida

and

D. L. DONOHUE
Analytical Chemistry Division, Oak Ridge National Laboratory, Oak Ridge, Tennessee

Contents

I. INTRODUCTION

Mass spectrometry is generally thought of as an analytical tool for organic chemical applications. Certainly, the early development and application of mass spectrometry resulted from the need to analyze complex mixtures in the field of petroleum and petrochemical industries. The structure elucidation of elaborate organic molecules represents an invaluable and continuing utilization. Inorganic materials, on the other hand often provide a set of intrinsic experimental difficulties for mass spectrometric analysis. In practice, gases, liquids, and even solids may be analyzed by mass spectrometry if a sufficient sample vapor pressure exists in the ionization source. The low operating pressure in a mass spectrometer source aids in the production of a vapor population from solution samples, and heated insertion probes allow many solids to be sufficiently vaporized for subsequent ionization and analysis. However, even a brief consideration of the relative volatility of various materials indicates that the mass spectrometric analysis of solids would be generally restricted to organics and to a relatively few inorganics. Furthermore, the ions produced by low-energy sources, such as electron impact, are predominantly of molecular fragments, rather than the elemental constituents that one might wish for inorganic analysis.

Therefore, a highly energetic vaporization–ionization source was needed to volatilize refractory materials and to provide atomic mass spectra for elemental analysis. A pulsed radio frequency spark has been the major ionization mode used for such applications and the field as a whole has come to be generally known as spark source mass spectrometry (SSMS). As shown in Fig. 1, a high-voltage spark volatilizes and ionizes a solid sample, and the ions are extracted into a double focusing mass spectrometer for analysis by electrical or photographic means (see Section III). The spark discharge allows the analysis for both metals and nonmetals to sub-ppm sensitivity of even such samples as stainless steel. The applicability of SSMS to the trace element analysis of many important materials has led to a substantial development of the technique (13), particularly in industrial laboratories.

A. HISTORICAL BACKGROUND

SSMS has had a relatively recent arrival. Although the fundamental concept of utilizing a spark discharge for the mass spectrometric analysis of solids goes back to the mid-1930s, the practical development of the technique was delayed until the availability of commercial instrumentation in the late 1950s and early 1960s. Dempster (46) in 1936 reported three different methods for producing ions from solid metals: (a) the low-voltage vacuum vibrator provided an inconstant beam of ions and excessive heating of the electrodes; (b) the condenser discharge "hot spark" produced large vapor densities, but no measurable ions; but (c) the inductively coupled spark discharge, using a Tesla coil-type circuit, yielded a satisfactory flux of singly and multiply charged ions for "solid elements and alloys." It is interesting that the spark circuit shown in Dempster's paper is basically the same as those used in modern instruments. The first complete instrument, which was the predecessor of today's solids mass spectrometers, was built by Shaw and Rall (126), who indicated that while they found qualitative analysis to be routine, quantitative analysis was "a more complicated problem." The same could be said 50 years later.

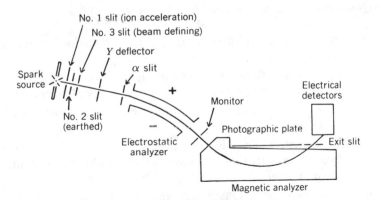

Fig. 1.   Schematic of MS-702 spark source mass spectrometer (108).

The analysis of stainless steel was demonstrated by Gorman, Jones, and Hipple (69) in 1951. They showed that electrical detection methods could be used with the fluctuating spark source to obtain quantitative results that were "better than it had been anticipated would be possible with a spark source." Yet it was almost 20 years later before commercial manufacturers would make electrical detection generally available for spark source instruments. Hannay (78) in 1954 constructed a spark source mass spectrograph of the Mattauch and Herzog geometry that yielded 0.1 ppm sensitivity for 3–10 min exposures. Hannay and Ahearn (77) in the same year used this instrument to analyze for trace elements in semiconductors for those applications where the emission spectrograph sensitivity was inadequate. The demonstration of SSMS capabilities was by then sufficient to persuade commercial instrument development. It was estimated that there were at its peak in the early 1970s approximately 150–200 spark source mass spectrometers active in world laboratories, about 60–70 of which were in the United States. Today, with the competition from more recently developed techniques, SSMS use has dropped significantly, but it still serves critical needs in many laboratories for solids analysis. Because of the special application nature of SSMS and the high instrumentation cost, the great majority of the instruments are located in industrial laboratories. This predominance of industrial users makes the open literature of SSMS somewhat misleading. The relatively small number of papers that appear each year (~ 25–30) would suggest limited new activity. However, many of the interesting industrial applications of SSMS are never published, other than perhaps as proceedings of national mass spectrometry conferences. It is difficult to obtain the proper perspective for the present importance of SSMS without considering these factors.

B.   ROLE OF SSMS IN ANALYTICAL CHEMISTRY

While SSMS can be used for the analysis of major elemental constituents, it is in the area of trace element analysis that it has its greatest importance. Here it has certain capabilities beyond those of most other analytical methods. The major advantages of SSMS can be summarized as follows:

1. *High sensitivity.* Detection limits for all elements are in the parts-per-billion (ppb) range.
2. *Broad response range.* Both metals and nonmetals are easily ionized in the spark discharge, thus allowing analysis for essentially all elements.
3. *Uniform sensitivity.* As a general rule, elemental sensitivities are within a factor of three.
4. *Minimal matrix effect.* The high energy available from the spark source produces a leveling effect that leads to relatively uniform sensitivities from one matrix to another.
5. *Results available rapidly.* Qualitative and semiquantitative data can be obtained from a mass scan of a few minutes.

Before comparison of SSMS to other methods, the limitations of the technique should also be considered:

1. *Only modest precision and accuracy.* Quantitative results are often no better than ± 20–30%.
2. *Complex and expensive instrumentation.* Large initial capital equipment costs plus significant maintenance expenses are involved.

The importance and potential of SSMS can best be seen by a brief (and very general) comparison with several other contemporary trace element analysis methods that are in popular use. Flame atomic absorption (FAA), for example, has sensitivity and selectivity comparable to SSMS for many elements, coupled with good quantitative accuracy, but FAA is normally a single element at a time method. Furthermore, many elements such as the nonmetals, are not analyzable by FAA. X-ray fluorescence is a rapid and accurate means of trace element analysis, but its sensitivity and total elemental response are not as favorable as in the case of SSMS. Spectrographic emission allows simultaneous multielement analysis, although the sensitivities range from very good to only modest. The inductively coupled plasma is an excellent solution technique for emission and mass spectrometry, but it is not generally suited to solids. Neutron activation analysis (NAA) shows excellent sensitivities for many metals and nonmetals and yields better precision than SSMS. Sample preparation is also often simpler for NAA. However, NAA will not allow trace analysis of as many elements as SSMS and usually requires more time for results to be obtained. Much larger interelement sensitivity differences are observed with NAA than for SSMS. Glow discharge mass spectrometry is the rising competitor for SSMS and has already supplanted it in many laboratories.

In summary, the overall capabilities of SSMS for rapid qualitative and quantitative simultaneous analysis of all elements present in a sample remain important for solids elemental analysis. These striking attributes have led to the utilization of SSMS in a wide variety of applications that will be documented in a later section of this chapter.

II. PRINCIPLES OF SPARK SOURCE MASS SPECTROMETRY

A. FORMATION OF IONS FROM SOLID SAMPLES

1. The rf Spark Discharge as a Solids Ionization Source

The standard electrical discharge used for solids analysis in mass spectrometry is the pulsed radio-frequency (rf) spark. In practice, a 500-kHz to 1-MHz rf oscillator is pulsed at a repetition rate ranging from single pulses to tens of thousands of pulses per second. The pulse length is variable from 25 to 200 μs. The peak-to-peak spark voltage may be as high as 100 kV, although voltages in the 10–30 kV range are more normally used. Under the 10^{-5}–10^{-8} torr vacuum conditions existing in the source, a spark discharge will occur between two conducting electrodes at a narrow gap (~ 10–$100\ \mu$m), thus vaporizing and ionizing the electrode components.

The rf spark is far from an ideal solids ion source. The ion beam is quite erratic, and the ions exhibit an energy spread of up to several kilovolts. The ion beam may be inhomogeneous. Furthermore, the rf voltage radiates spark noise that can cause serious interference in nearby electronic circuitry. Given these difficulties, why was the pulsed rf spark selected as the most suitable solids ionization source and why has it continued to maintain this primary position in a form that is basically unchanged from the original studies of Dempster (46).? Honig (89) suggests that the selection was dictated "on the basis of equipment and circuitry available." In addition, to quote Franzen (62): "Because of its commendable performance in the early work of Dempster, the rf spark enjoys the widest useage, although it is the least understood process, and the most difficult one to control." However much the rf spark may owe its preeminence to historical developments, the fact remains that there are several positive features about this ionization mode that made its selection and continued popularity altogether rational. Dempster (46) reported the advantage of significant quantities of multiply charged ions in such spectra. Hannay and Ahearn (77) reported the relative independence of the rf spark ion source from such factors as elemental volatility, ionization efficiencies, and sample matrix. The ability of the rf spark to ionize semiconductors (77) also contributes to its popularity in many industrial applications.

a.   THEORY OF THE VACUUM SPARK DISCHARGE

High-voltage discharges in vacuum have been the subject of much research, and many explanations have resulted, attempting to explain the observed phenomena. However, no unified theory has been clearly established. A high-voltage vacuum discharge, such as the rf spark, is generally subdivided into three distinct discharge regions (62, 89), as shown in Fig. 2. A "prebreakdown" region is shown for that part of the discharge process wherein the voltage rises to some high value, with attendent low currents, before entering the "breakdown" region. The voltage attained before breakdown may reach tens of kilovolts at normally submilliampere currents. The fully reversible process of prebreakdown suddenly converts to an irreversible breakdown mode of falling voltage coupled with rapidly increasing current (negative resistance). The discharge path leads to a low-voltage, high-current "arc discharge" region of short duration.

The translation of this qualitative outline of a high-voltage vacuum discharge to the practical rf spark source commonly used for solids ionization suggests the following:

1.   The rf voltage applied to the sample electrodes builds up in successive cycles to a breakdown value determined by the electrode gap. At a wide gap, fewer breakdowns per pulse are observed than for a narrow gap.

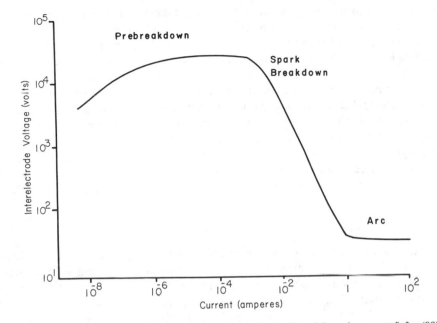

Fig. 2.   Display of voltage as a function of current for an electrical breakdown in vacuum [after (89)].

2.   The duration of the arc phase of the discharge probably differs from one type of spark source circuit to another, depending on power capacity and intrinsic circuit impedance. Therefore, the characteristics of the overall spark discharge may also vary similarly, as one circuit may contribute a greater low-voltage arc component than another.

3.   The discharge cycle is basically a violent, uncontrolled process that is not a closely reproducible phenomenon. This contributes to the erratic precision associated with SSMS.

b.   IONIZATION MECHANISMS IN THE SPARK

The process by which ions are formed in the rf spark discharge is not well understood. No doubt there are several different ionization modes occurring sequentially or even concurrently. In the initial stages of the discharge cycle (prebreakdown), electron release can occur by field emission, probably from small whisker protrusions (62, 89). The high-voltage buildup between the sample electrodes can create very high fields at needlelike surface features and significant localized heating effects at the anode electron bombardment spot. Significant material transport can occur (125) and ionization by electron bombardment, as well as by thermal and Townsend discharge effects, may occur.

The subsequent high-voltage breakdown and arc phases of the discharge follow and create the bulk of the ion flux. Bombardment of the vaporized anode material in the electrode gap by very high energy electrons causes extensive ionization to form singly and multiply charged species. The ions produced may be the cause of further ionization by sputtering interaction at electrode surfaces with the release of secondary ions and additional neutrals. From the species formed in the spark discharge, it is clear that low-energy electrons are also active in ionization as well. Furthermore, given the initial high pressure (62) that may exist at or near the electrode surface, localized discharges may occur in which the ionization mode is open to question. Thus, a nonequilibrium situation exists in which the net ionization results mainly from varying contributions of (a) electron impact, (b) thermal, and (c) sputter release of secondary ions. Other modes probably also exist to some extent. The primary ionization mode for a given analysis will be affected by such factors as spark voltage, electrode gap width, and electrode material. Net spark on-time can also affect thermal considerations.

c. CHARACTERISTICS OF SPARK SOURCE SPECTRA

The rf spark has been used to ionize many different types of samples, including organics. However, the more normally encountered application is in a survey mode for elemental analysis. The spectra produced are relatively simple for inorganic materials, particularly when compared to emission spectra. Both negative and positive ions are formed, although the latter predominate. The interpretation of spark source spectra is basically a quite straightforward evaluation of isotopic patterns and magnitudes, but an awareness of the broad range of species produced is necessary when considering possible spectral interferences.

(1) Positive-Ion Formation

It is here that SSMS has gained its well-deserved reputation as a premier qualitative analysis technique. The formation of singly charged ions of each isotope according to its relative abundance provides a characteristic spectrum wherein multiisotopic elements form easily recognizable patterns. A typical example is shown in Fig. 3. In addition to the predominant +1 ions, multiply charged ions are also formed and may be quite valuable in identification. The abundance of such species will normally decrease by a factor of 5–10 (88) with each increasing charge, but for major constituents such as a metal matrix, species with 10–15 electrons removed are not an unusual occurrence. Monoisotopic elements at odd nominal masses, such as arsenic, manganese, phosphorous, and aluminum, depend heavily on the appearance of spectral +2 lines for positive spectral identification.

The high energy of the rf spark does not, unfortunately, assure that all the positive ions formed are in the elemental state. Molecular ions are observed, particularly in certain types of matrices such as geological and biological samples, where both organic and inorganic combinations are possible. Combinations of major sample components with oxygen, hydrogen, nitrogen, or the electrode matrix (graphite, silver, etc.) can yield molecular species that may interfere with the detection of elemental ions. Both the occurrence and intensity

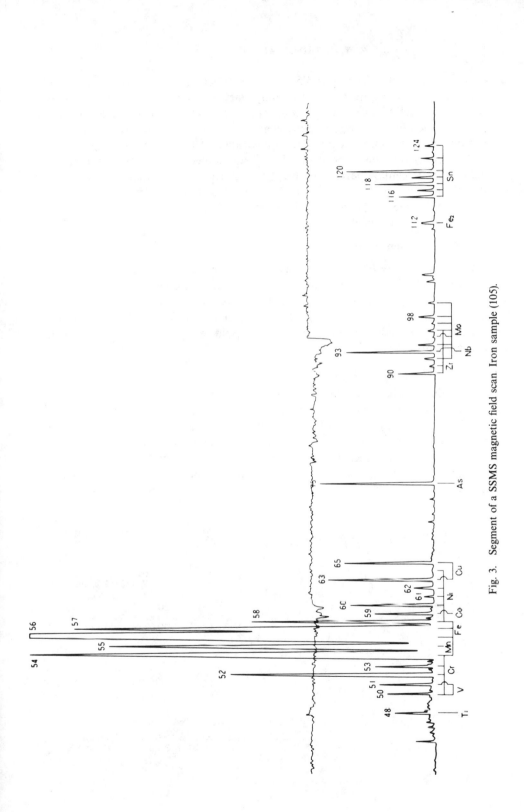

Fig. 3.   Segment of a SSMS magnetic field scan Iron sample (105).

of these molecular ions are difficult to predict with any confidence, and, depending on sparking conditions, may vary even between similar samples. Molecular cluster ions are also encountered. Carbon and silicon are particularly prone to such formations, but many elements form detectable cluster ions when present in high concentration. The appearance of C_n^+ and Si_n^+ throughout the normal inorganic mass range should be anticipated when these elements are present in percentage level concentrations. A significant difficulty in SSMS is that the normally used resolution of 1000–3000 may not be sufficient to separate all of these molecular ions from the elemental ions at the same nominal mass. This is a particular problem in the electrical scanning mode (see Section 4.b).

(2)   Negative-Ion Formation

Little work has been reported concerning negative ion SSMS. The high energies available in the spark create, for most elements, a much larger population of positive ions than negative ions. Modification of a commercial mass spectrometer to achieve negative-ion analysis (8) is not difficult, however. Elements with high electron affinities may yield reasonable sensitivities in the negative-ion mode.

(3)   Energy Spread of Ions

The range of energies obtained in electron bombardment sources is 0.2–0.5 eV (120), approaching the spread imposed by Boltzmann effects. By contrast, the rf spark shows energy spreads of 1–3 keV, thus necessitating an initial energy filter before the magnetic analyzer. The energy distribution depends (139) on spark gap width, the rf voltage, and the electrode material. Figure 4 shows the effect of spark voltage at a narrow gap width. The distribution width widens as the gap is increased. For singly charged ions, a generally uniform distribution is observed

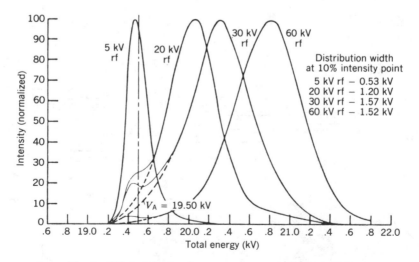

Fig. 4.   Effect of spark voltage on energy spread of ions (139).

(89), thus producing little net instrumental discrimination between sample elemental ions. The large energy spread in SSMS can cause analytical difficulties when two ions, usually one elemental and one molecular, are not entirely separated at a nominal m/e value.

2. Alternative Discharge Sources for Solids Ionization

Although this report is directed toward the rf spark as an ion source, there are other discharge sources that are presently used for solids mass spectrometric analysis and should be briefly noted here. The arc discharge sources were known and tested by Dempster (46), but the gas discharge is of more recent use as a solids ionization source for analysis.

a. VARIATION OF AN ARC DISCHARGE

Schuy and Hintenberger (124), among others, have shown that a simple low-voltage vibrating arc works well with conducting electrodes. A 6-V car battery provided currents of up to 15 A. Large ion currents are produced with a high concentration of $+2$ species. Although the simplicity of the source is attractive, it has not developed into widespread use because of its difficulty in handling brittle, compacted samples.

Honig and coworkers at RCA Laboratories (89) have studied pulsed dc arcs as solids ionization sources. They describe a short trigger discharge in which a trigger signal is coupled through a pulse transformer to initiate a high-current (5–50 A) discharge of 1–4 μs duration. Initial pulse voltage may be as high as 60 kV. The same investigators have also used a triggered low-voltage discharge to ionize solid samples. They report that the low-voltage source produces more stable ion currents and smaller ion energy spreads than does the rf spark.

The arc-type sources have never offered serious competition to the spark source for several reasons. Commercial arc sources are not conveniently available to couple to a solids mass spectrometer. The necessity of designing or building a suitable arc source has been enough to deter many investigators, particularly when the rf spark is readily available as an integral part of present commercial instruments. The spark does, of course, have the further advantages of handling poorly conducting samples and also of creating a predominance of singly charged species. There is enough promise, however, to justify a further look at arc sources, particularly pulsed mode, and compare their performance to the rf spark for a variety of samples relative to sensitivity and precision. The elimination of rf spark noise alone in the arc sources can improve detection limits in electrical readouts by improving the signal-to-noise ratio.

b. GAS DISCHARGE SOURCES

Gas discharges as ionization sources are certainly among the oldest known, but more recently ion bombardment in a gas discharge has been used to sputter and ionize solids samples for elemental analysis. Coburn and coworkers (33–35) have shown that an rf glow discharge in argon (Fig. 5) can sputter-etch thin films

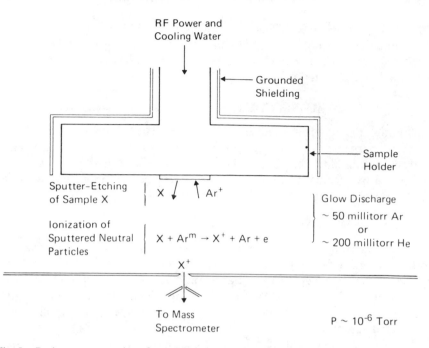

Fig. 5.   Basic processes and configuration utilized in glow discharge mass spectrometry (35).

or bulk solids and provide sensitivity to low $\mu g/g$ levels. The discharge generates a very stable ion flux with an energy spread for extracted species of about 1 eV. This source appears quite attractive as a means of obtaining elemental composition profiles in solids.

The dc glow discharge (83) also offers promise as a possible alternative to the rf spark. Again, sputtering by rare-gas ions creates ionization of solids (37, 84) or of surface films deposited from solution (41, 42, 48). Pahl and coworkers (86, 90, 91) have investigated many fundamental aspects of the discharge in their plasma diagnostics studies.

B.   ION SEPARATION

The basic fundamentals of ion separation have been described in Chapter 1 and will not be repeated here. The crucial difference between SSMS and most other forms of mass spectrometry lies in the mode of ion production; the subsequent extraction and separation are essentially routine with certain exceptions imposed by the nature of the generated ions. This section will address only those distinctions.

1.   Special Resolution Requirements for SSMS

As indicated previously, the energy spread, ΔE, of the spark produced ions is quite large. To minimize this, the kinetic energy, E, of the ions entering the magnetic field is made large by use of a high accelerating voltage. Thus, although the ΔE between ions remains constant during acceleration, $\Delta E/E$ is now reduced to an acceptable level for subsequent mass separation. Dempster, in his original studies, used 20 kV as the acceleration voltage, and this has been the nominal value reported for much of SSMS investigations to date. However, a 25-kV acceleration potential has also been used on commercial instruments, further improving the attainable resolution. Cleanliness of the source and acceleration regions of the mass spectrometer becomes particularly crucial at these higher voltages to avoid breakdown to ground. Because of the emphasis on trace element analysis in SSMS, the high acceleration voltages normally used are also useful in increasing ion extraction efficiency and net sensitivity.

Another special resolution requirement imposed upon SSMS by nature of the wide energy spread of the ions is double focusing (see Chapter 1). The need to have an initial energy filter for the ions, even given the high acceleration voltage, before they enter the magnetic analyzer has contributed significantly to the cost of spark source mass spectrometers, which in turn has limited, to some extent, the widespread use of the technique.

C. ION DETECTION

Standard types of ion detectors (Chapter 1) are used for SSMS. Both photographic plates and electron multipliers are available on commercial instruments. Because of the fluctuating nature of the spark ion beam, photoplates were the detector of choice almost exclusively up to the late 1960s. Gorman and coworkers (69) had shown in 1951 that electrical detection was feasible for SSMS, but the integrating nature of the photoplate led to its favored use. Because of the survey role that SSMS often plays, the plates were also useful as a full mass display, including the more volatile elements that might be rapidly evolved and go undetected in an electrical scan mode. Since about 1970, however, electrical detection has become more and more adopted, with the advent of commercial availability.

III. EXPERIMENTAL ASPECTS OF SPARK SOURCE MASS SPECTROMETRY

A. BASIC INSTRUMENTAL COMPONENTS

Of particular interest here are the instrumental components in a typical spark source mass spectrometer that are quite different from a more conventional organic instrument. These are (a) the ion source configuration, including the sample manipulation controls; (b) the rf spark circuitry; and (c) the data collection–readout system.

1. Ion Source Configuration

Typically, SSMS utilizes two small sample electrodes that are mounted before an ion exit slit and positioned by external manipulators. Figure 6 shows one such configuration. The electrode holders, ion exit slit plate, and back plate are all

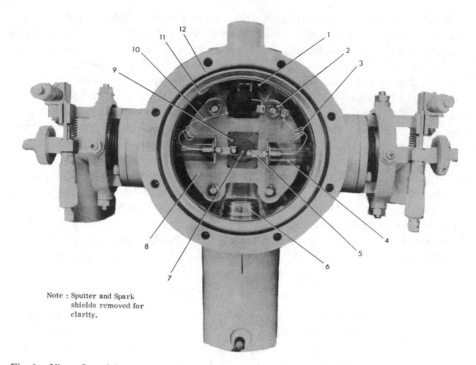

Note : Sputter and Spark
        shields removed for
        clarity.

Fig. 6.   View of spark ion source and electrode manipulators. (Courtesy of AEI, Ltd., Manchester, England) (1) Bracket for sputter shield, (2) 4BA screw and tantalum shield, (3) 20-kV lead to right-hand electrode holder, (4) right-hand glass insulator, (5) right-hand electrode holder, (6) source inspection lamp window, (7) sample electrodes, (8) large tantalum plate, (9) No. 1 slit plate, (10) rf lead to left-hand electrode holder, (11) stainless steel liner, (12) rubber gasket (for use with lead glass front plate).

constructed from tantalum to minimize spectral contribution from ion sputtering. A spark shield, not shown in Fig. 6 for clarity (see Fig.7), normally fits over the electrodes and holder to reduce source contamination from the sparking sample. A source inspection lamp, allows external viewing of the mounted electrodes through a small port in the spark shield box. Both insulators are mounted on flexible bellows vacuum feedthroughs, which then permit manipulation of the sample in each of three degrees of freedom by the micrometer controls.

Figure 7 shows a top-view pictorial of the ion source and associated voltages. The rf voltage is connected by wire leads to the electrode holders, which are isolated from the chassis earth by quartz insulators. The electrodes, spark shield, and No.1 plate are all held at the accelerating voltage. The earthed No. 2 slit acts as the negative appearing attraction site, and the No. 3 plate contains a small beam defining slit, typically 0.05×1.0 mm

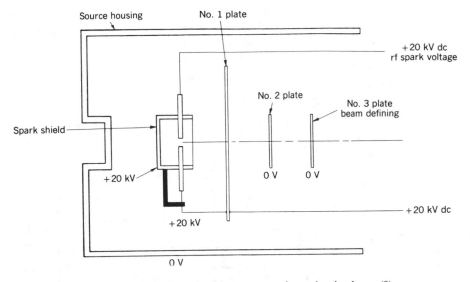

Fig. 7.   Pictorial of a SSMS ion source and associated voltages (9).

2. rf Spark Circuitry

The spark circuitry consists basically of a triggered rf oscillator feeding an rf power amplifier that is then Tesla coupled to the sample electrodes through an evacuated high-tension glass bushing. This is shown in block diagram form in Fig. 8. The trigger circuit allows the application of rf voltage to the sample electrodes for selected, reproducible time intervals. This is accomplished by turning on the rf oscillator to produce pulses of variable pulse length and repetition rate. These pulses are then translated through the power amplifier and Tesla transformer to the electrodes. By varying the pulse length and repetition rate, the net rf power applied to the sample can be selected. The spark variac also determines the discharge power by controlling the magnitude of the applied rf voltage, which can be as high as 80 kV after being stepped up by the Tesla transformer.

The beam suppress section of Fig. 8 provides for electrostatic deflection of the ion beam to the housing wall when the operator wishes to interrupt charge accumulation while still maintaining sparking conditions. This technique is also used in the ion beam chopper (IBC) (11) by applying the output of a pulse generator–amplifier to the deflection plates. Thereby, the ion beam is period-ically deflected as set by the IBC pulse length and repetition rate. In this manner, a range of ion charge accumulations can be conveniently obtained at a fixed set of spark conditions by varying the pulse parameters. It is particularly useful for controlling short exposures.

A number of investigators (17, 63, 118) have attempted to achieve more reproducible sparking conditions by modifying the high-voltage spark circuit. It

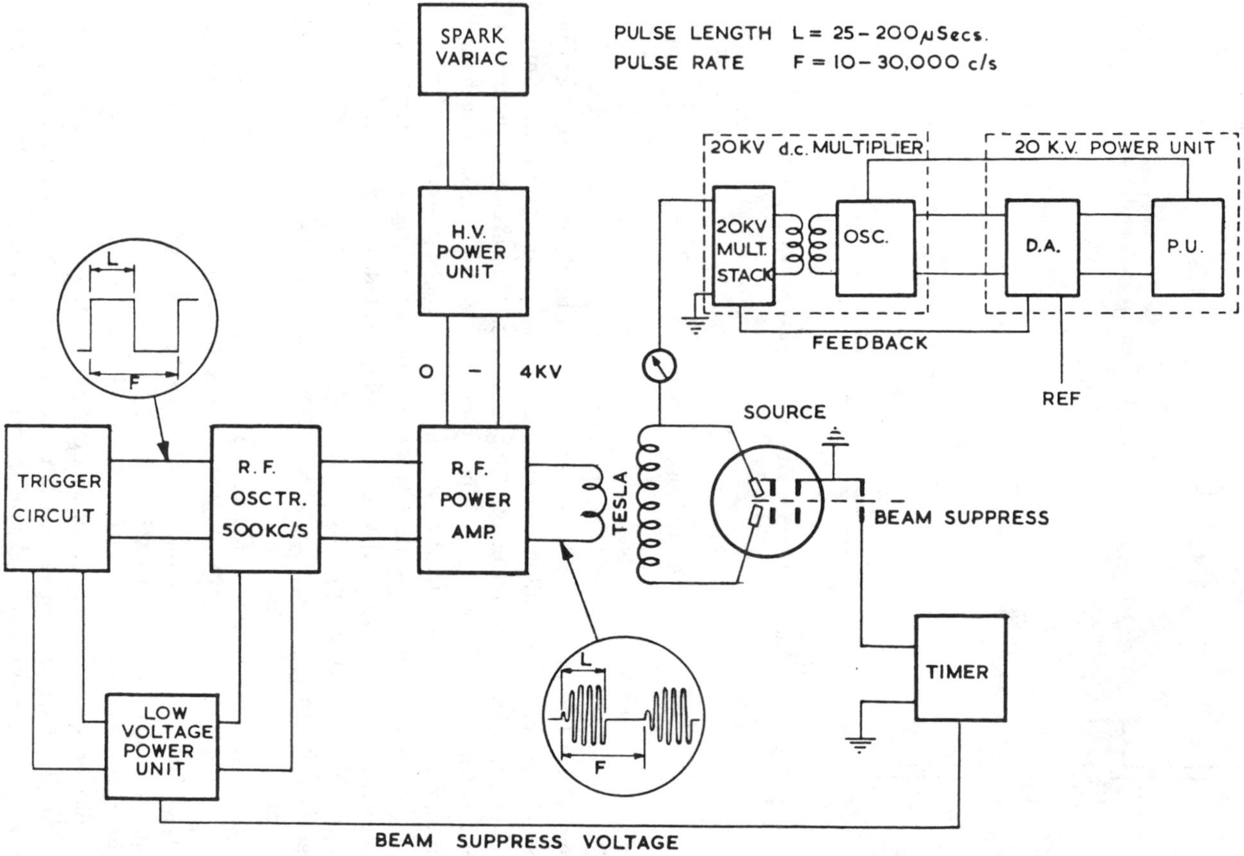

Fig. 8. Block diagram of circuitry for production and extraction of ions by a spark source mass spectrometer (9).

204

is known that the relative abundance of singly and multiply charged atomic ions as well as the abundance of molecular ion species are sensitive to the voltage and current conditions of the spark discharge. These conditions can be adjusted with appropriate circuit elements to give more or less of an arc character to the discharge. Related studies (56, 135, 136) have shown that the abundances of certain ionic species change during the breakdown period (about 1 μs). These time-resolved studies allow the experimenter to choose conditions that give a maximum signal for the analytically useful species in the spectrum relative to interferences.

3. Spark Gap Control

The interelectrode spark gap should be maintained as constant as possible while sparking because changes in this parameter can affect such critical factors as ion flux, ion energy, ion ratios, electrode shielding, and electrode temperature. Many investigators (21, 36, 104) have recognized the need for close gap control. A device to maintain automatically the spark discharge for SSMS may be current or voltage controlled. With the current control mode, the current flowing in the spark circuit is sensed and used as the governing signal while in the voltage type, the rf breakdown voltage between the electrodes acts as the control signal. Only one electrode is moved. For the current modes, as the gap is decreased starting from an open circuit with zero current, the discharge current rises to a maximum at a closest setting just before the electrodes begin to make contact. Further decreasing of the gap causes the electrodes to become shorted, with the current falling to zero. A current-activated spark control functions by seeking out the maximum current with regard to gap width. For a specific current-optimized gap width, any subsequent change in gap width will result in a decrease in discharge current. This makes it necessary for the device to continually cycle on each side of the current maximum to adjust the electrode position in the proper direction. This is the principle of the Autospark (21), a commercial gap controller, which then uses the current-induced signal to adjust electrode position.

In the voltage control mode (36, 103) a coil or capacitive probe is placed near the spark circuit, producing a voltage that varies linearly with gap width. The directly proportional relationship between rf voltage and gap width eliminates any ambiguities with respect to a change in control signal. The control element can correct the gap width to a preselected value (rf voltage) by moving one electrode as the control signal dictates.

Figure 9 shows the corrective action and resultant gap width for both types of spark gap control devices. The cyclic nature of the current controller is clear. However, it should also be noted that the *mean* gap is maintained quite well with time. For long exposures, the net gap variations will tend to average out. Voltage control, however, allows closer gap control and is particularly useful for short exposures (or integrations).

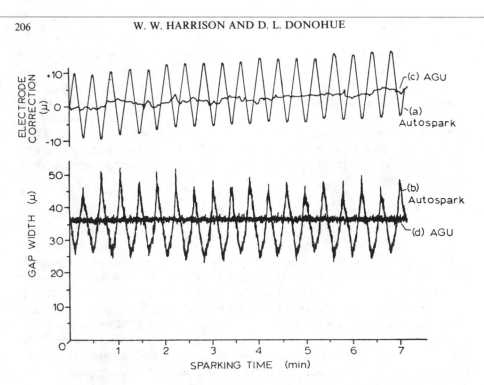

Fig. 9.   Correlation between gap width and electrode position for current control (Autospark) and voltage control (AGU) (stainless steel electrode using low vibration) (103). (a) Electrode position—Autospark, (b) gap width—Autospark, (c) electrode position—AGU, (d) gap width—AGU.

4.   Electrostatic and Magnetic Analyzers

These instrumental components for SSMS are rather conventional, normally of the double-focusing Mattauch–Herzog type (see Chapter 1).

5.   Ion Detection and Spectral Display

a.   PHOTOGRAPHIC

Until about 1970, essentially all SSMS analyses were carried out with ion-sensitive silver bromide plates as detectors, and even today the use of electrical detection has not removed the ion-sensitive plate from its role as the major detector in SSMS. The Mattauch–Herzog geometry of spark source mass spectrometers, with its simultaneous focusing of all ions at a plane, gives particular advantage to the use of such plates. A broad elemental mass spectrum extending from lithium to uranium can be simultaneously recorded for qualitative and quantitative analysis.

Although these detection plates are often referred to as "photoplates," there are significant differences between silver bromide emulsions prepared for the detection of photons versus those intended for ion detection. The modest penetrating ability of 15–20 keV ions dictates that the gelatin layer surrounding each silver bromide grain be very thin, no more than a few hundred angstroms. Also, for the high-vacuum conditions required in a mass spectrometer, low outgassing is desired with thin emulsions of minimal water and gas content. The Ilford Q-2 ion-sensitive plate became a popular ion detector by combining good sensitivity with adequate resolution and vacuum properties.

Use of the ion-sensitive plate as an ion detector in SSMS is accomplished by positioning the plate at the focal plane of the ions, with the impinging ions then producing a latent image for subsequent development. A series of exposures is normally taken, ranging from very short exposures that record the major constituents to long exposures where even the trace elements appear. As in all photographic processes, the relationship of plate darkening to incident flux is linear over only a limited range. As exposures increase, the lines from the major constituents saturate and may affect adjacent lines by formation of a surrounding "halo," apparently produced by a combination of primary ion scattering plus the release of secondary ions and neutrals from the emulsion. This can seriously affect detection limits for ion lines located near matrix lines.

Exposures in SSMS are not timed accumulations, such as those taken in emission spectrography. Rather, the random nature of the spark-generated flux (including intervals when the spark may be entirely out for seconds) necessitates the use of integrated charge accumulations. A monitor plate at the exit of the electrostatic analyzer (see Fig. 1) intercepts a fixed fraction of the ion beam that is then integrated. Exposures may range from 10^{-4} to 10^3 nC or larger, at approximately factor-of-3 increments.

Approximately 3000 ions are required to produce a barely perceptible line on an ion-sensitive plate. The plates are not uniformly sensitive for all the elements, however; there is a mass dependence term that should be considered. While there is incomplete agreement as to the exact quantitative relationship, sensitivity appears to decrease linearly with the square root of the isotopic mass. Ion energy also affects plate sensitivity: 25-keV ions should produce greater sensitivity than the more generally used 20-keV ions.

b. ELECTRICAL

Electrical detection has become of general application to SSMS. High-sensitivity electron multipliers, particularly the 20 stage Allen type (121), are now commonly used. This detector and its operation have been considered in Chapter 1. The aspects of electrical detection to be discussed here pertain to signal manipulation of the electron multiplier output. The fluctuating nature of the spark discharge and the application of SSMS to multielement survey analysis have led to specific readout circuitry designed to compensate for these factors.

(1) Scanning Mode

Perhaps the most powerful advantage of SSMS lies in this aspect: the ability to provide survey elemental analysis data in minutes. By scanning the magnet current or magnetic field, the various isotopes may be brought sequentially into register onto the electron multiplier. Because the spark discharge is unstable, a fluctuating ion beam results. Thus, a simple display of detector signal alone is not satisfactory. Instead, ratio circuitry is used to present a scan output signal that represents the ratio of the analytical ion (the electron multiplier signal) to the monitor ion beam. One example of this is shown in Fig. 10, from the work of Bingham and Elliott (21), who then use a three-decade logarithmic scale output display, as shown in Fig. 11. A wide dynamic range of elemental concentrations can be analyzed in each scan and by running three scans to cover the general

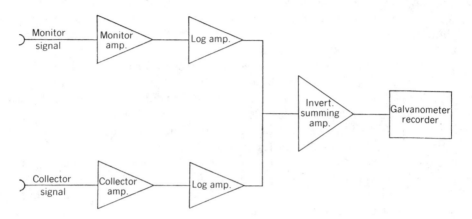

Fig. 10.   Block diagram of a ratio circuit for spark source mass spectrometry scans [after (21)].

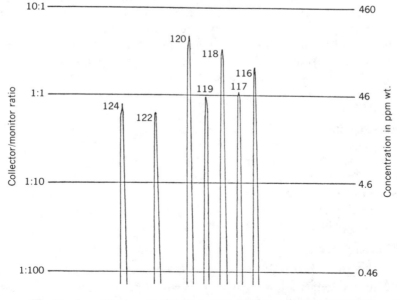

Fig. 11.   Logarithmic scan of tin isotopes in NBS low-alloy steel 462 (105).

ranges 0.1–100, 10–10^4, and 10^3–10^6 ppm, the entire concentration range from trace elements to majors may be examined.

Svec and Conzemius (130) have used a step function approach to electrical scanning. Signals from the electron multiplier and the monitor are fed into a ratiometer, the output of which is gate controlled by a staircase signal that also steps the electrostatic field. This mode allows convenient use of a computer to control and plot the scan.

Although the sensitivity of SSMS by electrical scan methods is quite good, the precision attainable has generally been no better than 25–40% for a single scan. It is recognized that precision increases if several scans are taken and the results averaged (21), but this is both time consuming and tedious. However, if repetitive scans are accumulated and averaged automatically, the improvement in ion statistics can be obtained with a minimum of operator intervention and effort. This has been accomplished (85, 107) by use of a multichannel analyzer as a spectral accumulator. An asymmetric triangular wave generator drives the magnet current in a cyclic mode until the desired number of mass scans are obtained, depositing each m/e species reproducibly in specific channels. Either full mass scans or selected mass regions may be programmed. The improvement in precision attainable by such signal averaging is shown in Table 1.

(2) Peak Switch Integration

The poor precision associated with SSMS electrical scanning can be improved by integration techniques, in which at a fixed magnetic field and electrostatic voltages a desired m/e is focused onto the electron multiplier. The detector current is then integrated for an appropriate charge accumulation at the monitor. Depending on ion flux, this may provide a significant improvement in

TABLE 1

Isotopic Abundance for Molybdenum Isotopes Showing the Effect of Signal Averaging from Multiple Scans (Sample: NBS No. 468 Steel) (85)

No. of Scans	Isotopes							
	100 (9.63%)	98 (23.78%)	97 (9.46%)	96 (16.53%)	96 (15.72%)	94 (9.04%)	92 (15.84%)	Av rel % error
1	8.83	23.76	12.91	18.07	16.87	4.91	14.65	16.38
5	12.43	20.60	7.51	16.99	17.01	9.95	15.51	12.31
10	9.58	22.83	8.17	16.15	18.49	9.21	15.57	5.95
15	9.66	23.42	8.71	16.65	16.89	9.27	15.44	3.28
20	9.66	23.41	9.33	15.61	16.56	9.56	15.88	2.88
Log recorder								
1	9.23	21.03	12.56	21.03	16.92	5.13	14.10	19.65
5	12.30	20.85	6.89	17.33	18.42	10.59	13.63	17.19
10	9.35	21.86	7.81	17.01	19.31	10.09	14.57	10.54
15	9.73	22.95	8.21	16.61	17.66	9.55	15.29	5.67
20	9.77	22.79	8.70	16.09	17.53	10.00	15.11	6.15

ion statistics over scanning methods. However, perhaps a greater factor in improving the precision may be the averaging of time-dependent phenomena (such as sample inhomogeneity and spark fluctuation) over the integration period, which may be tens of seconds to minutes, versus the corresponding scan residence time, which is often less than one second. Precision may be improved to better than 5% (21) by such integration methods.

The "peak switch" terminology arises from switching either (a) the accelerating and electrostatic analyser voltages or (b) the magnetic field to bring the next desired m/e rapidly into register for integration. The selection parameter may be preprogrammed such that several different isotopes may be successively selected from a multiposition switch.

(a) Electrostatic Peak Switch

Electrostatic peak switching has generally been favored (38, 54) over magnetic methods due to hysteresis problems associated with the latter. The prime advantage of electrostatic switching is the ability to switch rapidly between selected ions in a convenient and highly reproducible manner. Figure 12 shows a typical integration–readout system for SSMS peak switch analysis (105). The analytical ion current is integrated for a fixed charge as collected at the monitor. A digital voltmeter provides a net integrated voltage readout. Voltage-to-frequency converters have also been used (85) to convert the analog signal to a voltage-dependent frequency for accumulation on a count meter. Bingham and Elliott (21) found that the standard deviations between successive integrations could be reduced to 1.6%; the standard deviation for variations between sample loadings was 4.1%. However, the differences between experimental and reported

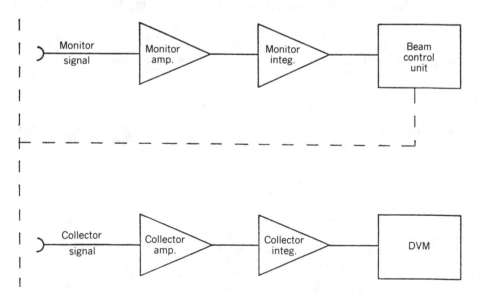

Fig. 12.   Typical integration readout system for spark source mass spectrometry (105).

elemental values for standard samples was found to be 7.9%. They attribute better than 4% of this error to uncertainty in the reported values for the standards.

(b)   Magnetic Peak Switch

Electrostatic peak switching may introduce several problems, including possible changes in ion extraction efficiency, variation of energy band pass into the magnetic analyzer, and electron multiplier gain dependence on ion energy. The general electrostatic peak switch mass limitation of M to $2M$ before required readjustment of the magnetic field is also a distinct disadvantage when carrying out a survey analysis for many elements. Magnetic peak switching eliminates these difficulties because the electrostatic voltages remain unchanged.

A simple and inexpensive manual magnetic peak switch system (104) is shown in Fig. 13. A Hall probe is used to produce a voltage analogous to the magnetic field. A calibration plot of mass versus Hall voltage as read from the digital voltmeter allows the analyst to move rapidly and reproducibly from one ion species to another for integration measurements. The Hall voltage is not used to control the magnetic field in this application.

Hull (93) has shown that a magnetic peak switch system can attain a precision of 0.5 parts per thousand using the system shown in Fig. 14, wherein the

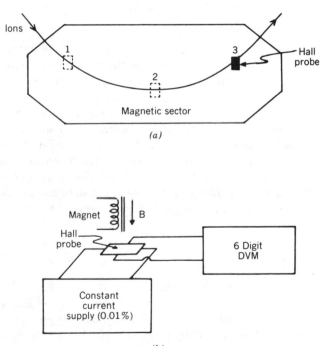

Fig. 13.   Example of Hall probe circuitry that simplifies manual peak switching. (a) Placement location of the Hall probe; 3 is best. (b) Hall probe measurement system (104).

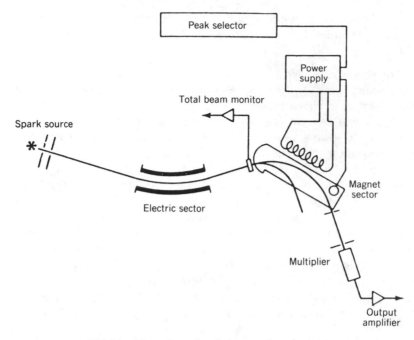

Fig. 14.   Magnetic peak switch control system (106).

electromagnet is controlled by a precision magnetic field regulator. The ion of interest is focused onto the detector by setting the proper voltage from the peak selector. The Hall probe provides a continuous field monitoring and acts as an error signal if the field drifts from the preset value.

Magee and Harrison (106) constructed a 10-channel magnetic peak switch and control circuit. An integrating operational amplifier sensed and corrected for the voltage differences between a Hall probe and a mass reference voltage. Arai and coworkers (7) also developed a Hall probe-controlled magnetic field switching circuit for a commercial instrument. The reproducibility of peak setting was rated as 0.02 mass unit. Degreve et al. (45), constructed and evaluated an efficient magnetic peak selector using a Hall probe sensor for an MS-702 mass spectrometer.

c. POSITION-SENSITIVE DETECTORS

The use of channel electron multiplier arrays (CEMAs) in SSMS has been studied by Donohue et al. (50, 51, 146) and Radermacher and Beske (119). The CEMA consists of one or two thin plates of semiconducting glass perforated by small channels of 10–25 μm diameter in a close packed array. Energetic particles or photons that strike the surface of such a plate will produce secondary electrons that are accelerated inside each channel by an applied voltage of several hundred volts per plate. These electrons collide many times with the wall of the channel, providing a multiplication effect similar to that of a discrete dynode electron multiplier. Thus, each incident ion may produce a swarm of up to 10^6 electrons leaving the back face of the channel plate. The electrons can be collected with an appropriate anode or allowed to strike a phosphor screen, giving a visible image that retains the spatial characteristics of the incident ion beam—a form of "electronic photoplate." The advantages of such a detector are high sensitivity, simultaneous detection of many mass-resolved ion beams, and real-time readout as compared to the laborious and time-consuming development and readout required for photoplates. A disadvantage is that the pulsed nature of the spark can saturate the output of the channels for large ion beams limiting the dynamic range. The spatial resolution provided by commercial channel plates is intermediate between that of standard electrical detection $(R = 1000)$ and photoplates $(R \geq 10{,}000)$.

B. DATA REDUCTION

1. Qualitative Analysis

In general, the spectra that result from SSMS are relatively simple and easy to interpret, particularly when compared to the multiplicity of lines observed in emission spectrographic analyses. The technique of qualitative analysis is basically the same whether the mode of detection is photographic or electrical. In each case, one obtains a display of the elemental isotopes plus certain molecular combinations. An example of a photoplate spectrum is shown in Fig. 15. A segment from an electrical scan was previously shown in Fig. 11. An experienced analyst can quickly identify patterns of isotopes as indicating specific elements, particularly if he or she is familiar with the sample type. For photoplates, a light box is useful for initial cursory examination, but a comparator allows more careful identification by displaying on a split viewing screen both the sample plate and a standard mass-marked plate. For electrical detection, the oscillographic recorder scan is compared to a mass-marked scan. A table with a built-in light box allows convenient superposition of the sample and reference scans.

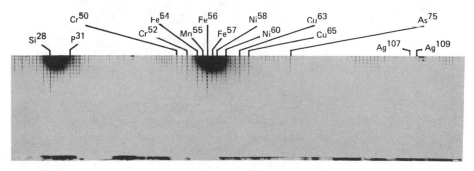

Fig. 15. SSMS photoplate spectrum (partial) of NBS 461 stainless steel. (Courtesy Nuclide Corp., State College, Pa.)

By obtaining a progression of photoplate exposures or electrical scans, all of the sample elements, from matrix to trace, may be displayed at a suitable sensitivity. For electrical detection, this often takes the form of three scans at low, medium, and high sensitivity. On a photoplate, approximately 15 exposures are taken, each differing by approximately a factor of three. The predominant species in the spectra are singly charged ions of each elemental isotope. In addition, multiply charged elemental ions are observed at concentrations $\frac{1}{5}$ to $\frac{1}{10}$ that of the singly charged parent ion. Molecular ions may also be present from hydrocarbons, or from molecular combinations such as FeO^+ or Si_2^+. These normally arise only from major constituents. Table 2 shows examples of the several types of molecular ions that may occur. Although the analyst must always be aware of possible interferences from molecular ions, in practice they are in most sample types a minor spectral contribution in SSMS.

Qualitative analysis depends on the systematic evaluation of the mass spectrum. Characteristic isotopic patterns should be present for each multi-isotopic element, reflecting the relative isotopic abundances. Where the abundance distributions appear out of line, the possible presence of interferences from isotopes of other elements or from molecular ions must be considered. If the element is present in sufficient concentration, multiply charged ions should be located for further verification. In the case of monoisotopic elements, such as arsenic, the As^{+2} species at m/e 37.5 may be the only sure means of identification.

TABLE 2

Molecular Interferences on Certain Isotopes (112)

Isotope obscured	A. From the Sample and Conductor Source
$^9Be^+$	$^{18}O^{+2}$, $^{27}Al^{3+}$
$^{12}C^+$, $^{13}C^+$	Graphite matrix and $^{24}Mg^{2+}$, $^{26}Mg^{2+}$
$^{14}N^+$, $^{15}N^+$	$^{28}Si^{2+}$, $^{30}Si^{2+}$
$^{32}S^+$	$^{16}O_2^+$, $^{64}Zn^{2+}$, $^{64}Ni^{+2}$
$^{33}S^+$	$^{16}O^{17}O^+$, $^{66}Zn^{2+}$
$^{34}S^+$	$^{16}O^{18}O^+$, $^{68}Zn^{2+}$
$^{45}Sc^+$	$^{13}C^{16}O_2^+$, $^{29}Si^{16}O^+$
$^{60}Ni^+$, $^{61}Ni^+$, $^{62}Ni^+$	Ni blank in silver and $^{12}C_5^+$, $^{12}C_4$ $^{12}C^+$, $^{50}Ti^{12}C^+$
$^{75}As^+$	$^{59}Co^{16}O^+$
$^{79}Br^+$, $^{81}Br^+$	$^{63}Cu^{16}O^+$, $^{65}Cu^{16}O^+$
In	Internal standard

Element obscured	B. From the Instrument Source
Ta	Ta sputtered from electrode chucks
Pt	TaN^+, TaC^+ sputtered from electrode chucks
Au	TaO^+ sputtered from electrode chucks
Bi	TaO_2^+ sputtered from electrode chucks
Hg	Hg diffusion pumps (intermittent)

A molecular ion does not normally show a +2 line in spark source spectra. Clearly, the best aid in spectral interpretation is experience, both with SSMS in general and with the specific sample type.

2. Quantitative Analysis

SSMS is rather unusual compared to other analytical methods in that quantitative analysis is most often performed using a single internal standard; that is, some isotope of known concentration in the sample is used as a reference against which all the elements to be analyzed in that sample are calculated. The unorthodox method is made possible by the relatively constant sensitivity that all elements exhibit in SSMS and made necessary by the incompletely controllable spark conditions that make use of an external standard somewhat hazardous; experimental conditions may change between successive samples. It is also true that reliable standards for survey analysis containing 30–40 elements are not readily available. External standards may be used (61, 81), however, provided that (a) a suitable standard is available and (b) proper control is exercised over critical electrode and spark parameters. The use of multielement isotopic dilution SSMS (29) offers interesting promise as a means of using multiple internal standards with excellent precision and accuracy (± 2–5%).

a. PHOTOGRAPHIC

For quantitative analysis, the density of an ion line is related to concentration by use of a densitometer, which can translate this line blackening into a recorded peak height or an analogous integrated area. Conversion of the peak heights or areas into corrected intensities can then be done by using the classical Churchill method (32) or by more empirical techniques such as those proposed by Hull (92) or Franzen (60). A working plot of ion intensity versus exposure allows the calculation of concentration for the analysis element, taking into account the several correction factors shown below:

$$C_x = C_s \cdot \frac{E_s}{E_x} \cdot \frac{A_x}{A_s} \cdot \frac{I_s}{I_x} \cdot S_{s/x} \tag{1}$$

where x and s refer to the analysis element and standard, respectively, C is concentration by weight, E is exposure at equivalent line areas or densities, A is the gram atomic weight conversion from atomic to weight percent, I is the isotopic abundance of the specific isotopes used, and S is the experimental relative sensitivity coefficient. Normally S is defined (55) as

$$S_{(s/x)} = \left[\frac{(C_s/C_x)_{\text{exp}}}{(C_s/C_x)_{\text{true}}} \right] \tag{2}$$

where the measured (experimental) concentration ratio of the standard and analysis element are divided by the true ratio of these elements. Also S includes other correction factors, such as mass response of the photoplate (54). In the use of external standards, in which case the standard and analysis element are the same element and isotope, expression (1) simplifies to

$$C_x = C_s \cdot \frac{E_s}{E_x} \tag{3}$$

For multielement analysis, however, the use of external standards is considerably more time consuming (as opposed to measuring only one standard) and has been used very little.

b. ELECTRICAL

Quantitative analysis by electrical detection should be subdivided into scanning versus peak switch methods (see Section 4.b). An oscillographic recording of a mass scan provides a spectrum (Fig. 11) from which peak heights may be taken for analysis elements and compared to the corresponding peak height for an internal standard, correcting for isotopic abundance and relative sensitivity, and if desired, atomic weight (to convert ppma to ppmw).

Peak switch quantitative analysis offers improved precision and accuracy, at the expense of time. Considerably more time per element is involved than in scanning methods. Therefore, broad survey analyses are not normally done by peak switch. Instead, samples in which a smaller number of elements need to be accurately determined are particularly appropriate for peak switch techniques. To take full advantage of the precision capabilities of this method, external standards are normally used, with a resultant relationship:

$$C_x = C_s \cdot \frac{i_{int(x)}}{i_{int(s)}}$$

where i_{int} refers to the integrated analytical ion multiplier current from the standard and unknown samples for a given total ion beam collection at the monitor. Precisions of better than 5% can be obtained by careful attention to critical parameters. For example, reproducible positioning of the electrodes is very important, requiring the use of a telescopic alignment system.

3. Automatic Data Reduction

With any analytical technique that produces as much information as does SSMS, the use of computer methods for data handling is a natural consideration. Both electric and photographic detection allow advantageous applications of computer data acquisition and processing. Little work has been reported toward computer control of the prior analysis steps. Given the broader scope of this review, the reader is directed to more detailed descriptions of computer applications in SSMS. A good survey by Woolston (141) summarizes the several approaches to interacting computers with SSMS. Specific programs were described by Woolston (140), Franzen and Schuy (60), and Degreve and Champetier de Ribes (43) for treating SSMS photographic plates. Burdo et al. (27) described a variation of the Hull method that is applicable to peak height, line width, or integrated intensities. Haney (74) evaluated a commercially available microdensitometer–computer package for acquisition of SSMS data from plates. Millett and coworkers (110) developed a computer system on-line to a microdensitometer to evaluate and intepret spark source spectra on plates. Computer techniques for electrical detection have been reported by Bingham (21) and Brown (24). Morrison et al. (114), described an on-line computer-controlled electrical detection system.

The interaction of a computer with a developed SSMS photoplate may simply involve correcting manually obtained line densities or areas to "intensities" (from stored calibration information) and then carrying out the arithmetic in the previously described quantitative expression. As such it may be quite useful, but this does not utilize the full capabilities of a computer to aid the analyst. A more frequent and more valuable application is automated plate reading. The analyst faced with evaluating a series of SSMS photoplates, each with 10–15 exposures, and each exposure consisting of from dozens to perhaps hundreds of lines, rapidly comes to appreciate an automated procedure. Computer evaluation of data taken directly from a densitometer scan of a photoplate allows decision making by the computer for qualitative analysis, based on the presence of proper isotopic distributions, possibly doubly charged lines, and the lack of inter ferences. Quantitative analysis is also performed.

Electrical detection is easily adapted to computer data acquisition and treatment. Scanning methods produce analog information that can be treated in a generally similar manner to densitometer scans taken from photoplates. Peak switch integration allows convenient translation of the electrical signal into a digital form for data reduction. The ability to convert ion current rapidly into an analytical result is particularly useful here because it allows operator interaction to study the sample or the effects of experimental parameters on readout values.

C. COMMERCIAL INSTRUMENTS

Compared to the many commercially available organic mass spectrometers, there have been relatively few manufacturers of spark source mass spectrometers, and today only one remains. This is not very surprising considering the high cost and more limited application that is made of inorganic mass spectrometry. Description of equipment no longer available is included so the reader has a sense of what instrumentation has been used in developing the SSMS field.

It is always hazardous to compare commercial instruments, particularly for instruments no longer available. Technical data from company literature may be misinterpreted and unfair extrapolations made between various instruments. The spark source mass spectrometers that have been used are, however, very similar both in design and performance. The three units discussed here are perhaps the best known and most highly developed instruments. All use a pulsed rf spark source for ionization and the Mattauch–Herzog spectrometer geometry. The three have 25 kV or better acceleration voltage, upward of 10,000 resolution, and sensitivities approaching 1 ppb. All were available with both electrical and photographic detection. Even the optional accessories were quite similar for each instrument.

Some specific comments on each model may be useful. The instruments are considered in alphabetical order by company name.

1.   AEI Scientific Apparatus, Ltd (Model MS-702R)

Figure 16 shows the MS-702R. A 500-kHz spark is provided with pulse lengths of 25–200 μs, repetition rates of 10–30,000, and 0–80 kV voltage. A 25-kV maximum accelerating voltage is used with a 380-mm radius ESA (31° 50′) and a 16,000 G magnetic analyzer to allow 10,000 resolution (50% density definition) at mass 208. Source slits are available as 0.05, 0.025, and 0.012 mm; an externally adjustable α slit (2 mm maximum) allows an appropriate compromise between transmission and resolution. Eight 250 × 50 mm photoplates may be loaded into the magazine with 15 exposures available per plate. Electrical detection modes include both peak scanning (500 resolution at 10% valley definition) and peak switching. Precisions for the latter are given as 1% for matrix level and 5% at the 1 ppm level. The Autospark (21) controls spark gap continuously.

Accessories include a cold finger sample cooler for low-melting-point samples, a cryogenic source pumping kit to reduce hydrocarbon and molecular species,

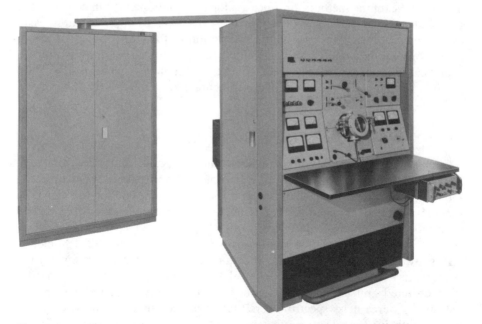

Fig. 16.   AEI MS-702 spark source mass spectrometer. (Courtesy AEI, Ltd., Manchester, England).

an ion beam chopper, an ion counting system, and a data reduction package. An ion microprobe attachment for the 702R was developed but did not become widely used.

2.   Jeol, Ltd. (JMS-01BM-2)

Figure 17 shows the JMS-01BM-2, which with variations in the ion source and detector can be converted to other modes, such as ion bombardment and field ionization. JEOL is the only supplier available today for a new spark source mass spectrometer. For SSMS, the pulsed rf (~ 1 MHz) allows maximum output of 100 kV and 2 kW at pulse widths of 20–80 μs and repetition rates of 10–10,000. Acceleration voltage is variable from 15 to 30 kV. This unit uses a spherical lens ESA (200-mm radius) for maximum ion transmission and sensitivity. An interesting feature is the feedback use of the ion monitor signal to control and maintain a preset beam intensity. Both the α and β slits are externally adjustable. The ability to vary the β slit, which defines energy dispersion, allows interesting experiments relative to the effect of ion source parameters. The magnetic field is 13,000 G with a 10,000 or better resolution. Sensitivity is quoted as 10^{-9} g (atomic) at ~ 2000 resolution. Available mass range is 1–500. The JMS-01BM-2 uses larger photoplates than the MS-702R. Eight 380×50 mm plates are accepted by the magazine at 30 exposures (1 mm height) per plate. The

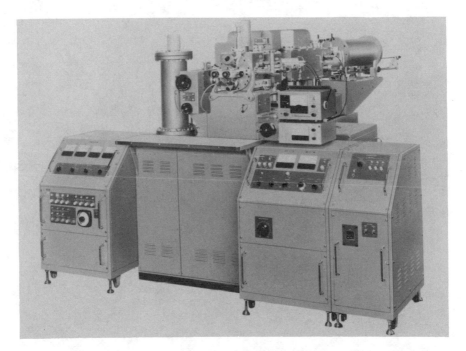

Fig. 17.   JEOL JMS-01BM-2 spark source mass spectrometer (98) (Courtesy JEOL, Inc., Cranford, N.J.).

electrical detection system relies upon an automatic spark gap control that may be either voltage *or* current controlled. In addition, both peak scanning and peak switching are available. The Hall probe control of the scanning unit is noteworthy: 500 resolution is reported for peak scanning with a 0.03 ppma detection limit; 0.01 ppma is reported for peak switch. A data analysis system is available for data acquired by electrical detection.

3. Nuclide Corporation (Graf-3S)

Figure 18 shows the Graf-3S. As for the previously described spectrometers, other types of sources are possible, but only the pulsed rf spark will be considered here. The spark voltage (0–100 kV) can be pulsed at 10–30,000 pulses per second with pulse lengths of 3.2–10,000 μs. The accelerating potential is continuously variable from 0 to 32 kV at 3-mA output current. An externally adjustable α slit is variable to a maximum of 0.200 in., allowing control over angular divergence of ions entering the ESA, which is the spherical type, 380-mm radius. A variable β slit is also a standard feature. The magnetic field is variable from 200 to 14,000 G. Both field and current regulation are available. A temperature-compensated Hall probe is used to sense and control the field mode. A *pair* of ion beam

Fig. 18. Nuclide Graf-3S spark source mass spectrometer. (Courtesy Nuclide Corp., State College, Pa.).

monitors located between the ESA and magnetic analyzers allows measurement of the instantaneous current and total exposure.

D. SAMPLE PREPARATION AND HANDLING

1. Contamination Problems

Given the extreme sensitivity shown by SSMS for all the elements, it is clear that contamination possibilities pose a serious problem. Mizuike (111) has discussed loss and contamination considerations for trace element analysis, and most of these apply to SSMS. Fortunately, chemical separations are rarely necessary in SSMS, eliminating one major problem area. However, several other critical contamination aspects exist.

Airborne elemental contamination is a very real problem that is often insufficiently guarded against. In many modern analytical laboratories, air entering the building is filtered and passed through electrostatic precipitors to remove dust particles. These measures are not totally successful, however, and may even give the analyst a false sense of security. Air-conditioning coils in air ducts may create condensation accumulations, which upon drying may become powdery or flaky and be picked up by the strong air flow. To prevent objectionable bacteria and fungi growth, it is common to spray "slimicides" on air-conditioning coils; these chemicals are laden with elemental additives, such as chromium, which may then be distributed throughout the building in the air supply. Lead compounds from automobile exhausts are almost always to be found in the air intake system. It is therefore necessary to take particular precaution by not exposing the SSMS samples to direct room "fallout" during treatment steps such as drying, mixing, or electrode formation. A glove box is effective for certain steps, but awkward to work in. Controlled-atmosphere drying chambers have been described by Thiers (133) and Chow and McKinney (31).

Analysts attempt to restrict the contact of the sample with other metals or even glass wherever possible. Polyolefin ware is now readily available to replace glassware for many trace element applications. Polyethylene volumetric flasks are reputed to be as accurate as standard glass flasks (53). Teflon, although expensive, allows the use of temperature up to about 250°C (22).

Ultra-pure reagents are desirable wherever possible. Blanks run on standard reagent-grade chemicals, such as the common acids, may show unacceptable impurity levels. Redistilled (129) and special-purity (13) acids are available. Other reagents of extra-high purity (134) may also be useful, possibly as added internal standards.

2. Removal of Organics

The presence of organics in a SSMS sample intended for elemental analysis can greatly complicate the resultant spectrum and create interferences at nominal elemental isotopic lines. The removal of these organics is thus necessary, but in such a way as to prevent loss of the desired inorganic species. A monograph by Gorsuch (70) has reviewed the many oxidation (ashing) methods that are available to the analyst. The commonly used techniques are (a) high-temperature dry ashing (HTDA) by furnace, (b) low-temperature dry ashing (LTDA) in an excited oxygen atmosphere, and (c) wet ashing by oxidizing acids.

All of these have been used in the oxidation of organic based samples for SSMS analysis. HTDA is the most rapid and simple of the three. A muffle furnace at 500–600 °C provides thorough ashing and requires no intervening operator attention. Volatile elements, such as mercury, are of course lost, and there are various reports, not always in agreement, that several other elements may suffer losses. This can vary with the sample matrix and chemical combination of the element. Some biological samples, such as tissues, create certain problems in HTDA by forming a melt of the alkali salts that upon cooling form a hard glassy residue often difficult to remove from the sample container. Quartz boats are attacked and soon ruined by this action. The advantage of HTDA for SSMS, apart from the aforementioned difficulties, is the rapid production of an inorganic residue that may be mixed directly with a conducting matrix for electrode formation.

LTDA (66) uses atomic oxygen species, produced by passing molecular oxygen through an rf discharge, to oxidize organic material. The reactivity of atomic oxygen allows this to occur at temperatures estimated at 100–200°C. Thus, the danger of elemental loss (other than mercury) is substantially reduced. For tissue samples, LTDA yields an easily handled powdery residue, thus avoiding the hard, glassy residue typical of HTDA. Ashing times are usually longer, however, for LTDA.

Wet ashing with mixtures of HNO_3, H_2SO_4, or $HClO_4$ can produce efficient oxidation of organic materials at temperatures that minimize elemental loss. Smith (128) has reviewed wet ashing methods, particularly the use of $HClO_4$. HNO_3–$HClO_4$ is a common combination that uses the HNO_3 to remove easily oxidized components, after which the $HClO_4$ oxidizes the more resistant species, such as lipids in tissues. Once experience is gained with wet ashing procedures, many samples can be run concurrently. Disadvantages include the contamination possibilities from the acid reagents and from the digestion vessels (111) and also the experience necessary to carry out the acid digestions safely and efficiently.

Which of these methods is best for SSMS? This will depend on the sample type and the mode of electrode preparation preferred. For those samples that contain no volatile elements, HTDA is often used, particularly if the residue is to be mixed directly with a conducting matrix in the dry state. LTDA allows the same procedure, but with less chance for elemental volatilization. If solution sample preparation is preferred, wet ashing has some advantage in that the sample is already in solution form prior to electrode preparation.

3. Shaping of Electrodes

a. CONDUCTING SOLIDS

Conducting solids, such as metals, are ideal samples for SSMS in that they require little or no pretreatment. The shaping step may simply involve cutting or machining a pair of small electrodes from the conducting material. The size will depend on the particular sample holders, but a 10–15 mm length, 1–2 mm diameter is generally suitable. Smaller sizes can also be accommodated. A single sample electrode can be run against a pure metal counter electrode in those cases where sample size is limited. Small sample particles can be mounted on the tips of conducting rods for analysis.

b. POWDERS

Powdered samples must be compacted into electrodes. Several types of pressing techniques have been utilized; one popular approach is through the use of a sample moulding die (10). A cylindrical polyethylene slug, $\frac{7}{16}$ in. diameter $\times \frac{5}{8}$ in. length is frozen in liquid nitrogen, and a small shaft is drilled to accommodate the powder sample. The slug is then placed in a stainless steel die and compressed at 10–12 tons pressure, forming an electrode that, after release of the pressure, can be gently tapped out of the slug.

Conducting powders, such as metals, can be compacted directly or perhaps with the addition of an internal standard (95). Nonconducting powders must be mixed with a conducting matrix of high purity; graphite, gold, and silver powders are commonly used.

c. SOLUTIONS

Solution sample residues can be analyzed by SSMS. The solution must be evaporated onto some suitable matrix for subsequent analysis. The preparation may be as simple as soaking a porous graphite rod in the analysis solution, followed by drying under an infrared lamp. Ahearn (2) has described the evaporation of a solution onto the surface of a germanium support electrode for sensitive qualitative analysis. A similar procedure (3, 12) involved the evaporation of a few drops of solution onto a disk, which was rotated and sparked against a counter electrode.

Better quantitative results are obtained by mixing the solution sample and any required internal standard with powdered graphite to form a homogeneous slurry (142). After infrared drying, the graphite, now containing the solution residue, is thoroughly mixed in a plastic vial using a vibrator, such as a Wig-L-Bug (39). Electrodes are then pressed for analysis. This approach has been successful even for metal samples (71) placed in a solution matrix by acid dissolution and treated as above. Improved quantitative results were reported over running the metals directly.

d. INSULATORS

Insulators require the presence of some auxiliary electrode or conducting medium if they are to be run directly in solid form. Insulating powders, of course, can be run as described previously. Ahearn (2) and James and Williams (94) showed that insulators could be sparked by mounting them onto conducting electrodes whereby the spark discharge strikes both. The components of the auxiliary electrode appear in the mass spectrum and may create interferences. Smaller pieces of insulators may be packed into conducting tubes or foil for analysis. All of these procedures provide adequate qualitative analysis, after background correction, but quantitative analysis is made difficult by the inconsistent sampling rate of the insulators.

IV. APPLICATIONS

SSMS has been applied to essentially all kinds of trace element sample types. The greatest daily use has been probably in the metals and semiconductor industries because of the concentration of instruments and needs in these areas, but many of these are routine applications that are never published. The variety of sample types of analyses described in the literature are often those that concern new or specialized uses, and thus may not accurately reflect the general utilization of SSMS in industry.

This section will concentrate on selective rather than exhaustive coverage of SSMS applications, with the idea of merely illustrating the capabilities of this technique. The recent review (13) and the monographs by Ahearn (4, 5) are recommended for additional coverage.

A. METALS

SSMS can be advantageously applied to a range of metal matrices (55). Degreve and Figaret (44) analyzed impurities in aluminum, using electrical detection. The analysis of gaseous elements in metals has been carried out by a number of workers (1, 15, 16, 52, 116, 123, 137). The accurate measurement of these elements in most SSMS samples is made more difficult by interferences arising from the residual gases in the ion source and by the desorption of gases and water vapor from surfaces inside the ion source chamber during the analysis. However, with improved vacuum pumping and attention to cleanliness in the ion source, it is possible to make meaningful measurements. Beske and coworkers (15, 16) used liquid helium cryopumping to improve the detection limits for oxygen in copper and silicon.

Konishi (100) determined oxygen, carbon, and nitrogen in iron, nickel, and copper using photographic detection. The instrumental blank level was reduced to a satisfactory level by use of a gallium arsenide counterelectrode, which appeared to produce a getter action for residual gases.

Evans, et al. (54) analyzed for elements in the Johnson-Matthey CA copper standards and the NBS 460 steel series. Table 3 demonstrates the precision obtained ($\pm 2\%$, relative standard deviation) for five elements; an average accuracy of $\pm 7\%$ was reported. Chastagner and Tiffany (17) reported the quality control analysis of curium and californium for all elements above mass 10 (except carbon, nitrogen, oxygen; and noble gases). Microgram size, or smaller, samples were mounted on gold electrodes and analyzed from a single photographic exposure. A typical analysis showed 29 elements detected as impurities in curium. Hamilton and Minski (73) examined 19 elements in NBS 461 steel and compared sensitivity factors with data from other laboratories. Sulfur, titanium, and silver showed high sensitivity relative to the iron matrix. Konishi (99) reported attempts to improve precision and accuracy in the analysis of steel, aluminum, copper, and lead matrices. By controlling source parameters, particularly spark gap (50 μm), a relative standard deviation of 5–15% was obtained on repetitive photographic exposures. Magee and Harrison (106) used magnetic peak switching to analyze for 15 elements in NBS 461 steel taking nickel as the internal standard. Guidoboni and Evans (71) reduced matrix effects in SSMS by dissolving their samples (aluminum and steel standards), doping with an internal standard, and depositing them onto a silver powder, which was used to form an electrode. The silver then became the new matrix and was thus constant from one sample to another. The resultant deviation from NBS values was reported to be about half that observed when the standard samples were analyzed in solid form

B. THIN FILMS AND SURFACES

The spark source is not generally well suited to the analysis of thin films and surface layers. The energy expended in the spark discharge is not readily controlled and presents a rather coarse sample probe that forms abrupt craters, rather than smoothly etching its way through a film. Secondary ion mass

TABLE 3

Spark Source Mass Spectrographic Analysis of 460 Series Steels (54)

NBS No.	Cr (wt%)		Mn (wt%)		V (wt%)		Nb (wt%)		Mo (wt%)	
	SSMS	NBS	SSMS	NBS	SSMS	NBS	SSMS	NBS	SSMS	NBS
461	0.133	0.13	0.38	0.36	0.025	0.024	0.009	0.011	0.26	0.30
462	0.85	0.74	0.94	0.94	0.068	0.058	0.089	0.096	0.084	0.080
463	0.275	0.26	1.13	1.15	0.105	0.10	—	—	—	—
464	—	—	—	—	—	—	0.033	0.037	0.025	0.029
465	0.0038	0.004	0.0024	0.002	0.035	0.032	—	—	—	—
466	0.012	0.011	0.0075	0.007	0.107	0.113	—	—	—	—
468	0.58	0.54	0.47	0.47	0.18	0.17	—	—	—	—
Precision (% RSD)	1.83		1.23		1.55		1.70		2.13	

spectrometry and ion-scattering spectrometry perform much more ably in this respect. However, if favorable surface analysis techniques are not available, SSMS may be pressed into service, often with surprisingly good results, if handled properly.

Hickam and Sweeny (87) and Skogerboe (127) have reviewed the microprobe utilization of the spark source. Desjardins (47) discussed the problems of using SSMS for thin films, particularly insulating films. The use of a gold wire (see Fig. 19) around the film and substrate, coupled with a gold counter electrode, allowed determinations in a tin oxide film to about 10 ppma. Meeuwsen (109) used SSMS to conduct surface analysis of germanium and GaSe disks by scanning the disc with a high-purity graphite counter electrode. An average layer of 2 μm for the germanium disks and 4 μm for the GaSe materials was removed per scan. Analyses were reported showing low ppma impurities. Liebich and Mai (102) used a high-purity niobium electrode, tip diameter $\cong 0.07$ mm) to examine 0.005–2-μm thick copper films on glass substrates. Well-defined craters were shown for a single rf pulse onto the thicker films. For film thicknesses from 0.005 to 1 μm, the spark penetration could be almost totally restricted to the film.

C. BIOLOGICAL, GEOLOGICAL, AND ENVIRONMENTAL APPLICATIONS

These applications are grouped because each normally requires mixture with a supporting matrix and subsequent pressing into suitable electrodes. The biological samples must be ashed to remove organic interferences, as must many of the environmental samples. In fact, the biological–environmental interface is not sharp because animal (or human) tissue may be analyzed to investigate an environmental effect. The great advantage of SSMS in all three of these areas is its ability to yield a complete elemental survey and indicate the presence of any element in sample types where, indeed, almost any element is possibly present.

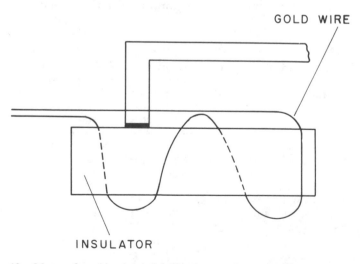

GOLD WIRE

INSULATOR

Fig. 19.   Means of sparking insulating films by use of a surrounding gold wire (47).

1. Biological Materials

Morrison and Roth (115) surveyed the biological application of SSMS to 1970. Since then, Brown et al. (25) have analyzed for selected elements in blood serum from cancer patients, reported 40 elements in a kidney stone, and found 43 elements in lung tissue of an ex-coal miner. Harrison and Clemena (82) determined 30 elements in human fingernail samples from 17 different donors. Some subjects were monitored monthly for one year. It is not surprising that the elements found were similar to those in human hair (142). Harrison et al. (80) determined 11 elements in human blood serum from 6 different donors and

TABLE 4
Analysis of Ashed Mammalian Blood (14)[a]

| Element | I.A.E.A.—Animal Blood 66/12 | | U.K.A.E.A.— Elementary Composition of Mammalian Blood |
	I.A.E.A. Range	SSMS Values	Mean Values
Na		1930	1990
Mg		59	41
Al		0.32	0.37
P		440	373
S		2150	2040
Cl		2780	2900
K		2250	1690
Ca		44	60
Cr	0.02–0.56	0.07	0.023
Mn	0.08–0.38	0.35	0.026
Fe	835–3270	1050	475
Cu	0.77–3.3	2.4	1.07
Zn	11.5–22	10.2	6.5
As		0.07	0.49
Se		0.24	<0.2
Br		4.3	4.6
Rb		4.7	2.7
Sr		0.06	0.0095
Y		0.004	
Zr		0.006	
Ag		0.02	0.19
Sn		0.009	0.13
Sb		0.002	0.0047
I		0.04	0.056
Cs		0.01	0.0028
Ba		0.24	
Hg		<0.006	0.0065
Pb		0.4	0.27

[a] All values in micrograms per gram wet weight.

analyzed up to 31 elements in human aortic tissue from 10 autopsy cases. SSMS data were compared to atomic absorption results for certain elements. Ball et al. (14) reported the high-resolution analysis of 28 elements (Table 4) in mammalian blood. Up to 19,000 resolution was used to prevent spectral interferences from such species as $NaCl^+$. Magee and Harrison (106) determined relative sensitivity factors for 7 elements in NBS Bovine Liver. Jaworski and Morrison (96) studied factors affecting sensitivity coefficients for 14 elements in NBS Orchard Leaves and Bovine Liver.

2. Geological Materials

A summary of geological applications through 1970 is available (115). Subsequently, Taylor (132) has discussed the problems of internal standards and plate processing relative to the analysis of lunar soils. By using lutetium as an internal standard, good agreement was obtained with neutron activation and isotopic dilution techniques. Andreani et al. (6) reported sample preparation techniques for geological materials and discussed corrections for interferences encountered in the complex mineral spectra. Data from 53 elements in 2 standard rock samples were compared with those obtained by ASTM methods. Morrison and Rothenberg (113) studied homogenization of geological samples with high-purity graphite, leading to an average precision of $\pm 7\%$ for 28 elements in meteorite material. Jaworski and Morrison (96) compared SSMS and neutron activation analysis data (Table 5) for lunar samples in a study centering around relative sensitivity coefficients.

Jaworski and Morrison (97) used 9 separated isotopes mixed in a graphite powder matrix to determine trace elements in geological samples. Donohue et al. (49) used a similar technique involving 15 separated isotopes in a high-purity

TABLE 5
Agreement of NAA and SSMS Lunar Analyses (96)

	Composition (ppm)					
	Sample 60315		Sample 62255		Sample 74220	
Element	SSMS	NAA	SSMS	NAA	SSMS	NAA
Cr	1600	1400	36	24	4900	4400
Co	83	95	1.4	1.3	75	65
Ni	1600	1400	ND	ND	90	70
Cu	9.5	10.8	0.82	1.09	30	34
Zr	800	1000	1.0	0.82	150	180
Yb	10.0	13.0	0.080	0.071	4.5	4.3
Lu	1.8	2.0	0.08	0.07	0.7	0.5
Av % dev	15.2		21.9		18.4	

Both using $W-1$ as comparative standard.

silver powder matrix to measure trace elements in coal ash and fly ash with a precision and accuracy of ± 5–10%.

3. Environmental Materials

The increased importance attached in recent years to environmental problems has produced greater interest in SSMS. Spark source mass spectrometers have been used in EPA laboratories for both air and water analyses, as well as for material of related interest, such as fuels.

Much of the introductory work in these areas have been demonstration studies by Brown and coworkers. Brown and Vossen (23) reported the analysis of 28 elements in particulates taken from New York City air (see Table 6). A

TABLE 6

Analysis of Air Particulates from New York City (23)

Elements detected	Concentration ($\mu g/m^3$)	Ratio (Sample/Blank)
Boron	0.004	ND[a]
Fluorine	0.11	ND
Sodium	> 5.5	10
Magnesium	11	ND
Aluminum	1.1	ND
Silicon	63	ND
Phosphorus	1.1	5
Sulfur	2.3	ND
Chlorine	0.28	ND
Calcium	2.8	5
Potassium	40	110
Titanium	0.23	100
Chromium	0.30	240
Vanadium	1.9	20,000
Manganese	0.07	100
Iron	2.4	22
Cobalt	0.007	ND
Nickel	0.32	114
Copper	0.26	250
Zinc	1.1	133
Arsenic	0.005	66
Bromine	0.12	250
Strontium	0.05	6
Zirconium	0.004	ND
Molybdenum	0.01	25
Tin	0.07	ND
Barium	0.02	57
Lead	4.3	5,000

[a] ND: Not detected in nitrocellulose blank.

nitrocellulose filter was used for a 9.5 hr pumping period, 19 liters of air per minute. Brown (26) discussed the use of SSMS for the analysis of air, water, and river sediments. Brown et al. (25) reported 45 elements in 1.65 mg of particulates taken from a miner's air sampler. Guidoboni (72) compared SSMS and atomic absorption data in the analysis of coal samples for 9 elements. Dry ashing versus wet ashing procedures were investigated. von Lehmden et al. (138) determined trace elements in coal, fly ash, fuel oil, and gasoline, comparing SSMS to optical emission spectrometry, neutron activation analysis, and atomic absorption. Table 7 shows a summary of the reported concentrations in coal by SSMS. Taylor (131) summarized the use of SSMS for water samples.

TABLE 7
SSMS Analysis of Fuel Samples (138)

Elements analyzed	Coal[a]	Fly Ash[a]	Fuel Oil[b]	Gasoline[c] (Premium)
Hg	<2	<1	NA	NA
Be	0.4	7	0.0005	<0.0001
Cd	6	<3	0.003	0.001
As	2	40	0.2	<0.02
V	10	250	High	<0.0001
Mn	20	300	0.4	0.005
Ni	<40	100	High	<0.01
Sb	0.6	10	0.003	<0.0004
Cr	<30	200	0.8	<0.001
Zn	<100	200	0.5	0.2
Cu	10	100	0.2	0.005
Pb	<4	200	2	High
Se	<15	<10	NA	0.001
B	15	500	0.002	<0.02
F	<2	30	0.004	<0.0003
Li	0.3	20	0.02	0.001
Ag	<2	<1	0.0006	<0.001
Sn	3	6	0.01	<0.003
Fe	2,000	High	10	0.1
Sr	100	150	0.4	<0.001
Na	600	2,000	0.4	<0.4
K	100	High	High	0.1
Ca	10,000	High	High	<0.05
Si	6,000	High	High	<1
Mg	2,000	10,000	2	0.02
Ba	400	200	2	0.001
P	—	—	—	0.2
S	—	—	—	NA

NA: not analyzed.
[a] ppm, by weight on sample direct.
[b] ppm, by weight, 450°C ashing.
[c] μg/ml, HCl extraction.

4. Nuclear Industry Applications

A number of workers, both in Europe and the United States have used specially shielded SSMS instruments to measure radioactive samples coming from the industrial applications of nuclear energy. Carter and Sites (28) described a Plexiglass glove box surrounding the ion source chamber. This instrument was able to measure samples with high α and β radioactivity, such as $^{233}UO_2$ (57) and diluted spent-fuel dissolver solutions (58). Similar instruments were used in Belgium (122) and France (18, 19) to measure impurities in samples of plutonium and mixed-oxide reactor fuels.

5. Forensic Applications

There have been a number of studies directed at cataloging trace elements in samples that are frequently encountered in forensic studies. Haney and Gallagher at the FBI (75, 76) have measured 26 commonly occurring elements in lead bullets with the goal of identifying the manufacturing lot. The determination of trace elements in window glass and container glass have been addressed by Scaplehorn and coworkers at Aldermaston (40, 64, 65). The forensic significance of certain biological samples such as human hair and fingernails has been studied by Harrison et al. (82, 142) and described in a more general review (79). Gooddy and coworkers (6, 68) used SSMS to study the elemental constituents of blood and cerebrospinal fluid as an aid in the diagnosis of neurological disease.

REFERENCES

1. Abelova, L. G., Andrikanis, E., and Kormilitsyn, D. V., *Ref. Zh., Metall.,* Abstr. No. 8K35 (1982).
2. Ahearn, A. J., *J. Appl. Phys.,* **32,** 1195 (1961).
3. Ahearn, A. J., *J. Appl. Phys.,* **32,** 1197 (1961).
4. Ahearn, A. J., Ed., *Mass Spectrometric Analysis of Solids,* Elsevier, New York, 1966.
5. Ahearn, A. J., Ed., *Trace Analysis by Mass Spectrometry,* Academic Press, New York, 1972.
6. Andreani, A. M , J. C. Brun, J. P. Mermoud, A. Fillot, and R. Stefani, *Method. Phys. Anal.,* 7, 258 (1971); through *Chem. Abstr.,* **76,** 529 (1972).
7. Arai, M., M. Narto, T. Takagi, and M. Takeuchi, 21st Annual Conference on Mass Spectrometry and Allied Topics, San Francisco, CA, May, 1973, paper V3.
8. Associated Electrical Industries, Ltd., Manchester, England, Technical Information Bulletin, "Conversion to Negative Ion Mass Spectrometry."
9. Associated Electrical Industries, Ltd., Manchester, England, Instruction Manual for MS-702.
10. Associated Electrical Industries, Ltd., Manchester, England, Technical Information Bulletin A-304, "A New Die for Bricketting Powders for Spark Source Mass Spectrographic Analysis."
11. Associated Electrical Industries, Ltd., Manchester, England, Technical Information Bulletin A 1506, "Ion Beam Chopper."
12. Associated Electrical Industries, Ltd., Manchester, England, Technical Information Bulletin A 2005, "A Spinning Disc Technique for the Analysis of Liquids."
13. Bacon, J. R., and A. M. Ure, *Analyst,* **109,** 1229 (1984).

14. Ball, D. F., M. Barber, and P. G. T. Vossen, *Biomed. Mass Spectrom.*, **1**, 365 (1974).
15. Beske, H. E., *Fresenius' Z. Anal. Chem.* **256**, 103 (1971).
16. Beske, H. E., G. Frerichs, and F. G. Melchers, *Fresenius' Z. Anal. Chem.*, **267**, 99 (1973).
17. Berthod, J., B. Alexandre, and R. Stefani, *Int. J. Mass Spectrom. Ion Phys.*, **10**, 478 (1973).
18. Billon, J. P., *Analusis*, **22**, 137 (1973).
19. Billon, J. P., M. Buffereau, and S. Deniaud, *Nucl. Sci. Abstr. 1976*, **33**, 14937 (1975).
20. Bingham, R. A., P. Powers, and W. A. Wolstenholme, Proceed. Int. Conf. on Mass Spectros., Kyoto, Japan, 1969.
21. Bingham, R. A., and R. M. Elliott, *Anal. Chem.*, **43**, 43 (1971).
22. Biolabs, Inc., Derry, N. H., Plastics catalog 74–1.
23. Brown, R., and P. G. T. Vossen, *Anal. Chem.*, **42**, 43 (1971).
24. Brown, R., P. Powers, and W. A. Wolstenholme, *Anal. Chem.*, **43**, 1079 (1971).
25. Brown, R., M. L. Jacobs, and P. G. T. Vossen, unpublished data.
26. Brown, R., Int. Symp. on Ident. and Measmt. Envir. Pollut., Ottawa, 1971, paper 79.
27. Burdo, R. A., J. R. Roth, and G. H. Morrison, *Anal. Chem.*, **46**, 701 (1974).
28. Carter, J. A., and J. R. Sites, *Trace Analytical Mass Spectrometry*, A. J. Ahearn, Ed., Academic Press, New York 1972, p. 347.
29. Carter, J. A., D. L. Donohue, J. C. Franklin, and R. W. Stelzner, Proceedings 23rd Annual Conference on Mass Spectrometry and Allied Topics, Houston, Texas, May, 1975.
30. Chastagner, P., and B. Tiffany, *Int. J. Mass Spectrom. Ion Phys.*, **9**, 325 (1972).
31. Chow, T. J., and C. R. McKinney, *Anal. Chem.*, **30**, 1499 (1958).
32. Churchill, J. R., *Ind. Eng. Chem., Anal. Ed.*, **16**, 653 (1944).
33. Coburn, J. W., *Rev. Sci. Inst.*, **41**, 1219 (1970).
34. Coburn, J. W., and E. Kay, *Appl. Phys. Lett.*, **18**, 425 (1971).
35. Coburn, J. W., E. Taglauer, and E. Kay, *J. Appl. Phys.*, **45**, 1779 (1974).
36. Colby, B. N., and G. H. Morrison, *Anal. Chem.*, **44**, 1263 (1972).
37. Colby, B. N., and C. A. Evans, *Anal. Chem.*, **46**, 1236 (1974).
38. Conzemius, R. J., and H. J. Svec, *Talanta*, **16**, 365 (1969).
39. Crescent Dental Mfg. Co., Chicago, Illinois.
40. Dabbs, M. D. G., B. German, E. F. Pearson, and A. W. Scaplehorn, *J. Forensic Sci. Soc.*, **13**, 281 (1973).
41. Daughtrey, E. H., Jr., D. L. Donohue, P. J. Slevin, and W. W. Harrison, *Anal. Chem.*, **47**, 683 (1975).
42. Daughtrey, E. H., Jr., and W. W. Harrison, *Anal. Chem.*, **47**, 1024 (1975).
43. Degreve, F., and D. Champetier de Ribes, *Int. J. Mass Spectrom. Ion Phys.*, **4**, 125 (1970).
44. Degreve, F., and R. Figaret, 31st G. A. M. S. Conference, Paris, France, 1972.
45. Degreve, F., R. Figaret, J. J. Le Goux, and L. Calavrias., *Int. J. Mass Spectrom. Ion Phys.*, **14**, 183 (1974).
46. Dempster, A. J., *Rev. Sci. Inst.*, **7**, 46 (1936).
47. Desjardins, M., *Advan. Mass Spectrom.*, **4**, 439 (1968).
48. Donohue, D. L., and W. W. Harrison, *Anal. Chem.*, **47**, 1528 (1975).
49. Donohue, D. L., J. A. Carter, and J. C. Franklin, *Anal. Lett.*, **10**, 371 (1977).
50. Donohue, D. L., J. A. Carter, and G. Mamantov, *Int. J. Mass Spectrom. Ion Phys.*, **33**, 45 (1980).
51. Donohue, D. L., J. A. Carter, and G. Mamantov, *Int. J. Mass Spectrom. Ion Phys.*, **35**, 243 (1980).

52. Dugger, D. L., and D. Oblas, *Talanta*, **24**, 447 (1977).

53. Dynalab Corporation, Rochester, New York, 1975 Catalog.

54. Evans, C. A., R. J. Guidoboni, and F. D. Leipziger, *Appl. Spectrosc.*, **24**, 85 (1970).

55. Farrar, H., IV, "Relating Mass Spectrum to Sample Composition," in A. J. Ahearn, Ed., *Trace Analysis by Mass Spectrometry*, Academic Press, New York, 1972, pp. 239–295.

56. Franklin, J. C., *Nucl. Sci. Abstr.*, **25**, 15474 (1971).

57. Franklin, J. C., L. Landau, D. L. Donohue, and J. A. Carter, *Anal. Lett.*, **A11**, 347 (1978).

58. Franklin, J. C., L. Landau, and J. A. Carter, *Radiochem. Radioanal. Lett.*, **41**, 217 (1979).

59. Franzen, J., K. H. Maurer, and K. D. Schuy, *Z. Naturforsch.*, **21A**, 37 (1966).

60. Franzen, J., and K. D. Schuy, *Z. Naturforsch.*, **21A**, 1479 (1966).

61. Franzen, J., and K. D. Schuy, *Z. Anal. Chem.*, **225**, 295 (1967).

62. Franzen, J., "Physics and Techniques of Electrical Discharge Ion Sources" in A. J. Ahearn, Ed., *Trace Analysis by Mass Spectrometry*, Academic Press, New York, 1972, pp. 11–56.

63. Fursov, V. Z., and M. S. Chupakhin, *Zh. Anal. Khim.*, **29**, 1677 (1974).

64. German, B., D. Morgans, A. Butterworth, and A. W. Scaplehorn, *J. Forensic Sci. Soc.*, **18**, 113 (1978).

65. German, B., and A. W. Scaplehorn, *J. Forensic Sci. Soc.*, **12**, 367 (1972).

66. Gleit, C. E., and W. D. Holland, *Anal. Chem.*, **34**, 1454 (1962).

67. Gooddy, W., T. R. Williams, and D. Nicholas, *Brain*, **97**, Pt. 2, 327 (1974).

68. Gooddy, W., E. I. Hamilton, and T. R. Williams, *Brain*, **98**, Pt. 1, 65 (1975).

69. Gorman, J. G., E. J. Jones, and J. A. Hipple, *Anal. Chem.*, **23**, 438 (1951).

70. Gorsuch, T. T., *The Destruction of Organic Matter*, Pergamon, Oxford, 1970.

71. Guidoboni, R. J., and C. A. Evans, Jr., *Anal. Chem.*, **44**, 2027 (1972).

72. Guidoboni, R. J., *Anal. Chem.*, **45**, 1275 (1973).

73. Hamilton, E. I., and M. J. Minski, *Int. J. Mass Spectrom. Ion Phys.*, **10**, 77 (1973).

74. Haney, M. A., Proceedings 23rd Annual Conference on Mass Spectrometry and Allied Topics, Houston, Texas, May, 1975.

75. Haney, M. A., and J. F. Gallagher, *Anal. Chem.*, **47**, 62 (1975).

76. Haney, M. A., and J. F. Gallagher, *J. Forensic Sci.*, **20**, 484 (1975).

77. Hannay, N. B., and A. J. Ahearn, *Anal. Chem.*, **26**, 1056 (1954).

78. Hannay, N. B., *Rev. Sci. Inst.*, **25**, 644 (1954).

79. Harrison, W. W., G. G. Clemena, and C. W. Magee, *J. Ass. Offic. Anal. Chem.*, **54**, 929 (1971).

80. Harrison, W. W., M. A. Ryan, L. D. Cooper, and G. G. Clemena, "Determination of Trace Elements in Biological Materials by Spark Source Mass Spectrometry," in W. Mertz and W. E. Cornatzer, Eds., *Newer Trace Elements in Nutrition*, Marcel Dekker, New York, 1971, pp. 391–420.

81. Harrison, W. W., and G. G. Clemena, *Anal. Chem.*, **44**, 940 (1972).

82. Harrison, W. W., and G. G. Clemena, *Clin. Chim. Acta*, **36**, 485 (1972).

83. Harrison, W. W., and C. W. Magee, 21st Annual Conference on Mass Spectrometry and Allied Topics, San Francisco, CA, May, 1973.

84. Harrison, W. W., K. R. Hess, R. K. Marcus, and F. L. King, *Anal. Chem.*, **58**, 341A (1986).

85. Harrison, W. W., and W. A. Mattson, *Anal. Chem.*, **46**, 1979 (1974).

86. Helm, H., F. Howarka, and M. Pahl, *Z. Naturforsch., Teil A*, **27**, 1417 (1972).

87. Hickam, W. M., and G. G. Sweeney, "Mass Spectrographic Microprobe Analysis," in A. J. Ahearn, Ed., *Mass Spectrometric Analysis of Solids*, Elsevier, New York, 1966, pp. 138–163.

88. Hickam, W. M., and E. Berkey, "Analysis of Low Melting and Reactive Samples," in A. J.

Ahearn, Ed., *Trace Analysis by Mass Spectrometry*, Academic Press, New York, 1972, pp. 323–345.

89. Honig, R. E., "The Production of Ions from Solids" in A. J. Ahearn, Ed., *Mass Spectrometric Analysis of Solids*, Elsevier, New York, 1966, pp. 16–55.

90. Howorka, F., and M. Pahl, *Z. Naturforsch., Teil A*, **27**, 1425 (1972).

91. Howarka, F., W. Lindinger, and M. Pahl, *Int. J. Mass. Spectrom. Ion Phys.*, **12**, 67 (1973).

92. Hull, C. W., 10th Annual Conference on Mass Spectrometry and Allied Topics, New Orleans, LA, 1962.

93. Hull, C. W., *Int. J. Mass Spectrom. Ion Phys.*, **3**, 293 (1969).

94. James, J. A., and J. L. Williams, *Advan. Mass Spectrom.*, **1**, 157 (1959).

95. Jaworski, J. F., R. A. Burdo, and G. H. Morrison, *Anal. Chem.*, **46**, 805 (1974).

96. Jaworski, J. F., and G. H. Morrison, *Anal. Chem.*, **46**, 2080 (1974).

97. Jaworski, J. F., and G. H. Morrison, *Anal. Chem.*, **47**, 1173 (1975).

98. JEOL, Inc., Medford, Mass., Brochure MS-OP70 for JMS-01BM-2 spark source mass spectrometer.

99. Konishi, F., *Shitsuryo Bunseki*, **19(4)**, 284 (1971); through *Chem. Abst.*, **76**, 80526h (1972).

100. Konishi, F., *Int. J. Mass Spectrom. Ion Phys.*, **9**, 33 (1972).

101. Krebs, K. H., *Fortschr. Phys.*, **16**, 419 (1968).

102. Liebich, V., and H. Mai, *Adv. Mass Spectrom.*, **6**, 655 (1974).

103. Magee, C. W., and W. W. Harrison, *Anal. Chem.*, **45**, 220 (1973).

104. Magee, C. W., and W. W. Harrison, *Anal. Chem.*, **45**, 852 (1973).

105. Magee, C. W., Ph.D. Thesis, University of Virginia, Charlottesville, Virginia, May, 1973.

106. Magee, C. W., and W. W. Harrison, *Anal. Chem.*, **46**, 474 (1974).

107. Mattson, W. A., and W. W. Harrison, *Anal. Chem.*, **47**, 968 (1975).

108. Mattson, W. A., Ph.D. Thesis, University of Virginia, Charlottesville, Virginia, May, 1975.

109. Meeuwsen, M., "Surface Analysis of Ge Discs," MS-702 Users Conference, Manchester, England, 1972.

110. Millett, E. J., J. A. Morice, and J. B. Clegg, *Int. J. Mass Spectrom. Ion Phys.*, **13**, 1 (1974).

111. Mizuike, A., "Separations and Preconcentrations," in G. H. Morrison, Ed., *Trace Analysis*, Interscience, New York, 1965, pp. 103–159.

112. Morrison, G. H., and A. T. Kashuba, *Anal. Chem.*, **41**, 1842 (1969).

113. Morrison, G. H., and A. M. Rothenberg, *Anal. Chem.*, **44**, 515 (1972).

114. Morrison, G. H., Corly, B. N., and Roth, J. R., *Anal. Chem.*, **44**, 1203 (1972).

115. Morrison, G. H., and J. R. Roth, "Insulators, Powders, and Microsamples," in A. J. Ahearn, Ed., *Trace Analysis by Mass Spectrometry*, Academic Press, New York, 1972, pp. 297–322.

116. Nyary, I., and I. Opauszky, *Fresenius' Z. Anal. Chem.*, **309**, 274 (1981).

117. Owens, E. B., and N. A. Giardino, *Anal. Chem.*, **35**, 1172 (1963).

118. Pustovit, A. N., and G. G. Sikharulidze, *Mikrochim. Acta*, **2**, 219 (1981).

119. Radermacher, L., and H. E. Beske, *Adv. Mass Spectrom.*, **8A**, 378 (1980).

120. Roboz, J., "Mass Spectrometry," in G. H. Morrison, Ed., *Trace Analysis*, Interscience, New York, 1965, pp. 435–509.

121. Roboz, J., *Introduction to Mass Spectrometry*, Interscience, New York, 1968.

122. Rymen, T., *Int. J. Mass Spectrom. Ion Phys.*, **47**, 299 (1983).

123. Saprykin, A. I., I. R. Shelpakova, and I. G. Yudelevich, *Izv. Sib. Otd. Akad. Nauk SSSR. Ser. Khim. Nauk*, **5**, 77 (1982).

124. Schuy, K. D., and H. Hintenberger, *Z. Anal. Chem.*, **197**, 98 (1963).

125. Schwabe, S., *Z. Angew. Phys.*, **12**, 244 (1960).

126. Shaw, A. E., and W. Rall, *Rev. Sci. Inst.*, **18**, 278 (1947).

127. Skogerboe, R. K., "Surface and Thin Films Analysis" in A. J. Ahearn, Ed., *Trace Analysis by Mass Spectrometry*, Academic Press, New York, 1972, pp. 401–422.

128. Smith, G. F., "The Wet Chemical Oxidation of Organic Compositions," G. F. Smith Chemical Co., Columbus, Ohio, 1965.

129. G. F. Smith Chemical Co., Columbus, Ohio.

130. Svec, H. J., and R. J. Conzemius, *Adv. Mass Spectrom.*, **4**, 457 (1968).

131. Taylor, C., 23rd Annual Conference on Mass Spectrometry and Allied Topics, Houston, TX, May, 1975.

132. Taylor, S. R., *Geochim. Cosmochim. Acta*, **35**, 1187 (1971).

133. Thiers, R. E., "Separation, Concentration, and Contamination," in J. H. Yoe and H. J. Koch, Jr., Eds., *Trace Analysis*, Wiley, New York, 1957, pp. 619–636.

134. United Mineral and Chemical Corp., New York, N.Y.

135. Van Puymbroeck, J., R. Gijbels, M. Viczian, and I. Cornides, *Int. J. Mass Spectrom. Ion Processes*, **56**, 269 (1984).

136. Viczian, M., I. Cornides, Van J. Puymbroeck, and R. Gijbels, *Int. J. Mass Spectrom. Ion Phys.*, **51**, 77 (1983).

137. Vidal, G., P. Galmard, and P. Lanuse, *Anal. Chem.*, **42**, 98 (1970).

138. von Lehmden, D. J., R. H. Jungers, and R. E. Lee, Jr., *Anal. Chem.*, **46**, 239 (1974).

139. Woolston, J. R., and R. E. Honig, *Rev. Sci., Inst.*, **35**, 69 (1964).

140. Woolston, J. R., *RCA Rev.*, **26**, 539 (1965).

141. Woolston, J. R., "Computer Techniques for Solids Analysis," in A. J. Ahearn, Ed., *Trace Analysis by Mass Spectrometry*, Academic Press, New York, 1972, pp. 213–238.

142. Yurachek, J. P., G. G. Clemena, and W. W. Harrison, *Anal. Chem.*, **41**, 1666 (1969).

Chapter 4

PLASMA CHROMATOGRAPHY

Roy A. Keller

Department of Chemistry, State University of New York,
College at Fredonia, Fredonia, New York

Contents

I. INTRODUCTORY REMARKS

The Plasma Chromatograph™ (Franklin GNO, West Palm Beach, Florida) (PC) is most appropriately described as an atmospheric pressure chemical ionization drift time spectrometer (23). It qualifies as a chromatograph only as differential migration from a narrow zone. It falls short of the resolution expected from common chromatographic techniques. Objection has been made to the term "plasma," which can mean a relatively stable cloud of gaseous ions. Gaseous electrophoresis has also been suggested (31). Historically the basic equipment (excepting the drift gas) and results resemble rudimentary mass spectrometry when high-vacuum technology was aborning (34). Incorporation of modern electronic techniques have given results far superior to those of more primitive days. It seems that most basic research of the PC is complete, and it has reached the applications stage, particularly in repetitive trace analysis for a particular component.

Initial great interest was produced by the lower limit of detection (LLD) of the instrument. The now out-dated but intensive review of Keller and Metro (25) reported results of 1.5×10^{-14} mol of a dioxin, a compound of much current interest, in 1 μl of solution in benzene. Seemingly 10^{-11} mol parts, depending on the species under study, is about the average experience. In some cases this would yield a LLD 20 times superior to the electron capture detector of gas chromatography and is much superior to the flame ionization detector. A second advantage is that the PC operates at atmospheric pressure, which detours the need for the mechanical pumping system associated with a vacuum mass spectrometer. A third advantage is that a simple reversal of the direction of the electric field allows the collection of either positive or negative ions at the collecting electrode. Ion–molecules are examined rather than charged fragments of shattered sample molecules common to mass spectrometry. This simplifies the spectrum and avoids computer reduction of the data. A bonus is the compactness of the instrument. Figure 1 shows the Model Beta/VII, an early version. We elect to discuss it rather than later modifications because we employed this configuration for a number of years. Quantitative studies are largely lacking.

II. INSTRUMENTATION

Figure 1 is the basic package. The focal point of the instrument is the Separate Ion Formation and Drift Tube (SIFAD)™ (Franklin GNO Corporation) shown in Fig. 2. Heated *carrier* or *reactant gas*, generally nitrogen or air, of very low water content (it is passed over a molecular sieve) passes through the sample injector and over a ^{63}Ni source of an electron capture detector. This produces *reactant* or *primary ion–molecules*, positive or negative, in the reaction region, and they move toward the drift region. The arrows indicate the flow patterns. An admitting electronic gate is pulsed to admit a "wedge" of these ions to the drift region, and it is here where differential migration from a narrow zone occurs depending on the "width" of the gate window. Counter to the reactant gas, *drift gas* is admitted at the other end of the tube. A second electronic gate is positioned at the end of the drift tube and can be pulsed to open at carefully regulated increasing delay times after the admitting gate opens. Each admitted pulse is sampled once by the second gate. The sampled pulse is sensed by an electrode, sent to a fast electrometer amplifier, and generally read out by an x–y recorder. Other readout options exist. Our experience has been that a 2-min scan of a 20-ms drift time range is about all that one can tolerate without peak distortion by the recorder.

The sequence of events is that the carrier gas, encountering radiation from the ^{63}Ni source, forms reactant ion–molecules. This is shown in Fig. 3 as the scans at the front of the diagram, the time scale being from front to back. At some point sample molecules are injected into the system, whereupon charge transfer reactions occur between reactant ions and sample to produce new species, the *sample* or *product ion–molecules*. Note that reactant ions may be completely depleted. As the sample is depleted, reactant ions again appear.

A. CARRIER GAS

The nature of the reactant ions depends on the identity and purity of the carrier gas. Air and nitrogen are popular, air for negative ion collection and nitrogen or air for positive ions. Our experience has been with ultra-pure nitrogen, further passed over a molecular sieve. Water content and temperature play an important part of reactant ion formation as they determine the extent of

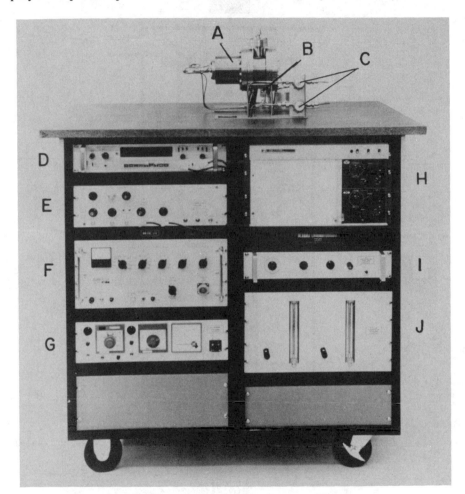

Fig. 1.   Model Beta/VII Plasma Chromatograph (25): (A) SIFAD™ Tube (Franklin GNO Corporation, West Palm Beach, Florida). Sample inlet on the left. The entire unit is encased in heating mantles. (B) Solid-state preamplifier. (C) Carrier and drift gas shutoff valves. (D) Hewlett-Packard Model 5325A Universal Counter. (E) GNO PC Controller. (F) Fluke Model 408B high-voltage power supply. (G) GNO temperature controller and sensor. (H) Hewlett-Packard 7035B x–y recorder. (I) GNO PC electrometer. (J) GNO PC flow controller. The rotameters shown here are replaced by Hastings Mass Flowmeters. Reprinted with permission from Keller, R. A., and M. M. Metro, *Sep. Purif. Methods*, **3** (1), 207 (1974). Copyright 1974, Marcel Dekker, New York.

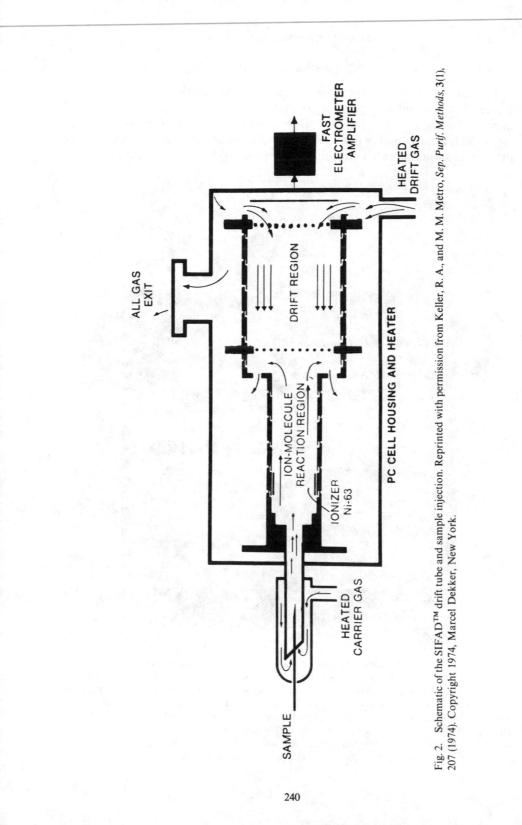

Fig. 2. Schematic of the SIFAD™ drift tube and sample injection. Reprinted with permission from Keller, R. A., and M. M. Metro, *Sep. Purif. Methods*, **3**(1), 207 (1974). Copyright 1974, Marcel Dekker, New York.

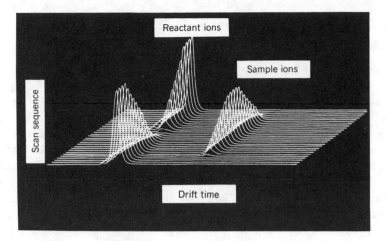

Fig. 3.   Composite of successive plasmagrams of Musk Ambrette. Reprinted with permission from Keller, R. A., and M. M. Metro, *Sep. Purif. Methods*, **3**(1), 207 (1974). Copyright 1974, Marcel Dekker, New York.

ion clustering. Values range from 1 to 100 ppm of water (25). The partial pressure of water, after passing over the sieve, has been estimated at 10^{-5} torr.

B.   RADIATION SOURCE AND REACTION REGION

The carrier gas entrance is followed by a repeller electrode (not shown) that collects ions of charge opposite to that of the ion–molecules of interest. This is followed by the ^{63}Ni source. Electron energies are as high as 60 keV and an approximate rating is 12 mCi. After consideration of inappropriate directions of radiation, source self-absorption, and so on, it is estimated that about 10% of the high-energy electrons are available to produce ionization. About 2000 thermal ion pairs are produced per source emitted electron. The current in the reaction chamber is about 1.5×10^{-9} A in a 1 cm diameter beam. It has been claimed that the formation of reactant ions occurs within a fraction of a millisecond and within 1 mm of the ^{63}Ni source. A licence is required for operation.

Much study has been devoted to the identity of the reactant ions. Keller and Metro (25) reviewed the results and speculations that were published prior to 1974. The following were reported as identified by mass number as studied by coupling a quadrupole mass spectrometer to the PC outlet. Species claimed to be so identified are indicated by an asterisk.

Positive reactant ions:

General $(H_2O)_nH^+$, $(H_2O)_nNO^+$

Specific $(H_2O)H^{+*}$, $(H_2O)_2H^{+*}$, $(H_2O)_3H^{+*}$, $(H_2O)_4H^+$, NO^{+*},

$(H_2O)NO^{+*}$, $(H_2O)_2NO^{+*}$, (11)

Negative reactant ions:

    General $(H_2O)_nO_2^-, (H_2O)_n(CO_2)O_2^-$

    Specific $O_2^-{}^*, (H_2O)O_2^-{}^*, (H_2O)_2O_2^-, CO_2^-, (CO_2)O_2^-{}^*, CO_3^-{}^*, N_2O^-$

Several important points must be made: (1) Results may be suspect because of changes arising at the interface between the SIFAD tube at atmospheric pressure and the vacuum of the mass spectrometer. (2) Although a single peak is observed for the reactant ion species, there is mass spectral data that indicates that there may be more than one ion–molecule species present in the positive reactant ions responsible for the response. A now almost ancient publication by Keller and Giddings (24) theoretically demonstrated that interacting species that maintained equilibrium at reaction rates very much faster than the rate of migration would produce a single peak, the position of which (retention time or volume) was not characteristic of the individual species present. This also must apply to product ions. A paper published by Carroll et al. (3) after the 1974 review supported this contention and went on to state that *"earlier literature identifications are based upon assumed mass and mobility equivalency and are incorrect"* (pg. 1959). They added NH_4^+ and $NH_4^+ (H_2O)$ to the list. Spangler and Collins (32) went on to treat negative ions adding $(H_2O)_nOH^-$ and NO_2^-.

The debate is nontrivial. The analyst is interested in (a) reproducibility of results, (b) LLD, (c) sensitivity (rate of variation of the signal intensity with concentration), and (d) linear range. The first point is questionable, the second has been fairly well answered, and the last two have not been attacked.

C. SAMPLING

Sampling is a severe problem. We defined *overload* as any sample size that leads to the disappearance of the reactant ions. Unlike most chromatographic systems, the LLD is not the problem but instead a component concentration sufficiently small so as not to swamp the detector. The power of chromatography is the ability to deal with a sample of an unknown number of components of unknown identity at unknown concentrations. In this sense the PC is very limited. There are just so many reactant ions available to the sample components, and the distribution of charge occurs before entering the differential migration path if the PC is to be used as a simultaneous separator/identifier.

At best the sampling procedure is crude. To avoid overload and peak distortion, the sample concentration in the carrier gas should not exceed 1 ppm and should remain below 1 ppb. This is difficult to achieve. One method is to flame a wire, cool it, and while still warm, expose it to the vapors of the sample, allowing it to cool, and then insert it in the injection port. Such a method must be rejected when dealing with unknown mixtures. The headspace vapor concentration must be related to the liquid solution composition, and further fractionation may occur on evaporation-desorption from the wire. One may also draw liquid into a 1-μl syringe, expel it, pull the plunger, and inject the residual vapor. The LLD coupled with the large internal adsorptive surface of the SIFAD tube combine to present a formidable problem. It has been observed that dilute aqueous solutions of some compounds generated data for several hours. Reports on one sample at 1 ppm to 1 ppb levels still allowed 15–30 min of data generation. This is not a happy state of affairs if the PC is interfaced with a gas chromatograph. A component does not clear the PC before the next one appears. The residence time can be reduced by silylation of the surfaces followed by bake-out at elevated temperature at reduced pressure for extended periods of time. Even then the suspicion has been expressed that the apparently inactive hydrocarbon surface umbrella may react with the sample to release artifacts. Various combinations of the PC with stream splitters where a particular gas chromatographic peak region is picked off, dilution flasks, and so on have been reported.

The preceding discussion deals with pulse sampling. *Steady-state sampling* has been employed; that is, the sample is continuously injected into the PC. A far more extensive discussion of sampling appears elsewhere (25).

D. PRODUCT (SAMPLE) ION–MOLECULES

On injection of the sample, charge transfer reactions occur between the reactant ions and the sample to produce new species. These will depend on competition of the sample ions with the reactant ions for the charge, the concentration of the sample, the temperature, and, for mixtures, the competition between component ion–molecules, which is also concentration dependent. Fast reactions between sample species may be as expected as that between reactant ion species.

Implication that no fragmentation occurs is not intended. It has been reported for the halogenated benzenes and Freons.

E. DRIFT GAS

The drift gas flows countercurrent to the carrier gas. It is generally the same as the carrier gas, purity included. Its purpose is to sweep uncharged sample molecules out of the drift region and prevent further ion–molecule reactions. This hypothesis does not exclude interaction of the ion–molecules with one another or with the drift gas, including its impurities. The paper by Cram and Chesler (5) is particularly significant on this point. They stated that *it is impossible to identify the ion–molecules from the drift time because an eluting peak may be due to ion–molecules which spent a significant part of their drift time as some other species.* They insist that identification can only be achieved by coupling a mass spectrometer to the PC outlet and that data are restricted exclusively to the ions passing the PC–mass spectrometer interface.

F. READOUT SYSTEMS

The most common readout system of opening a terminal electronic gate to the sensing electrode at regulated delay times has been described. This has been abbreviated as the PC-mg (moving gate) mode. The objection to this mode is that changes that occur in ion–molecule chemistry within the drift tube during the 2-min scan are not detected. The high-speed electronics makes it possible to leave the exit gate open and scan the ion–molecules in a single admitted pulse. The signal is read out on an oscilloscope, PC-os. The instantaneous signal may also be recorded by a signal averaging computer, stored, and an accumulation of them read out on a recorder, PC-sac. We once referred to this as a "sum-of-sins" because it can be a composite of rapidly changing behavior. The work of Benezra (1) is an excellent example of this technique. His work is also noteworthy because of his appreciation of the dangers of overload.

III. DRIFT TIME

Drift times, the time of transit through the drift tube, are reported in terms of the *reduced mobility*, K_0, the speed of an ion–molecule in an electric field of 1 V/cm in a gas at 273°C and 760 torr. It is given by

$$K_0 = \frac{1}{t} \frac{D^2}{V} \frac{p}{760} \frac{273}{T}$$

where t = observed transit time (s) at p, T, and V
         D = drift space distance (cm)
         V = potential across the drift tube (V)
         p = pressure (torr)
         T = drift space temperature (K)

Much early effort was devoted to relating the reduced mobility to properties of the ion–molecules thought to be involved. The first correlation was application of the Langevin mobility equation (25), which involved a mass relationship. The superb theoretical paper by Revercomb and Mason (31) challenged the simplicity of this even for a unique ion–molecule species. It was this group that suggested that gaseous ion electrophoresis be used because molecular size and shape and drift gas pressure must be considered [also see Lin et al. (27)]. From prior discussion it is clear that such treatments are academic if proof is lacking that the cloud of ionic species in a signal pulse is a single species that spent all of its drift experience as that species. This author strongly supports Benezra's attitude (1) (pg 123): "No attempt was made to identify any ion species through a mass mobility curve without mass spectrometric evidence."

IV. CHROMATOGRAPH/IDENTIFIER

The power of chromatography is its ability to separate and detect the members of a multicomponent mixture. Separation and detection most often involve separate instrument components or separate steps. It would be ideal if the two functions could be combined. It is expected that the chromatogram is an additive composite of the chromatograms of the individual components (once identified).

Constancy of behavior of a single component was examined by Metro and Keller (28). Their PC-os scans for di-*n*-propyl ether are shown in Fig. 4. The sample was very much at overload so that the early scans are rapidly evolving. One can, as has been suggested, only consider scans when the reactant ions begin to reappear (6 ms), which in this case, is at 1061 s (17.6 min). The figure is a series of unretouched photographs of the oscilloscope screen; drift times are inaccurate because the time scale is difficult to establish. The time of reappearance of the reactant ions varies with the nature of the compound for the same sample size. We also adjusted our sample size of a pulse sample of diethyl ether so that reactant ions were present on the initial PC-mg scan (appropriate sample). The results are shown in Fig. 5 (2). Reactant ions are indicated by *R*. Other peaks are

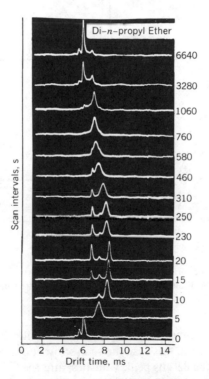

Fig. 4.   Successive PC-os scans of di-*n*-propyl ether. Reprinted with permission from Metro, M. M., and R. A. Keller, *J. Chromatogr. Sci.*, **11**, 520 (1973). Copyright 1973, Preston Publications, Niles, IL.

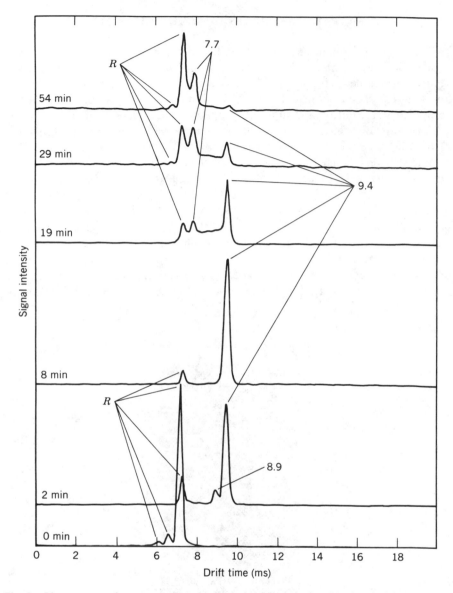

Fig. 5. Plasmagrams of an appropriate pulse sample of diethyl ether. Reprinted with permission from Bird, G. M., and R. A. Keller, *J. Chromatogr. Sci.*, **14**, 574 (1976). Copyright 1976, Preston Publications, Niles, IL.

identified by their drift times. The scan completed after 2 min ought to be the significant one. Yet, even in this case the scan shows an evolution over time. The 9.4-ms peak persists, the 8.9-ms peak vanishes, and the 7.7-ms peak appears. To puzzle this out Bird and Keller (2) injected diethyl ether under steady-state conditions by thermostating the ether in a tube and passing the vapor into the

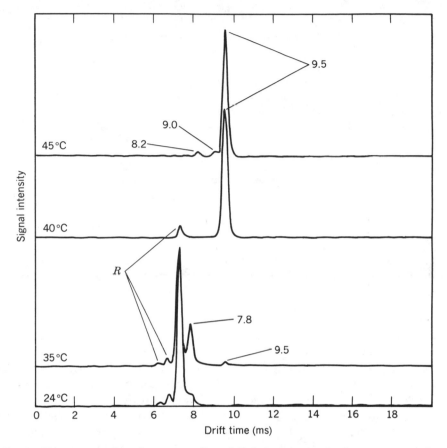

Fig. 6. Plasmagrams of steady-state sampling of diethyl ether at various vapor concentrations. Reprinted with permission from Bird, G. M., and R. A. Keller, *J. Chromatogr. Sci.*, **14**, 574 (1976). Copyright 1976, Preston Publications, Niles, IL.

sample injection port through a thermometer capillary. Figure 6 shows the results; temperatures of the thermostat are listed on the left. The lower the temperature, the smaller the sample concentration. We must conclude that the plasmagram, even for an appropriate sample, is highly concentration dependent. Such a change indicates that pulse sampling, even if appropriate, yields a concentration change as the sample is depleted from the system and an evolution of the plasmagram. To reiterate, one must question the PC-mg mode, as this evolution occurs over the scan time. This was examined by Metro and Keller (29). They recorded a PC-os scan, switched to a PC-mg scan, recorded a following PC-os scan, and compared the three. Results were too often ambiguous.

Because "chromatography" implies mixtures, we introduced binary mixtures of ketones (29) (Benezra used mixtures of aromatic ketones) and alcohols (26).

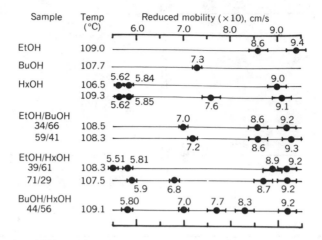

Fig. 7. Reduced mobilities of positive ion molecules from binary solutions of ethanol (EtOH), 2-butanol (BuOH), and 1-Hexanol (HxOH) (26). Reprinted with permission from Keller, R. A., and M. M. Metro, *J. Chromatogr. Sci.*, **12**, 673 (1974). Copyright 1974, Preston Publications, Niles, IL.

Figure 7 shows the reduced mobilities of the alcohols. The careful reader will note that the units of the reduced mobilities are incorrect. This was amended in a subsequent review (23). The error in units does not alter the conclusions. The bars represent the estimated error. The plasmagram mobilities of the pure alcohols are shown in the top three lines. Binary mixture compositions are in weight percent. Reduced mobilities beyond 8.3 have no value in discriminating between the alcohols. The 7.0–7.3 peak for BuOH might be taken as distinctive except that it is dangerously close to the 7.6 peak for HxOH, which, incidently, is missing for the 39:61 EtOH–HxOH mixture. Interpretation of mobilities between 6.8 and 8.3 is highly speculative and far too dependent on preknowledge of the composition. The 5.62 and 5.84 peaks for HxOH, one of which is missing for the 71:29 EtHO–HxOH and the 44:56 BuOH–HxOH mixtures.

We support the contention of Cram and Chesler (5) that 'the plasmagram of a binary mixture is very difficult to interpret by itself' and dispair the use of PC for qualitative fingerprinting.

Spangler and Collins (33) dealt with peak shape and plate theory for the PC.

Use of the PC as a detector coupled to a gas chromatograph is attractive because of the LLD and that either positive or negative ions may be sensed. The potentially long residence time in the system, coupled with the problem of overload pulse sampling, must always be kept in mind.

A very interesting modification by Carroll et al. (4) was to use the ion source of the PC as an atmospheric pressure ionization (API) source for a quadrupole mass spectrometer. Subpicogram detection was reported. One may speculate that this step led this group to later trace analysis of biological samples using API.

V. APPLICATIONS

In order to keep the length of this chapter within reason and avoid redundancies in the literature, this writer has referred to the review by Keller and Metro (35), which covers applications prior to 1974 and has elected not to repeat them here. The paper by Spangler and Collins (32) has an excellent bibliography. Karasek (8) also reviewed the early basic work.

Karasek and Kane (15) used the PC to elucidate the electron capture properties of aromatics, which is not directly applicable to PC as an analytical instrument. An interesting but nonanalytical application is that by Hau (6) who used the PC to determine densities.

The following analytical applications are of note:

Halogenated nitrobenzenes (17)

n-Alkanes and n-alkyl halides (13)

n-Alkyl halides (22). [This paper is of particular interest because the PC was coupled with a thermal conductivity (TC) cell of a conventional gas chromatograph to "pick off" the peak of interest and admit it to the PC. This is probably the best application. The vast difference in the LLD of the TC and PC should be kept in mind. Also see Karasek and Kim (19) for this modification.]

N-Nitrosamines (12)

Organic compounds in general with emphasis on the halogenated biphenyls with both positive and negative productions (16)

Nitrated toluenes (7, 10)

n-Alkyl acetates (21)

Pesticides and their metabolites (30)

Lysergic acid diethylamide and cannabinols (18)

Heroin and cocaine (14)

Karasek and Denney (9) used the PC as a liquid chromatography detector employing the much investigated moving wire (chain) interface. Karasek, Laub, and DeDecker (20) applied computer resolution of PC peaks, which seems presumptive because so many questions remain as to what is transpiring chemically within the system.

Acknowledgements   The author is grateful to the National Science Foundation, Grant GP-31824 and the State University of New York, Research Foundation, Small Grant-in-Aid for support of our projects. The author was very fortunate to be able to assist Mr. Michael M. Metro and Mr. George M. Bird in their studies. I also thank Dr. F. W. Karasek for drawing his attention to this interesting but chemically complex instrument.

ADDENDUM

In 1975 Franklin GNO Corporation ceased manufacture and the technology was continued by a new company PCP Inc., West Palm Beach, Florida. Abandoning identification with chromatography, referral was made to ion mobility spectrometry, IMS. However, as late as 1984 T. W. Carr edited a book *Plasma Chromatography* (35) which was critically reviewed by Karasek (36). Papers still appear with the first designation. The bewildered neophyte must consider both names. The name change is appropriate and not merely a polemic issue because the method is more closely related to mass spectrometry than it is to chromatography which is considered by most to be the separation of a mixture into its homogeneous components.

REFERENCES

1. Benezra, S. A., *J. Chromatogr. Sci.*, **14**, 122 (1976).
2. Bird, G. M., and R. A. Keller, *J. Chromatogr. Sci.*, **14**, 574 (1976).
3. Carroll, D. I., I. Dzidic, R. N. Stillwell, and E. C. Horning, *Anal. Chem.*, **47**, 1957 (1973).
4. Carroll, D. I., I. Dzidic, R. N. Stillwell, M. G. Horning, and E. C. Horning, *Anal. Chem.*, **46**, 706 (1974).
5. Cram, S. P., and S. N. Chesler, *J. Chromatogr. Sci.*, **11**, 391 (1973).
6. Hau, C. S., *Spectros. Lett.*, **8**, 583 (1975).
7. Karasek, F. W., *Res./Dev.*, **5**, 32 (1974).
8. Karasek, F. W., *Anal. Chem.*, **46**, 710A (1974).
9. Karasek, F. W., and D. W. Denney, *Anal. Lett.*, **6**, 993 (1973).
10. Karasek, F. W., and D. W. Denney, *J. Chromatogra.*, **93**, 141 (1974).
11. Karasek, F. W., and D. W. Denney, *Anal. Chem.*, **46**, 633 (1974).
12. Karasek, F. W., and D. W. Denney, *Anal. Chem.*, **46**, 1312 (1974).
13. Karasek, F. W., D. W. Denney, and E. H. DeDecker, *Anal. Chem.*, **46**, 970 (1974).
14. Karasek, F. W., H. H. Hill, Jr., and S. H. Kim, *J. Chromatogra.*, **117**, 327 (1976).
15. Karasek, F. W., and D. M. Kane, *Anal. Chem.*, **45**, 1210 (1973).
16. Karasek, F. W., and D. M. Kane, *J. Chromatogra.*, **93**, 129 (1974).
17. Karasek, F. W., and D. M. Kane, *Anal. Chem.*, **46**, 780 (1974).
18. Karasek, F. W., D. E. Karasek, and S. H. Kim, *J. Chromatogra.*, **105**, 345 (1975).
19. Karasek, F. W., and S. H. Kim, *J. Chromatogra.*, **99**, 257 (1974).
20. Karasek, F. W., R. J. Laub, and E. DeDecker, *J. Chromatogra.*, **93**, 123 (1974).
21. Karasek, F. W., A. Maican, and O. S. Taton, *J. Chromatogra.*, **110**, 295 (1975).
22. Karasek, F. W., O. S. Tatone, and D. W. Denney, *J. Chromatogra.*, **87**, 137 (1973).
23. Keller, R. A., *Am. Lab.*, **7**, 35 (1975).
24. Keller, R. A., and J. C. Giddings, *J. Chromatogra.*, **3**, 205 (1960).
25. Keller, R. A., and M. M. Metro, *Sep. Purif. Methods*, **3**(1), 207 (1974).
26. Keller, R. A., and M. M Metro, *J. Chromatogr. Sci*, **12**, 673 (1974).
27. Lin, S. N., G. W. Griffin, E. C. Horning, and W. E. Wentworth, *J. Chem. Phys.* **12**, 4994 (1974).
28. Metro, M. M., and R. A. Keller, *J. Chromatogr. Sci*, **11**, 520 (1973).
29. Metro, M. M., and R. A. Keller, *Sep. Sci.*, **9**, 521 (1974).

30. Moye, H. A., *J. Chromatogr. Sci.*, **13**, 285 (1975).
31. Revercomb, H. E., and E. A. Mason, *Anal. Chem.*, **47**, 970 (1975).
32. Spangler, G. E., and C. I. Collins, *Anal. Chem.*, **47**, 393 (1975).
33. Spangler, G. E., and C. I. Collins, *Anal. Chem.*, **47**, 403 (1975).
34. Tyndall, A. M., *The Mobility of Positive Ions in Gases*, Cambridge University Press, London, 1938.
35. Carr, T. W., ed., *Plasma Chromatography*, Plenum Press, New York, 1985.
36. Karasek, F. W., *Anal. Chem.* **57**, 336A (1985).

Chapter 5

LOW-ENERGY ION-SCATTERING SPECTROSCOPY

By A. C. MILLER

Zettlemoyer Center for Surface Studies, Sinclair Laboratory 7, Lehigh University, Bethlehem, Pennsylvania

Contents

I. INTRODUCTION

A. GENERAL REMARKS

The many techniques available to analyze solid surfaces can be divided into those that identify elements by their nuclear charge and those that identify elements by their mass. Techniques that fall into the first group include Auger electron spectroscopy (AES), soft X-ray appearance potential spectroscopy (SXAPS), and electron spectroscopy for chemical analysis (ESCA) or X-ray photoelectron spectroscopy (XPS). Two of those in the second group are secondary ion mass spectroscopy (SIMS) and low-energy ion-scattering spectroscopy (ISS). The basic requirements of low-energy ion-scattering spectroscopy for surface analysis are simple (see Fig. 1). A monoenergetic noble-gas ion beam (typically 0.5–5 keV) strikes the target surface, and the energy distribution of ions scattered at a fixed angle is recorded. The positions of peaks in the energy spectrum are used to identify the elements present on the surface, and the peak heights are a measure of the number of atoms of a given species present (see Fig. 2). For single-crystal targets, surface structure information can be derived from the position and relative size of different peaks as a function of the angles of incidence and scattering. Due to the nature of the interaction between the noble-gas ions and surface atoms, low-energy ion-scattering spectroscopy is sensitive to only the outer one or two monolayers. This can be extremely important for a number of applications. ISS can be used to analyze

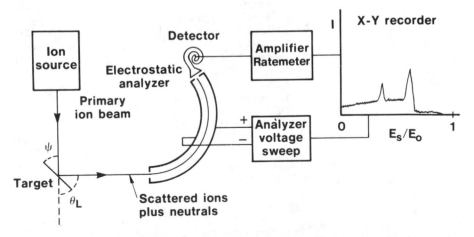

Fig. 1.   Schematic representation of a typical ion-scattering apparatus.

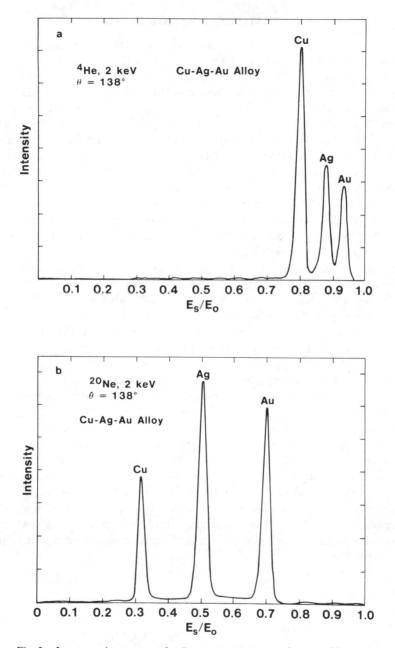

Fig. 2.   Ion-scattering spectra of a Cu–Ag–Au alloy using ⁴He and ²⁰Ne ions.

any vacuum-worthy material regardless of its conductivity. Depth profiling (obtaining elemental composition as a function of depth) is obtained by sputter removal of the surface atoms.

B. HISTORICAL BACKGROUND

The idea of using backscattered ions as a tool to determine the surface composition of solids has existed for many years (20, 65). Most of the early experiments, however, either used high energies (> 5 keV), reactive ions, or both. It will become evident later that these conditions placed restrictions on the usefulness of ISS for surface analysis. Early experiments did, however, point out a number of key features of ion-scattering spectra. For example, Brunee's (20) work with 0.4–4 keV alkali ions scattered from Mo showed that the maximum scattered energies could be predicted by single binary elastic collisions. Panin (65) also found that the spectra of H^+, He^+, N^+, O^+, and Ar^+ scattered from molybdenum and beryllium at high energies (7.5–80 keV) could be interpreted in terms of binary elastic scattering. Other early work by Mashkova and Molchanov (5), Fluit et al. (34), and Datz and Snoek (29) reconfirmed the earlier results and gave some evidence that the interaction potential could be described by a screened Coulomb potential.

The first clear documentation of low-energy (0.5–3 keV) ion-scattering spectroscopy as a surface analytical technique was reported by Smith (72) in 1967 when he used He^+, Ne^+, and Ar^+ ions to study molybdenum and nickel targets. He realized the importance of sharp spectral peaks in detecting multiple elements present on the surface. An example of this is shown in Fig. 3 where H_2^+ and He^+ have been scattered from copper (56). Smith interpreted the low-energy tail in the H_2^+ spectra as due to ions that had penetrated into the solid and experienced inelastic energy losses before and after the primary binary scattering event. These ions are not observed for noble gases because there is a high probability for subsurface neutralization. Using He^+ ions, Smith was able to obtain peaks corresponding to the surface atoms of a molybdenum substrate and from adsorbed oxygen and carbon. In later experiments with Strehlow (76), he was able to determine the cadmium and sulfur faces of cadmium sulfide single crystals by analyzing the relative peak heights. He was also able to show the quantitative nature of ISS by analyzing the oxygen and aluminum peak heights from aluminum oxide and pure aluminum. Thus, in a classic set of experiments Smith (73) demonstrated that ISS was surface sensitive, able to detect multiple elements present on a surface, and could yield quantitative information and determine crystal structure.

II. THEORY

A. SCATTERING PROCESSES

1. Binary Collisions

The fundamental principle of ISS is that the scattering process can be described as a binary elastic collision. The kinetics of such a scattering process are described by conservation of energy and momentum. Figure 4 depicts the

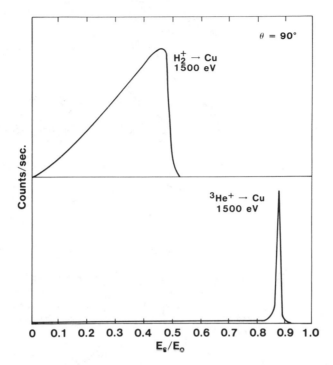

Fig. 3.   Energy spectra of H_2^+ scattered from Cu (56). (Reprinted with permission of the publisher, Academic Press, Inc.)

scattering of an ion of initial energy E_0 and mass M_0 by a stationary atom of mass M_s. Conservation of energy before and after the collision gives

$$\tfrac{1}{2} M_0 V_0^2 = \tfrac{1}{2} M_s V'^2 + \tfrac{1}{2} M_0 V_s^2 \tag{1}$$

where V_s is the velocity of the scattered incident particle and V' is the postcollision velocity of the surface atom. Conservation of momentum gives

$$M_0 V_0 = M_0 V_s \cos\theta + M_s V' \cos\phi \tag{2}$$

$$0 = M_0 V_s \sin\theta - M_s V' \sin\phi \tag{3}$$

where θ and ϕ are the scattering angles of the incident particle and surface atom, respectively. By combining these equations one obtains

$$\frac{E_s}{E_0} = \left(\frac{M_0}{M_0 + M_s}\right)^2 \left\{\cos\theta + \left[\left(\frac{M_s}{M_0}\right)^2 - \sin^2\theta\right]^{1/2}\right\}^2 \tag{4}$$

This is the basic equation for all ion-scattering experiments. By knowing the mass and energy of the incident ion and the scattering angle, and measuring the scattered energy, the mass of the target atom can be calculated from Eq. (4).

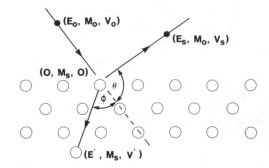

For binary elastic collisions

$$E_0 = E_s + E'$$

$$M_0 V_0 = M_0 V_s \cos \theta + M_s V' \cos \phi$$

$$0 = M_0 V_s \sin \theta - M_s V' \sin \phi$$

$$\frac{E_s}{E_0} = \left(\frac{M_0}{M_0 + M_s}\right)^2 \left[\cos \theta + \left(\left(\frac{M_s}{M_0}\right)^2 - \sin^2 \theta\right)^{1/2}\right]^2$$

For $\theta = 90°$

$$\frac{E_s}{E_0} = \frac{M_s - M_0}{M_s + M_0}$$

Fig. 4.   Idealized representation of ion–surface atom scattering.

Strictly speaking, Eq. (4) is applicable only for free atoms at rest. To be applicable for solid surfaces, a number of assumptions must be valid. First, it is assumed that the target atoms are free during the collision. This approximation is expected to be appropriate since the collision times for low-energy ions are on the order of 10^{-16}–10^{-15} s and are much shorter than the characteristic lattice vibrations ($\sim 10^{-13}$ s). Second, it is assumed the target atoms are stationary before the collision; that is, the lattice is cold. The vibrational energy of an atom in a room-temperature solid is about 0.05 eV and can usually be neglected when comparing the large energy transfer in the scattering process. Third, the energy losses of the primary particle during the collision are assumed to be completely kinetic; that is, the electronic energy transfer will be small.

All of these are good approximations in the energy range of interest for the present discussion (0.1–5 keV). Smith and Goff (74) experimentally tested the binary elastic model and found it to be valid in the range of 0.5–3 keV. Tongson and Cooper (84) have found that the binary collision approximation is valid for 90° scattering at energies of 40–1000 eV for the $He^+ \to Cu$ system and 20–1000 eV for the $Ne^+ \to Cu$ system.

Similar data were obtained by Hart and Cooper (41) for the $Ar^+ \to Cu$ system at energies of 25–100 eV. These investigators found no evidence of an effective mass of the target atoms as suggested by Veksler (87).

For some systems, deviations from the binary elastic interaction approach have been observed. For the scattering system $He^+ \to Ta_2O_5$, Wheeler (90) and

McCune et al. (51) found that the position of the tantalum peak, when referenced to the position of a $He^+ \rightarrow Ag$ peak, was shifted to progressively lower values of E/E_0 as the energy of the primary ion beam was reduced. These results could not be interpreted in terms of a binary elastic collision. Baun (4) has further examined the tantalum pentoxide and similar oxide systems and has suggested that the energy shift may result from inelastic collision exchange processes between the noble-gas ion and oxygen on or near the surface. Other investigators (11, 31) have also reported peak shifts due to inelastic energy losses. These shifts are usually on the order of 10–50 eV and may often go unnoticed when relatively large primary beam energies are used.

Some recent work (75a, 75b, 83a) has shown that an ion-scattering peak often contains two components. The first component arises from "true" ion scattering in which a primary ion is directly scattered by a surface atom without any changes in charge state. The second component results from an event in which an incident ion is neutralized near the surface on the incoming trajectory, is scattered as a neutral atom, and then is reionized after the collision. There is an inherent energy loss associated with the latter process that has been found to be on the order of the first ionization potential of the incident species (83a). With this two-component model for an ion-scattering peak, the actual peak position will depend on the relative intensities of the components. The relative intensities have been found to depend on the target material (83a). This model of an ion-scattering peak provides a partial explanation for the asymmetry that is frequently observed on the low-energy side of many ion-scattering peaks and for the apparent changes in peak position.

Helbig et al. (44) have shown that in the case of some insulators the positional deviations may be interpreted in terms of the electric field produced by charge buildup on the sample. This leads to a reduction in the primary ion energy and a change in the total scattering angle. In some cases, peak shifts will occur if there are changes in the relative positions of the sample surface and the energy analyzer.

2. Cross Sections

The elemental sensitivity and quantitative determination of the number of surface atoms are directly related to the probability that a scattering event will occur; that is, the differential scattering cross section $d\sigma/d\Omega$.

For a laboratory scattering angle θ, the differential scattering cross section is defined as

$$d\sigma(\theta) = -2\pi p \, dp \tag{5}$$

where p is the impact parameter and $d\sigma(\theta)$ is the annular area between impact parameters p and $p+dp$ for which incident particles scatter into a cone between the planar angles θ and $\theta - d\theta$. For rotational symmetry around the incident beam axis, the solid angle associated with $d\theta$ is $d\Omega = 2\pi \sin\theta \, d\theta$. Thus, the differential scattering cross section for scattering into a unit solid angle becomes

$$\frac{d\sigma(\theta)}{d\Omega} = \frac{p}{\sin\theta}\left|\frac{dp}{d\theta}\right| \tag{6}$$

To evaluate the cross section, it is necessary to specify $p(\theta)$. This is difficult in the laboratory system of coordinates. Calculations are simplified if the problem is solved in the center-of-mass (CM) system and then transformed back to the laboratory system.

In the CM system, the relation between p and θ_{CM} is given by the scattering integral (36),

$$\theta_{CM} = \pi - 2p \int_{r_m}^{\infty} \frac{dr}{r^2\left[1 - p^2/r^2 - V(r)/E\right]^{1/2}} \tag{7}$$

(See Fig. 5 for definition of the parameters.) It should be noted that the form of $V(r)$, the interaction potential, is required for the calculation.

At low energies the screening of the nucleus is important and, consequently, a screened potential must be used. Among those in current use is the Bohr-screened Coulomb potential

$$V(r) = \frac{Z_1 Z_2 e^2}{r}\exp\left(-\frac{r}{a}\right) \tag{8}$$

where r is the nuclear separation,

$$a = \frac{a_0}{(Z_1^{2/3} + Z_2^{2/3})^{1/2}} \tag{9}$$

is the electron screening length, $a_0 = 0.53$ Å is the radius of the first Bohr orbit in the hydrogen atom, and Z_1 and Z_2 are the nuclear charges. Tables of screened potentials have been published by Everhart et al. (33) and Bingham (13). Cross-

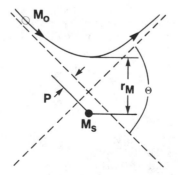

Fig. 5. Scattering parameters in the center-of-mass system: p is the impact parameter, r_m is the distance of closest approach, and θ is the center-of-mass scattering angle.

section tables for He^+, Ne^+, and Ar^+ scattered at 90° and 138° have been published by Nelson (57).

Born–Mayer potentials of the form $V(r) = A\exp(-br)$ have also been used, but since they do not contain a $1/r$ preexponential factor, they are not repulsive enough for close encounters and thus have a very limited range of applicability. Tables of the A and b parameters for the Born–Mayer potential have been published by Abrahamson (1).

The Thomas–Fermi potential $V(r) = (Z_1 Z_2 e^2/r)\phi(r/a)$, where ϕ is a given approximation to the screening function, is the potential believed to most closely represent the observed scattering. The Moliere approximation to the screening function is used most frequently (86). All of the cross sections resulting from these potentials increase with decreasing energy and scattering angle. This is illustrated in Fig. 6 (79) for the Born–Mayer potential. The other scattering potentials yield qualitatively similar curves. The cross sections also increase with increasing mass ratio. Figure 7 is a plot of the differential scattering cross sections as a function of atomic number of 2000 eV ^3He and ^{20}Ne ions scattered through 138° for the Bohr, Moliere, and Born–Mayer potentials. Since the exact potentials are not known, these cross sections can only be used as a guide in interpreting spectra and cannot be used for quantitative measurements.

3. Multiple Reflections

While single scattering peaks usually dominate the scattered energy spectrum, additional features due to multiple scattering can frequently be observed. This is especially true for heavy ions scattered in the forward direction. These features usually appear as shoulders on the high-energy side of a single scattering peak.

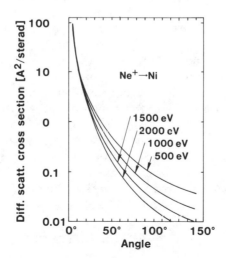

Fig. 6.   Variation of the differential scattering cross section for $Ne^+ \rightarrow Ni$ as a function of scattering angle and energy for the Born–Mayer potential (79). (Reprinted with permission of the publisher, Springer-Verlag.)

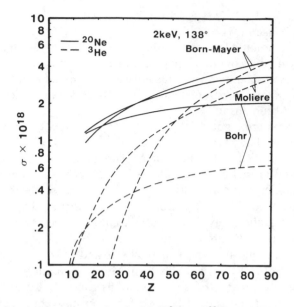

Fig. 7.   Differential scattering cross sections for 2 keV ^{3}He and ^{20}Ne at 90° and 138° (as a function of atomic number Z) for the Born–Mayer, Moliere, and Bohr potentials.

Though multiple scattering events are more pronounced for single-crystal targets, multiple scattering events have been observed for He$^+$, Ne$^+$, and Ar$^+$ scattering from polycrystalline materials (5).

Figure 8 (6) shows a spectrum of Ne$^+$ scattered from indium. The inset schematically depicts a double scattering event. The single scattering peak falls at the position predicted by the binary collision model. The shoulder on the high-energy side is due to ions that have scattered from two atoms with a total scattering angle of 90°. The maximum position of the peak can be calculated by applying Eq. (1) twice and using the condition that $\theta_1 = \theta_2 = 45°$.

A detailed calculation of the peak position, however, is not possible since it depends on the potential used through the scattering angle. Figure 9 is a plot of the energy for double scattering (scattering through 45° twice) as a function of target mass for a total scattering angle of 90°.

When scattering from single-crystal surfaces, the regular spacing of atoms in surface rows leads to a predictable and pronounced multiple scattering behavior. Kivilis et al. (48) introduced the chain model to account for this more complex scattering. The reader is referred to the work of Suurmeijer and Boers (77) for a more comprehensive review of multiple scattering.

B.   INTENSITY MODIFICATIONS

1.   Shadowing

Depending on the angle of incidence and the total scattering angle, a varying degree of shadowing of one atom by another may occur. This is demonstrated

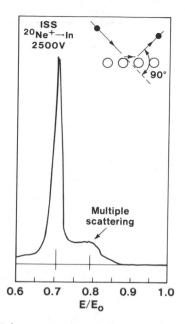

Fig. 8. ISS spectrum of $^{20}Ne^+$ scattering from indium showing the double scattering peak. A double scattering trajectory is pictured in the inset (6).

schematically in Fig. 10, which illustrates how m_2 lies within the shadow cone (3) of m_1 and is thereby shielded from the incident ion flux. Atomic shadowing on polycrystalline samples can lead to changes in the scattered ion yields and consequently can complicate the quantitative interpretation of data. Relatively little work has been done to understand this kind of shadowing quantitatively. These effects can be minimized by using high angles of incidence and reflection.

Atomic shadowing has been used to great advantage for determining the position of atoms on single crystals. In this case one wants to accentuate the shadowing by using a low angle of incidence and a small scattering angle. This technique has been successfully used by Heiland and Taglauer (42) and Brongersma and Mul (15) and will be discussed in greater detail in Section IV.

In addition to low-angle atomic shadowing, it is possible to have shadowing of one part of a sample by another, that is, surface roughness effects. This has been modeled by Nelson (58) for 90° scattering with a 45° angle of incidence using a model that depicts the surface as an array of cubes whose faces are oriented at various angles with respect to the surface normal. The surface distribution of cube angles was assumed to be Gaussian. Figure 11 is a plot of the signal ratio from a rough surface to that of a smooth surface as a function of the root-mean-square (rms) slope of the surface. By making measurements on a smooth surface and two surfaces that had been characterized for surface roughness, it was possible to show excellent agreement between the experimentally measured intensity ratios and those predicted by the model.

A. C. MILLER

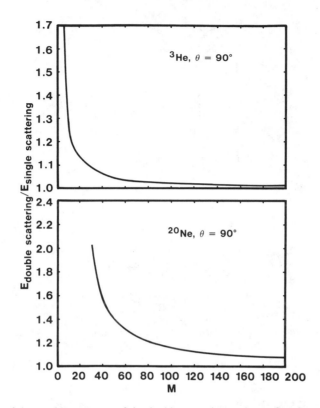

Fig. 9.   Plot of the maximum energy of the double scattering peak as a function of target mass.

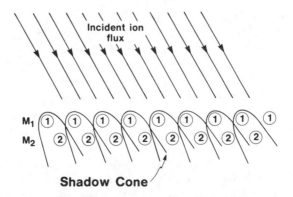

Fig. 10.   Schematic representation of the shadowing of atoms in the "second" atomic layer by atoms in the "first" atomic layer. The relative sizes of the atoms and shadow cones have been exaggerated for clarity.

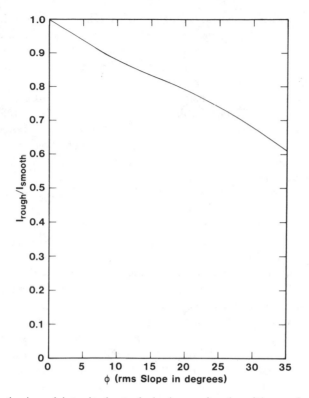

Fig. 11.   Reduction in peak intensity due to shadowing as a function of the rms slope of the surface (58). (Reprinted with permission from the publisher, The American Institute of Physics.)

2.   Vibrational Motion of Surface Atoms

The surface atoms under study are not at rest as assumed by the simple binary collision theory, but are actually vibrating around some mean position. If the atoms are vibrating in the same plane as the incident ions, it can be shown that the scattered peak will be broadened by approximately $(E_0 E_k)^{1/2}$ where E_k is the kinetic energy at the vibrating atom. Due to the large energy spread in the scattered peak, it is usually difficult to observe such a small change in the peak width. For further discussions on thermal vibrational effects the reader is referred to the work of Poelsema et al. (66).

3.   Ion Neutralization

Ion neutralization is the most important phenomenon in low-energy ion-scattering spectroscopy. Due to the fact that only ions are collected by the electrostatic analyzer used for the energy analysis, an understanding of how and to what extent the neutralization varies with energy, chemical composition, and angle of incidence is necessary for both qualitative and quantitative interpretation of low-energy ion spectra. Ion neutralization has been given by Smith (72) as the reason for the high (one to two monolayer) surface sensitivity of low-energy ion scattering. The ions that penetrate to the third or fourth layer have such a high probability for neutralization that if they emerge from the lattice, they do so as neutral particles. This is fortunate not only for the surface sensitivity, but it is also responsible for the sharp peaks observed with noble gases. However, the effects of different beam energies (i.e., different lattice penetration) can be observed at the energies normally associated with low-energy ion scattering. This can be seen in Fig. 12, which shows spectra of $^4He^+$ scattering from aluminum at 1 and 4 keV. The decrease in the low-energy tail with the decrease in incident energy is clearly evident.

Another possible explanation for the surface sensitivity is a beam attenuation effect for low-energy ions. That is, a large fraction of the incident beam is scattered out of the original direction after passage through the first one or two monolayers due to the large scattering cross sections. Buck et al. (21) have studied this problem by using a time-of-flight mass spectrometer that measures both neutral and ion-scattering yields. Their data indicate that both beam attenuation and neutralization effects contribute to the characteristic shape of low-energy scattering spectra. They find that the ratio of the scattered ion yield to the total scattering yield of ions plus neutrals is about 1–10%, depending on the ion–atom system. Eckstein et al. (31) have reported similar results for the ion/(neutral + ion) yield ratios, though they used electron-stripping techniques to reionize the scattered neutrals rather than time-of-flight techniques.

The neutralization process is not understood in detail, though it seems clear that several processes are involved. For example, there is evidence that incoming ions can be neutralized at the surface and then reionized on the outward path (22, 75b). Data have also been reported that indicate the neutralization probability can be a function of the ion trajectories in the vicinity of the surface atoms (35).

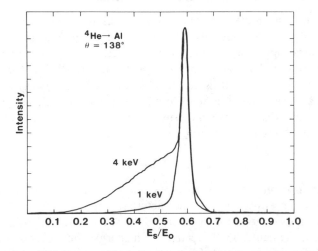

Fig. 12.   ISS spectra of $^4He^+$ scattered from Al at 1 and 4 keV showing the change in the low-energy tail with as a function of incident beam energy.

It is suggested that this is of particular significance when ISS is used to determine the positions of adsorbed atoms on the surface.

The most widely accepted neutralization processes are the Auger and resonance neutralization mechanisms first described by Hagstrum (38). Both of these have an exponential dependence of the ion fraction on the scattered ion velocity of the form

$$P \propto \exp \frac{-V_0}{V_\perp} \tag{10}$$

where P is the probability the ion will *not* be neutralized (ion escape probability), V_0 is a characteristic velocity, and $V_\perp$ is the velocity component normal to the surface. The characteristic velocity depends on the ion–atom system. This implies that the neutralization will be different for different atoms. This has been shown by Taglauer and Heiland (80) for sulfur and nickel where the neutralization is seven times higher for nickel than for sulfur. In addition, data from Leys (49) suggest that neutralization for a given atom can change when that atom is chemically bound to others. Using He^+ as a probe gas, he compared the magnitudes of the zinc signals in zinc metal and zinc sulfide and found that the magnitude of the zinc peak in the zinc sulfide spectrum was much smaller than would be predicted by considering only the areal density of zinc atoms on the two surfaces. When the experiment was repeated using Ne^+ as the probe gas, the ratios of the zinc peaks agreed with the predicted values. Nelson (59) has measured the experimental variation with energy of the copper peak height for copper metal and cuprous oxide. By dividing the experimental data by the screened Coulomb differential scattering cross section, he was able to obtain the probability of ion neutralization. By plotting this as a function of $1/V$, V_0 can be obtained. These data indicate that the neutralization parameter changes in going from copper to cuprous oxide when He^+ is used as the probe gas (in agreement with the observation of Leys for zinc sulfide) and does not change when neon is used. In addition, data from graphite (Fig. 13) (60) show that the neutralization parameter is much higher than for higher atomic number elements, such as for copper (59) as indicated by the data that have been plotted for comparison.

For other systems ($He^+ \rightarrow Ni$, O, and S and $Ne^+ \rightarrow Ni$ and S) (81), the results agree qualitatively with this theory but not in detail; that is, a semilogarithmic plot of P versus $1/V$ does not give a straight line.

An additional neutralization mechanism has been reported by Erickson and Smith (32). They observed oscillations in the scattered ion yield as a function of incident energy for He^+ scattered from lead. A similar behavior has been reported for several other elements (67). For example, Fig. 14 (56) shows the scattered ion yield curve for germanium, which has several oscillations between 500 eV and 3 keV, but the behavior of copper is more typical with no oscillations. This phenomenon is attributed to oscillations in neutralization probability like those studied by Ziemba et al. (93) in gases. The peaks are more or less evenly spaced when plotted as a function of $1/V$, which is proportional to the interaction time. The qualitative explanation given by Erickson and Smith (32) is that

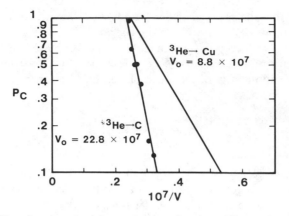

Fig. 13. Probability of an ion remaining an ion as a function of inverse ion velocity for $^3He^+$ scattered from C and Cu. The vertical scale is arbitrary (59, 60). (Reprinted with permission from Ref. 59. Copyright 1974, The American Chemical Society.)

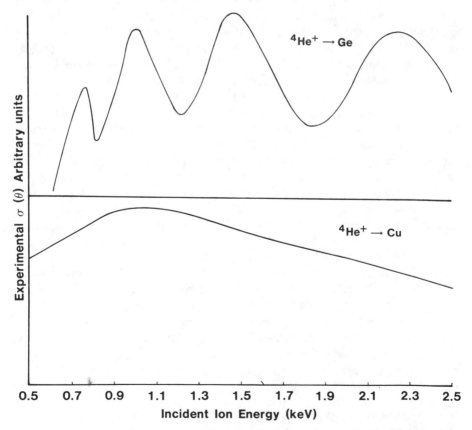

Fig. 14. Scattered ion yield curves for $^4He^+$ scattered from Ge and Cu. The oscillations in the Ge curve are the result of resonance neutralization (56). (Reprinted with permission of the publisher, Academic Press.)

resonant charge exchange occurs and that at high velocities the interaction time is so short that an electron from a neutral target atom has only time to jump to the ion and not back again. As the velocity is lowered, additional transfers of the electron back and forth occur.

Rusch and Erickson (67) have carried out a survey of the ion yield characteristics for most of the metallic elements using He, Ne, and Ar ion beams. According to their data, the yield curves can be separated into four classes. Elements exhibiting oscillatory yield curves for the scattering are designated Class II. This group includes Ga, Ge, In, Sn, Sb, Pb and Bi. Rusch and Erickson (67) have suggested a correlation of the occurrence of these oscillations with two criteria. First, the target atom must have an electronic state with a binding energy within approximately ± 10 eV of the first ionization potential of the bombarding species. Second, if the first criterion is satisfied, what are the orbital symmetries of the electronic state and the vacant ionic state?

These yield curves contain information about the electronic states of surface atoms. In addition, changes in the surface atom chemical environment can be manifested as changes in the oscillatory yield curve. This topic has been studied in detail by Christensen (24, 25) who has used Fourier transform techniques to clarify the differences in the ion yield curves obtained from pure Class II elements and their compounds.

In addition, because the oscillatory behavior of lead and bismuth is different, one can use this technique to distinguish between these elements, even though their mass difference is too small to produce a detectable difference in scattered energy.

Baun (7) reported an interesting result obtained during a study of HgCdTe. He observed that tellurium in HgCdTe exhibited an oscillatory yield behavior similar to Class II elements, whereas the yield curve for pure tellurium did not. This suggests that ion yield curves might be used to study bonding and the formation of molecular orbitals.

Though ion neutralization is recognized as a major parameter in the quantification of ISS, the subject has generally received little systematic investigation from both the experimental and theoretical viewpoints. However, a renewed interest in neutralization phenomena has produced several quantum-mechanical formulations of the neutralization processes (14, 54, 55, 70, 85). These calculations have been able to reproduce some of the experimental results qualitatively.

III. PRACTICE

A. LOW-ENERGY ION-SCATTERING SPECTROMETER

1. Ion Beam

The experimental apparatus needed for low-energy ion-scattering spectroscopy varies with the type of experiment being conducted, but for the majority of activities it is quite simple. As shown schematically in Fig. 1, an ion-scattering spectrometer consists primarily of an ion beam source, a target, an energy analyzer, and the electronics necessary for energy analysis and signal detection. Because of the ion beam requirements and the necessity for maintaining a clean surface, it is necessary that the experiment be conducted in an ultra-high vacuum system.

Instruments that have been designed by individual laboratories may be more elaborate in that they may be capable of orienting the target for surface structure studies on single-crystal targets. In addition, they may mass filter the primary ion beam to make certain only one species of ion is striking the target. Finally, they may have an analyzer that can be rotated so that spectra can be obtained at more than one scattering angle.

Typically, the ions are produced in an electron bombardment source at a pressure of 1.3×10^{-3} to 1.3×10^{-2} Pa $(1 \times 10^{-5} \times 10^{-4}$ torr). The source pressure is often achieved by backfilling the entire vacuum chamber. This method may not be acceptable because outgassing from the vacuum system or the sample could lead to undesirable levels of residual gases in the analysis chamber. This problem can be minimized by the use of differentially or dynamically pumped ion sources, which permit continuous pumping of the vacuum system during the analysis.

The ions are electrostatically extracted from the source and are focused to beams that typically range from 50 μm to 2 mm in diameter (full width at half maximum, FWHM). The radial ion density within the beam is usually Gaussian. Total beam currents vary from a few nanoamperes to several microamperes. The maximum current densities available from commercial ion guns are on the order of 600 μA/cm^2. A typical energy spread of the ions is $\sim 0.5\%$ FWHM. Most ion guns incorporate deflection plates, which are used to position the ion beam or raster the ion beam over an area on the sample. Rastering is desirable for a number of reasons. First, it minimizes the surface damage by the beam. Second, when sputter depth profiling, it enables one to accept electronically (gate) only that portion of the signal that arises from the center of the sputtered area. This helps reduce the edge effect of the sputtered crater. A third advantage of rastering is that it enables imaging of the sample if the ion beam diameter is sufficiently small. Not only is this useful for elemental mapping of a surface, but it is also a useful technique for locating the beam on a given spot on the sample.

The ion beam current can be determined either by using a Faraday cup or by applying a positive voltage of 25–100 V to the target to suppress the secondary electron contribution. The beam profile can be determined by measuring the current striking a small wire (whose diameter is less than the beam diameter) as the beam is scanned across the wire. Details of this procedure are reported in ASTM Standard Practice E684-79 (2).

2. Energy Analyzer

Electric sector analyzers are most frequently used to obtain the energy/charge spectra of the scattered ions. As the voltage between the analyzer plates is stepped (or scanned), the scattered ions of energy E_s/q are transmitted through the analyzer and counted by a detector. A number of different analyzers and scattering geometries have been used in ISS instruments. The four configurations that either have been or are currently used in commercial instruments are shown in Fig. 15.

Historically, the first low-energy ion-scattering experiments were carried out using the 127° sector analyzer illustrated in Fig. 15a. For a sector analyzer, the energy that passes through the analyzer is given by $E_s/q = VR/2d$, where V is the voltage between the two plates ($+V/2$ and $-V/2$) so that the center between the plates is at ground potential, R is the mean radius of curvature, and d is the plate separation. These analyzers are useful if one wants to be able to vary the angle of incidence or scattering angle or perform angular emission experiments because the acceptance angle of the analyzer relative to the beam spot on the sample surface is well defined. However, for a given scattering angle a large fraction of scattered ions are lost due to the small azimuthal angle subtended by the

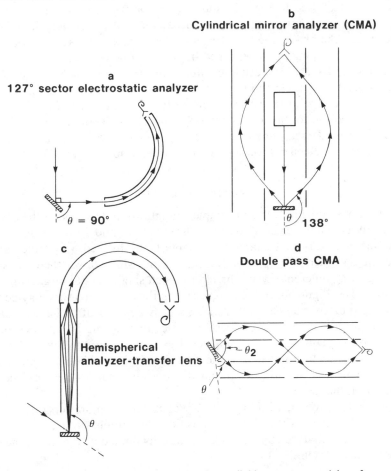

Fig. 15. Scattering geometries that are currently available on commercial surface analysis instruments.

analyzer entrance slits. The cylindrical mirror analyzer shown in Fig. 15b takes advantage of all of these ions. This results in an increased intensity of a factor of ~100 that leads to reduced surface damage for a given sensitivity and more rapid accumulation of data. This analyzer also has a larger acceptance area on the sample, which is essential for imaging. In this case the ion gun and detector are mounted coaxially in the energy analyzer. It is necessary to reach a compromise between the energy resolution $\Delta E/E$ of the analyzer (to reduce peak width) and sufficiently high acceptance to keep the counting rate acceptably high. A resolution of about 1% is usually sufficient to meet these criteria.

One of the features of electrostatic analyzers is that particles of either charge can be energy analyzed by reversing the potentials on the various analyzer elements and the detectors. This has prompted several electron spectrometer suppliers to provide the capability for doing ISS with systems that were originally designed for AES and XPS. The two principle geometries of these systems are illustrated in Fig. 15c and 15d. In the transfer lens–hemispherical analyzer combination, the scattering angle is defined by the angle between the ion beam and the axis of the transfer lens. The angular acceptance around the primary scattering angle is determined by the acceptance cone of the lens. The double-pass CMA uses a mechanical aperture in front of the entrance slit to define the scattering angle and the angular acceptance. By rotating the aperture around the CMA axis, the scattering angle can be continuously varied from the forward scattering to the backscattering directions. Though both of these geometries provide reasonable ISS spectra, they do not achieve the overall performance of the coaxial ion gun–CMA combination in Fig. 15b.

3. Sample Configuration

The type of sample holder and sample configuration depends on the electrostatic analyzer. Taglauer et al. (82) and Brongersma and Mul (16) have used sample manipulators that enable the sample to be rotated about the surface normal at the incidence spot of the ion beam. This allows them to study scattering along different crystal directions when single-crystal targets are used as samples. The alignment of the samples using these manipulators may be time consuming. For routine elemental analysis, it is convenient to have a sample holder that can accommodate several samples during one pumpdown of the vacuum system. This can be accomplished by using a carousel assembly that allows samples to be rotated into the correct position while maintaining the crucial alignment of the ion beam and the analyzer, but it often restricts the sample size rather severely.

Sample holders for use with the CMA are considerably more versatile because the only requirement is the ability to position the sample surface close to the focal point of the analyzer. Sample size is primarily limited by the size of the port on the vacuum system.

In addition, special sample holders that allow heating or cooling of the sample can be conveniently used with a CMA geometry.

4. Detection and Spectral Display

The detectors most commonly used for ISS are electron multipliers such as the channel electron multiplier and the copper–beryllium dynode types. Due to a drastic drop in efficiency for low-energy ions, the ions from the ESA are often accelerated by 1–3 keV before they strike the detector. The CEM is usually operated in the saturated mode to avoid gain shifts. Although this detector is sensitive to high currents, and care must be taken to avoid them, it is nonetheless quite rugged and withstands vacuum to atmosphere cycling with little degradation. The pulses from the electron multiplier pass through a preamplifier, an amplifier, and into a recording device. This is usually a count rate meter but can be a digital scaler and timer. The data can be displayed on an x–y recorder, but automatic data collection, either by a multichannel analyzer or computer, offers many advantages and is used in most laboratories. The use of an automated system has the advantage that the data are available for later manipulation such as smoothing, background subtractions, or peak integration, and no real time decisions are necessary regarding scale sensitivity factors. This can be extremely important for samples in which the ion beam on-time must be minimized.

A further advantage of automated data collection is that it permits the analysis of experimental data by suitable addition or subtraction of "standard" spectra. It has been shown (8) that spectrum subtraction techniques improve quantitative analysis, determination of background shape, and detection limits of minor constituents. For example, McCune (52) has shown that the experimental spectra of $MgAl_2O_4$ and Mg_2SiO_4 could be synthesized from the spectra of MgO, Al_2O_3, and SiO_2.

With the advent of mini ion beams ($< 100 \ \mu m$), it is possible to obtain some spatial elemental information. These elemental maps can be extremely helpful in solving practical problems.

5. Vacuum System

Due to the large mean free paths necessary for ISS experiments and the necessity of keeping the surface free of contamination during the measurements, base pressures of 1.3×10^{-7} Pa (10^{-9} torr) or lower should be achieved in the scattering chamber. This can be done by using adequate pumping and baking of the scattering chamber wall. It may also be desirable to heat the target in situ (although this is not usually required for "practical" samples). The most common pumping system consists of sorption pumps for roughing and sputter ion pumping along with titanium sublimation and liquid nitrogen cryopumping to achieve ultra-high vacua. The titanium sublimation pump and the liquid nitrogen grap are also used to remove active gases during the time the chamber is backfilled with the noble gas required for the ion beam.

B. QUALITATIVE ASPECTS

1. General Features of Spectra

Figure 16 shows spectra of helium and neon scattered from gold. The spectra are dominated by large, sharp, binary elastic peaks and a low background down to zero energy. This is desired for good elemental analysis and should occur for binary elastic scattering from the top monolayer. However, in some cases a number of other features can be observed. Among these are (1) double scattering, (2) charge exchange peaks, (3) sputtered ions, and (4) inelastic scattering.

Double scattering has already been discussed. An example is shown in Fig. 8 where $^{20}Ne^+$ is scattered from indium. The principal concern associated with double scattering is that the shoulder may mask another peak, or conversely, the shoulder could be mistaken for another element. Charge exchange peaks occur when the beam contains doubly charged ions, as is often the case for neon and argon ion beams. These doubly charged ions may lose a single charge during a scattering event and pass through the analyzer as a singly charged ion. An example of this is shown in Fig. 17. Since the energy of a doubly ionized primary particle is twice that of a singly ionized species when accelerated through a given potential, many charge exchange scattering events will have energies greater than E_0. Consequently, they do not usually interfere with normal single scattering peaks though there are exceptions. This problem does not occur when the ion beam is mass analyzed (actually m/q analyzed). However, few commercial instruments used for ion scattering have mass-analyzed ion beams. Therefore, it is necessary to minimize the number of doubly charged ions in the beam. This can be done by properly choosing the ionization voltage.

For insulating samples or "dirty" samples, there are large yields of secondary ions that can be observed as a broad, poorly defined peak in the spectrum. This peak occurs at low energies and interferes only with the detection of elements that fall at low E_s/E_0 values. Figure 18 is a spectrum obtained from a thin dirty oxide layer showing the sputter peak and the hydrocarbon contamination.

ISS spectra obtained with He^+ and Ar^+ may have long tails on the low-energy side. A typical sample is given in Fig. 19. This tail is believed to be due to inelastic scattering and reionization. The reason they occur for only some elements and are not usually observed when Ne^+ is used as the probe gas is not yet understood. The practical significance is that the tail can mask the presence of other elements that fall within its region. This problem can be overcome somewhat by reducing the energy of the ion beam, as can be seen in Fig. 20 where the peak shapes and the relative O/W signal are enhanced by lowering the energy of the beam. This is the result of increased neutralization and the shallower penetration depth of lower-energy ion beams.

2. Surface Sensitivity

The sensitivity of ISS to only the outer one or two monolayers is one of its major advantages. This feature is especially important for studies of certain

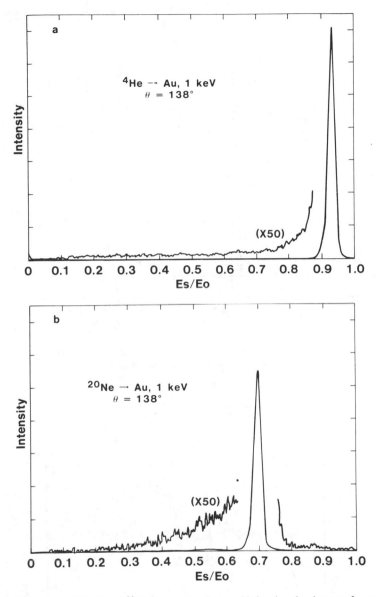

Fig. 16.   ISS spectra of $^4He^+$ and $^{20}Ne^+$ scattering from gold showing the sharp surface scattering peak and low background levels.

phenomena, like catalysis or surface passivation, in which the outer monolayer governs the surface activity. The extreme surface sensitivity is due primarily to the large probability of subsurface neutralization of the incident ion beam and the large surface atom–incident ion scattering cross sections. The surface sensitivity of ISS has been demonstrated by the adsorption of carbon monoxide

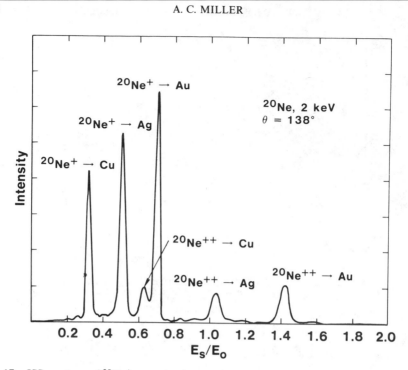

Fig. 17. ISS spectrum of $^{20}Ne^+$ scattering from a Cu–Ag–Au alloy. Note that the energy range of the analyzer was extended to $2E_s/E_0$ to show that charge exchange peaks that have an energy greater than E_0.

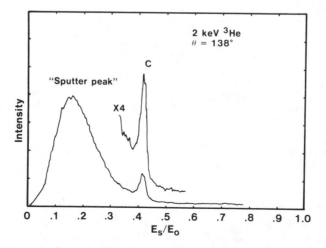

Fig. 18. ISS spectrum of a surface heavily contaminated by hydrocarbons. The broad "sputter peak" is attributed to sputtered ions.

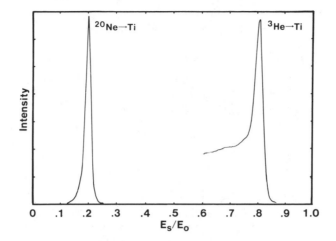

Fig. 19.   A comparison of ^{3}He$^+$ and ^{20}Ne$^+$ scattering from Ti illustrating the low-energy tail that is often observed when He$^+$ is used as a probe beam.

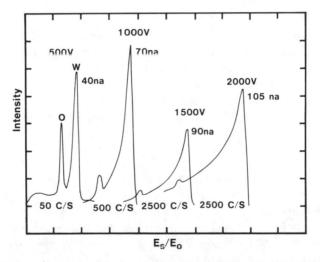

Fig. 20.   ISS spectra of tungsten oxide showing the changes that occur in the spectral shape as the beam energy is changed.

on nickel (72) and bromine on silicon (17). In both cases, the substrate signal essentially disappeared when one monolayer had been adsorbed. Surface sensitivity has also been demonstrated in studies of the polar faces of non-centro-symmetric crystals such as CdS, CdSe, and ZnS. Strehlow and Smith (76) were able to identify the (0001)Cd and the (000$\bar{1}$)S faces of CdS by comparing the Cd–S ratio obtained from spectra of the two different crystal faces (see Fig. 21). If several layers were analyzed, the Cd and S signals would average out and no difference would be observed in the two faces. Bongersma and Mul (15) have

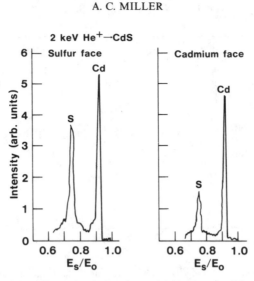

Fig. 21.   ISS spectra of the sulfur and cadmium faces of CdS. Note the difference in the Cd–S peak height ratio for the two faces (76). (Reprinted with permission from the publisher, The American Institute of Physics.)

done similar experiments on zinc sulfide. Also, Taglauer and Heiland (80) have looked at sulfur and nickel and have concluded the nickel signal disappears after one monolayer of sulfur is adsorbed. Of course, this fine surface sensitivity is only useful if spectra can be recorded before the layer has been sputter-removed by the beam. When the 127° analyzer is used, this may be a problem. However, with the cylindrical mirror analyzer, the sensitivity is such that sputtering need not be a problem.

3.   Resolution

For a mass-sensitive technique, the mass resolution determines whether neighboring elements can be separated from each other (actually neighboring masses). Equation (4) shows that E_s/E_0 is not a linear function of target mass and that differences in E_s/E_0 for adjacent masses become progressively smaller as the target mass increases. This means that for a given energy analyzer it becomes more difficult to separate elements with large Z values and nearly equal masses. In general, the mass resolution is given by

$$\frac{M_s}{\Delta M_s} = \frac{E_s}{\Delta E_s} \frac{2M_s/M_0}{1 + M_s/M_0} \frac{M_s/M_0 + \sin^2\theta - \cos\theta(M_s^2/M_0^2 - \sin^2\theta)^{1/2}}{M_s^2/M_0^2 - \sin^2\theta - \cos\theta(M_s^2/M_0^2 - \sin^2\theta)^{1/2}} \qquad (11)$$

where $E_s/\Delta E_s$ is the energy resolution of the analyzer. This resolution is plotted in Fig. 22 as a function of target mass M_s for scattering angles of 90° to 138° and constant energy resolution of $E_s/\Delta E_s$ of 60. The best mass resolution is obtained if the ion and target masses are nearly equal. This is clearly demonstrated in Figs. 2 and 23, which are spectra of a Cu–Ag–Au alloy and a Ni–Zn alloy

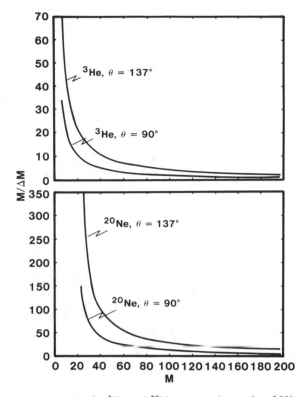

Fig. 22. Mass resolution for ^{3}He and ^{20}Ne at scattering angles of 90° and 138°.

obtained with both helium and neon. The increased resolution is quite evident in both cases. However, there is a drawback to improving the resolution by using a heavier probe ion. As the mass of the probe gas is increased, the ability to detect low-mass elements is lost. For example, using neon scattered at 90°, elements that have a mass below 20 amu cannot be detected. By using the 138° scattering angle of the CMA, the resolution for a given primary ion is increased somewhat. The poor mass resolution for high-mass elements (above ∼ 40 with helium, 60 with neon, and 80 with argon) is the major limitation of a low-energy scattering technique. In some cases this inability to separate peaks can be overcome by a knowledge of special techniques such as those presented by Erickson and Smith (32) for lead and bismuth, where they used the difference in scattering cross section to identify these elements. However, the best way to overcome the mass resolution problem is to combine ISS with another technique that does not, in general, have problems with mass resolution. Since an ion beam is being used to obtain the spectra, the most logical complementary technique is secondary ion mass spectroscopy (SIMS). This combination is now available commercially and has proven useful for cases where ISS has resolution problems. Baun et al. (9) and Grunder et al. (37) have discussed these complementary techniques.

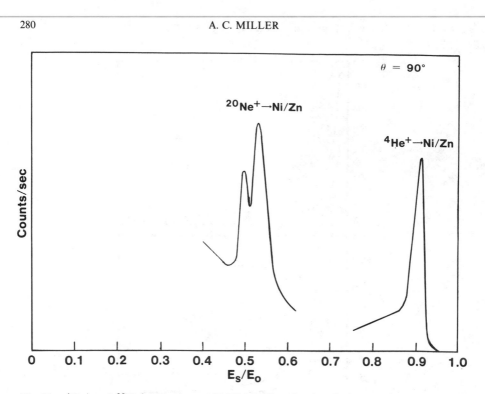

Fig. 23. ^{4}He$^+$ and ^{29}Ne$^+$ ISS spectra of a Ni–Zn alloy showing the improved mass resolution obtained when ^{20}Ne is used as the probe ion (56). (Reprinted with the permission from the publisher, Academic Press.)

C. QUANTITATIVE ASPECTS

1. Elemental Sensitivity

The scattered intensity at a given scattering angle and energy is given by

$$I_i = I_0 N_i \frac{d\sigma_i(\theta)}{d\Omega} TP_i G_i \Delta\Omega \tag{12}$$

where I_0 is the incident ion beam intensity, N_1 is the areal density of target atoms of type i, P_i is the probability the ion will *not* be neutralized (ion escape probability), $d\sigma/d\Omega$ is the differential scattering cross section, G_i is a geometric factor to account for shadowing, T is the transmission factor for the energy analyzer and detector, and $\Delta\Omega$ is the acceptance angle of the analyzer. As pointed out above, little is known about the factor G, and it is, in most cases, taken to equal 1. The scattered intensity is then related to the scattering cross section and the neutralization probability. Both of these factors favor the detection of high-atomic-number elements (i.e., the cross section increases with Z and the neutralization probability decreases with Z). Elemental sensitivity curves for He$^+$ and Ne$^+$ at 2 keV are shown in Fig. 24. These curves should only be used as

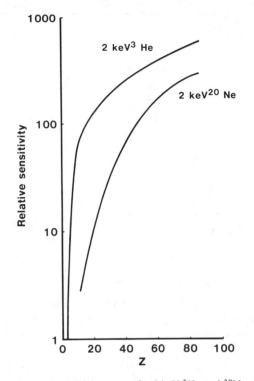

Fig. 24.   Approximate elemental sensitivity curves for 2 keV ^{3}He and ^{20}Ne as a function of atomic number Z.

a general guide for quantitative measurements. They do, however, indicate that the elemental sensitivity varies by nearly three orders of magnitude across the periodic table (for He$^+$). The absolute elemental sensitivity is a function of a number of other parameters, among which are scattering geometry and ion velocity. The increase in signal (and, therefore, elemental sensitivity) obtained with the CMA over the 127° ESA is a good example of optimizing the scattering geometry. The dependence on ion velocity is clearly demonstrated in Fig. 25 for ^{3}He$^+$ and ^{4}He$^+$ scattered from carbon. The increased signal strength for ^{3}He for a given energy is thought to result from the higher velocity of the ^{3}He ion, which in turn spends less time near the surface and thus reduces the neutralization. The absolute sensitivities vary from about 0.3 to 10^{-4} monolayers in going from lithium to gold.

2.   Standards

Due to the unknown quantities of scattering cross sections and neutralization probabilities, it is necessary to calibrate the technique using standard samples and comparisons with other techniques to obtain quantitative information. One would like linear dependence of the scattered yield on the target atom density

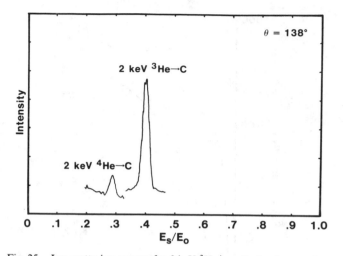

Fig. 25.   Ion-scattering spectra for 2 keV $^3He^+$ scattering from carbon.

(i.e., the surface density). Smith (73) found a linear relationship between scattered ion yield and composition for a series of Au–Ni alloys. Ball et al. (3) measured yields from gold on silicon that had been calibrated by neutron activation. Taglauer and Heiland (81) studied sulfur and oxygen adsorbed on nickel at submonolayer coverage and found a linear dependence on surface coverage. Detection limits for sulfur and oxygen were reported to be 10^{-3}–10^{-4} mono-layer.

Leys (49) found that pure element standards could be used as sensitivity standards for compounds when Ne was used as the probe gas but not He or Ar. Cu–Ar, Cr–Au, Ag–Au, and Pt–Rh peak height ratios obtained with He and Ne show linear curves when plotted as a function of the atomic concentrations of the bulk samples (see Fig. 26 for example).

Nelson (61) has shown that pure silver and gold can be used for sensitivity standards for determining the silver and gold surface concentrations for both He^+ and Ne^+. The major problem in obtaining quantitative information from standards is obtaining a surface standard because one must be certain that such factors as surface roughness are the same for both the standard and the unknown sample.

3.   Calculated Correction Factors

In principle, it is possible to calculate a relative sensitivity factor for a specific instrument from Eq. (2). However, factors other than those included in the equation must be considered. As pointed out previously, surface roughness can have a significant effect on the scattered ion intensity. In addition, selective sputtering (alteration of the surface due to higher sputtering rates for a given element) can influence the results. Selective sputtering will be discussed in Section IV.

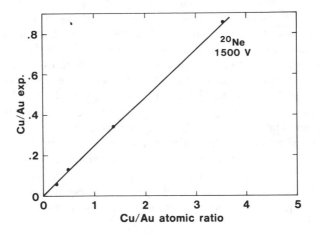

Fig. 26. Copper–gold peak height ratios obtained with ISS as a function of bulk alloy composition (56). (Reprinted with permission of the publisher, Academic Press.)

The usual approach to quantitative measurements is to write Eq. (12) for two elements and divide them to eliminate conversion terms:

$$\frac{I_1}{I_2} = \frac{N_1 P_1 G_1 (d\sigma_1/d\Omega)}{N_2 P_2 G_2 (d\sigma_2/d\Omega)} \tag{13}$$

The unknown terms for $d\sigma/d\Omega$, G, and P are usually lumped into a sensitivity factor S, and Eq. (13) becomes

$$\frac{I_1}{I_2} = \frac{N_1 S_1}{N_2 S_2} \tag{14}$$

The sensitivity factors can be obtained from standards using pure elements. In this case, N_i is replaced by A_i, where the A_i are atom sizes, that is,

$$\frac{S_1}{S_2} = \frac{I_{01}}{I_{02}} \frac{A_1}{A_2} \tag{15}$$

where I_{01} and I_{02} are the measured intensities from pure elements. However, changes in the neutralization probability with the chemical state (e.g., Ref. 49 and 59) suggest that care must be taken when employing this approach to the quantification of ISS results.

IV. APPLICATIONS

Low-energy ion-scattering spectroscopy has been applied to a wide variety of problems. Among these are adhesion, failure analysis, corrosion, catalysis, friction and wear, coatings, semiconductors, biochemistry, monitoring chemical treatment and cleaning, electrochemical effects, and surface contamination. These studies may range from a short, simple experiment to find whether element X exists on a surface to a long-term detailed study to find, for example, how X arrived at the surface, its spatial distribution, and, if desired, how it can be removed. As with all techniques ISS has its strong and weak points in its ability to solve such problems. Among its advantages are the ability to analyze the top atomic layer, simple mass identification for all masses above helium, structural information by shadowing and multiple scattering, quantitative analysis by using standards, ability to analyze conductors and insulators, and relatively low beam damage. Its major disadvantages are poor resolution for high-mass elements, a good vacuum is required, peak tailing can reduce elemental sensitivity, and the inevitable sputtering of the sample surface.

A. SURFACE COMPOSITION ANALYSIS

1. Conductors

A majority of the surfaces analyzed are conductive. This results partially from the fact that charge buildup on insulating surfaces can cause analysis problems. In addition, a number of interesting insulators have a relatively high vapor pressure (e.g., organics) and have been avoided because they were deemed not to be vacuum worthy; but, as will be pointed out later, this has changed in the past few years.

For conducting surfaces, one field of interest is the study of the surface versus bulk composition of alloys. Simple thermodynamic arguments suggest the surface of a binary alloy should be enriched in the component that has the lowest surface free energy (75). Because of its high sensitivity to the outermost monolayer, ISS is ideally suited for studies of surface segregation. For example, Brongersma and Buck (18) studied the Cu–Ni system and found that at 450°C the surface was enriched in copper as predicted by the thermodynamics models. Nelson (61) has made similar measurements for the Ag–Au systems and again found qualitative and quantitative agreement with the thermodynamic predictions. This is demonstrated in Fig. 27, where the silver and gold peak heights are shown for the surface and bulk of an alloy with a 30/70 bulk composition. Figure 28 contains plots of the surface versus bulk composition as predicted by the regular solution model (91) and the measured enrichment values. As can be seen, agreement is quite good.

ISS is also well suited for the study of thin-film diffusion. An example is the work of Nelson and Holloway (62) where the diffusion of chromium through thin gold films was studied. Hybrid microcircuits that consist of a TaN_x resistor layer, a chromium adhesion layer, and a gold conducting layer are used extensively in the electronics industry. During the processing of these circuits they are heat treated in air at 300°C for two hours. During this time some of the chromium diffuses through the gold layer and forms Cr_2O_3, which interferes with bonding

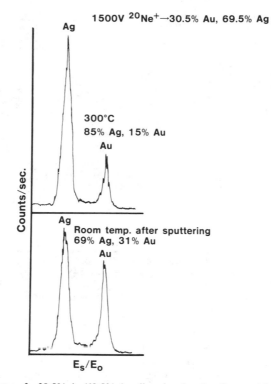

Fig. 27.   ISS spectra of a 30.5% Au/69.5% Ag alloy showing the silver enrichment that is observed at a sample temperature of 300°C (60, 61). (Reprinted with permission of the publisher, North-Holland Publishing Company.)

to the gold layer. By measuring the amount of surface chromium as a function of time and temperature, it was shown that the diffusion mechanism was due to grain boundary transport. Figure 29 is a plot of the surface chromium as a function of heat treatment time for several temperatures. From these data it was possible to determine the activation energy and preexponential factor for the diffusion coefficient. Results have also been reported for Au–Ni (23), Group VIII–Au and Group VIII–Sn (27), and Be–Cu and Sn–Cu (12) systems.

2.   Insulators

As pointed out, until the past few years, the surface analysis of insulators has been avoided. But it is now known that by using charge neutralization this important class of materials can be successfully studied. In addition, organic samples are routinely analyzed; Baun (10) has reported a number of examples where ISS has been used for studying bonding by organic adhesives, and Thomas et al. (83) have reported measurements on a number of polymer surfaces. Figure 30 shows an ISS spectrum of a copper oxide film that was prepared in $^{18}O_2$. It was demonstrated that the ratio of the ^{16}O and ^{18}O peak heights is an accurate

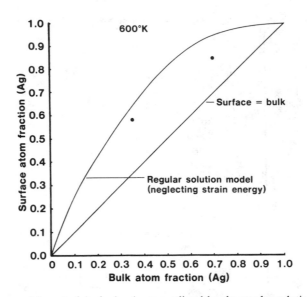

Fig. 28.   Surface enrichment of Ag in Ag–Au as predicted by the regular solution model and as measured by ISS (60).

measure of the relative amounts of the two isotopes (28). By using these labeled oxide films as substrates for polymer films, Miller et al. (53) were able to elucidate the mechanism associated with the copper-catalyzed oxidative degradation of polypropylene by following the changes in the isotope ratios within the thin films. The idea of using an ^{18}O tracer to follow reactions with ISS was first suggested by Czanderna et al. (28) and is applicable to a wide variety of problems.

Measurements made on more conventional insulating surfaces have been reported by Harrington (39) who studied the surface composition of glasses. Figure 31 shows spectra obtained from discolored Al_2O_3 before and after treatment with a gold etchant. These data confirm that the discoloration is due to the presence of residual gold on the surface. They also show that there is residual tantalum on the surface.

3.   Catalysts

Because of its high surface sensitivity and ability to analyze insulators, ISS is ideally suited for catalytic studies. Examples of these are the work of Shelef et al. (71), Wu and Hercules (92), and Swartzfager (78). Shelef et al. studied the surface composition of spinel catalysts. They were able to show that $CoAl_2O_4$ and $ZnAl_2O_4$ do not have cobalt or zinc in the top surface layer while $NiAl_2O_4$ and $CuAl_2O_4$ do contain nickel and copper, respectively. These results were shown to be in good agreement with the observed differences in the catalytic activity of these materials. Wu and Hercules used ISS to characterize the surfaces of nickel-supported catalysts. From the ISS data, they were able to infer that nickel moved

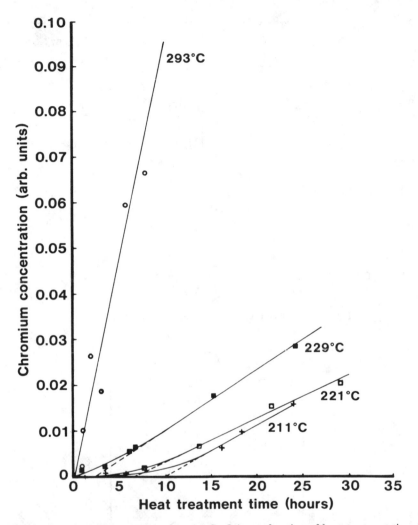

Fig. 29.   Surface chromium concentrations (as Cr_2O_3) as a function of heat treatment time and temperature as experimentally determined by ISS (62). (Reprinted by permission of the publisher, American Society for Testing and Materials.)

from octahedral sites on the γ-alumina to tetrahedral sites as the calcination temperature was increased. Swartzfager has used ISS to investigate the chemisorption of carbon monoxide on nickel and Cu–Ni alloys. He was able to show that at low coverages in the alloy surfaces, the carbon monoxide was bound at sites that are nickel-rich with respect to the clean surface composition. In addition, from the observed C–O ratio it is possible to confirm that carbon monoxide is bound with the carbon atom down and the oxygen atom up.

A. C. MILLER

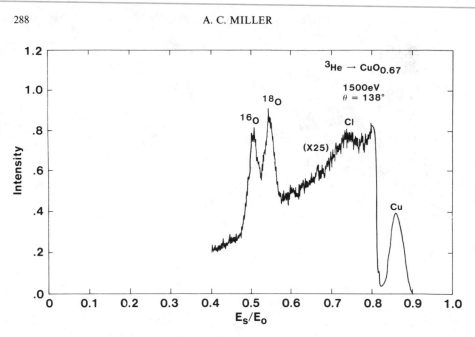

Fig. 30. ISS spectrum of isotopically labeled copper oxide.

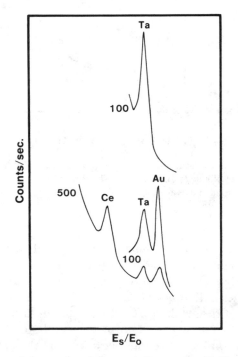

Fig. 31. ISS spectra of glass surfaces before and after treatment with a gold etchant (39).

B. IN-DEPTH CONCENTRATION PROFILES

The ion beam used for the ISS experiment can also be used for obtaining in-depth information by sputter removing the outer layers of the sample. This can be important for determining the extent to which a contamination or enrichment occurs. However, there are a number of factors connected with the sputtering process that need to be understood in order to interpret these data properly. This section will deal with a number of these parameters.

1. Sputtering Rate

The rate at which a sample is sputtered is given by S (angstroms/hour) $= 0.207\ Y\bar{V}I/d^2$ (47), where Y is the sputter yield (atoms/ion), I is the incident current (nA), d is the full width at half maximum of the beam (mm), and $\bar{V}$ is the average atomic volume (angstroms^3). Much information is available in the literature on sputtering yields. However, since these yields are a function of energy, ion species, and angle of incidence, one must be careful in applying them to a given sputtering geometry. Very little information is available on sputter yields between 0.6 and 5 keV, which is the energy range most commonly used for ion scattering. The data most often used are those of Wehner (88), which are normal incidence and energies below 600 eV. These data have been plotted as a function of atomic number in Fig. 32. As can be seen, there is considerable

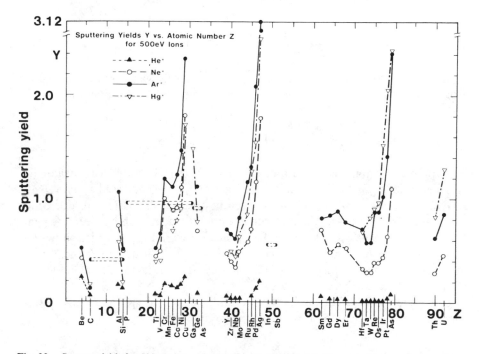

Fig. 32. Sputter yields for 500-eV ions, after Wehner's data (88, 47). (Reprinted by permission of the publisher, Elsevier Sequoia S.A.).

variation from element to element. In addition, published data indicate that "pure element sputtering yields" may not always be applicable to compounds or alloys. However, they can be used as a guide when no other information is available. To determine sputtering depths accurately, one should determine the sputtering rate under the experimental conditions to be used by measuring a "standard" film whose depth has been determined by some other technique.

2. Differential Sputtering

The rate that an atom of a given species is removed from a surface is a function of its concentration and sputtering yield. For a binary system whose atoms are of equal size, one can write the change in surface concentrations as (63)

$$\frac{dF_1}{dt} = - Y_1 I_0 F_1 + F_{10} I_0 (F_1 Y_1 + F_2 Y_2) \tag{16}$$

where F_1 and F_2 are the fractional concentration of type 1 and 2 atoms at time t, Y_1 and Y_2 are the sputter yields, I_0 the incident ion intensity, and F_{10} the bulk atom fraction of type 1 atoms. Using $F_1 + F_2 = 1$, this expression can be integrated to yield

$$\frac{F_1}{F_{10}} = \frac{Y_2}{Y_1 - F_{10} Y_1 + F_{10} Y_2} + \left\{ \left(1 - \frac{Y_2}{Y_1 - F_{10} Y_1 + F_{10} Y_2} \right) \right.$$
$$\left. \times \exp\left[-I_0 t (Y_1 - F_{10} Y_1 + F_1 Y_2] \right\} \tag{17}$$

This equation assumes that the ratio of the sputtering yields remain constant. For a large ion dose, that is, long sputtering times, Eq. (17) reduces to

$$F_1 = \left(1 + \frac{Y_1}{Y_2} \frac{F_{20}}{F_{10}} \right)^{-1} = \left(1 + \frac{Y_1}{Y_2} \frac{1 - F_{10}}{F_{10}} \right)^{-1} \tag{18}$$

This predicts a nonlinear relationship between the absolute peak height (or surface composition) and the bulk atom fraction for the sputtered surface. Data have been obtained where this is apparently observed (see Fig. 33). For other cases, such as the Ag–Au system, there is a linear relationship between the peak heights and the bulk atom fraction. This is illustrated in Fig. 34, which shows the experimental results and curves generated from Eq. (18) using Wehner's pure element sputtering yields for silver and gold. The data imply that selective sputtering does not occur in this case.

However, once equilibrium has been established on the sputtered surface (i.e., long sputtering times), the ratio of the elements is given by

$$\frac{F_1}{F_2} = \frac{Y_2}{Y_1} \frac{F_{10}}{F_{20}} \tag{19}$$

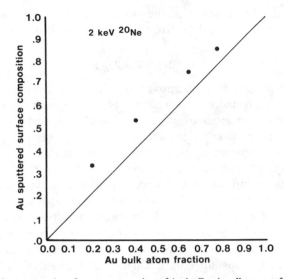

Fig. 33.   Plot of the sputtered surface concentration of Au in Cu–Au alloys as a function of the bulk Au composition showing the nonlinear relationship between the surface and bulk concentrations.

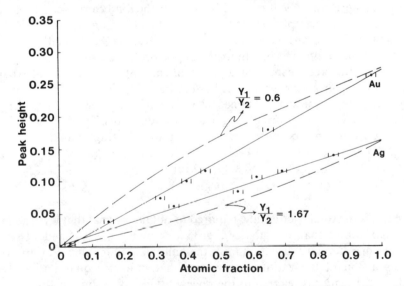

Fig. 34.   ISS Au and Ag peak heights plotted as a function of bulk atomic fraction. The solid lines are least squares fits to the data. The dashed lines are what would be predicted based on pure element sputtering yields (63). (Reprinted with permission from the publisher, The American Institute of Physics.)

so that a linear dependence is expected for the ratios of the two components even if preferential sputtering is occurring. This has been observed for the Cu–Au system (see, e.g., Fig. 26). Another sputtering effect that has been reported is an altered surface layer several angstroms thick (26). This is due to the strong interaction of the ion beam with the surfaces. Differential sputtering is not completely understood and definitive experiments are just being started.

3. Depth Resolution

Since ion sputtering does not remove atoms in an ideal atom by atom fashion, one expects to see a broadening of what starts out as a perfectly sharp interface. This may be due to preferential sputtering, knock-ons, surface roughness, and crater edge effects. Preferential sputtering has already been discussed. Surface roughness tends to obscure the definition of the surface layers. This may be due to the sputtering process itself, that is, what starts off as a smooth surface ends up very rough. Wehner and Hajicek (89) have shown that the extreme is a case where sharp cones are formed. Knock-ons occur when collisions move an atom from one layer into another layer. This has been observed particularly at high energies and is minimized by lowering the energy to below 1 keV. Crater edge effects are due to the fact that the ion beam used for the sputtering is not uniform in current density but is nearly Gaussian. If the beam used for the analysis is the same as the sputter beam (as it is in ISS), then the sides of the crater are also sampled. This effect can be reduced by using a large beam for sputtering and a small beam for analysis. A more convenient method is to raster the ion beam over a large area and electronically accept only those ions that come from the center of the rastered area. This is one technique used for ISS. Rusch (68) has mathematically determined the optimum conditions for depth profiling using a stationary or static beam, rastered beam, and a rastered and gated beam. An alternative approach to raster/gating techniques has been developed by Hoffman (45), who analytically treats the data obtained from a stationary ion beam to remove artifacts caused by signals originating from the crater walls. The depth resolution or interface width has been defined by Honig and Harrington (47) as

$$W = \frac{t(0.84) - t(0.16)}{T(0.50)} \times 100\% \qquad (20)$$

where $t(x)$ is the total sputter time required to obtain a signal that is a fraction x of the maximum signal strength (see Fig. 35 for details). The 0.84 and 0.16 limits represent one standard deviation from the half-maximum value. For an extensive discussion of the various parameters that influence quantitative depth profiling, the reader is referred to the review article by Hofmann (46).

Harrington et al. (40) have used the profiling technique with ISS to examine the interface between silicon and thermally grown silicon oxide. They found that the oxide composition is stoichiometric to within 15–20 Å of the interface. Their depth profiles clearly established a region of excess silicon that extended from the interface into the oxide for 15–20 Å. The excess silicon was calculated from the experimental data to be about 20%.

Another example of profiling by ISS is shown in Fig. 36 (69). This is a profile of equiatomic electroplated Sn–Ni film. These films have been shown to have a thin

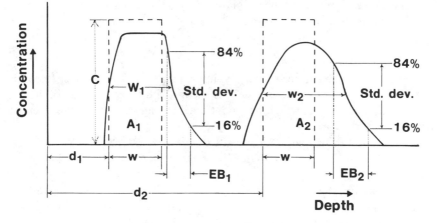

Fig. 35. Parameters for concentration depth profiles: (1) d_1, d_2 layer locations, (2) w layer width (as deposited), (3) w_1, w_2 layer width, as measured (FWHM), (4) $A_1 = A_2 =$ CW areas, (5) EB, EB_2 edge broadening (47). (Reprinted by permission of the publisher, Elsevier Sequoia S.A.)

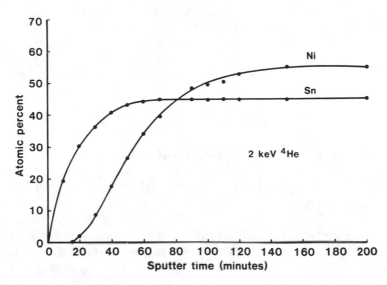

Fig. 36. Depth profiles of SnNi electroplate as determined by ISS (69). (Reprinted by permission of the publisher, The Electrochemical Society, Inc.)

passive film on the surface. The nature of the passive surface has been of considerable interest but had not been completely defined due to the sampling depth of other surface-sensitive techniques. It had been speculated that the film consisted entirely of tin oxide, but that had not been clearly shown. When examined by ISS, it is clear that the surface is void of nickel. The rise in the tin peak is due to the removal of carbonaceous material and tin oxide. The delay in the observation of the nickel signal clearly indicates that nickel is not present in the surface layer. From suitable estimates of the sputtering rate the thickness of the surface layer was found to be 5–10 Å thick.

C. SURFACE STRUCTURE

Because of its high surface sensitivity, low-energy ion-scattering spectroscopy is well suited for surface structure measurements. The determination of the polar faces of cadmium sulfide, which was mentioned earlier, was the earliest reported surface structure work.

This type of surface work has been extended by combining ISS with low-energy electron diffraction (LEED). LEED has been used for a number of years to determine the spacing and symmetry of surface atoms but has had some difficulty in giving the location of foreign atoms relative to the substrate atoms. Recent complex LEED calculations have had some limited success in suggesting atom location. Heiland and Taglauer (81) used He^+ scattering along with LEED to study oxygen on the (110)Ni surface. Two different surface structures had been proposed from the LEED measurements. The ion-scattering data from the shadowing of the nickel atoms by oxygen indicated that the model where every second place in the top rows of nickel is occupied by an oxygen atom is the correct one. Brongersma and Theeten (10) performed similar measurements and computer simulations for oxygen on (100)Ni and determined the oxygen is situated 0.9 Å above the nickel. Similar measurements have been reported for oxygen on (110)Ag (43), oxygen on tungsten (64), and oxygen on lead (47).

V. CONCLUSIONS

Low-energy ion-scattering spectroscopy is a highly surface-sensitive technique that is useful for qualitative and, in some cases, quantitative compositional analysis of the surface of a wide variety of materials. It has its greatest advantage in cases where this surface sensitivity is utilized. Examples of these that have been cited are the surface versus bulk composition of alloys, atom adsorption sites determined by shadowing, and adsorbed atom positions on single crystals. Additional areas where it can contribute significantly are those that use its mass analysis capability (e.g., oxidation by using ^{18}O) and its ability to analyze insulators (e.g., analyze Li, Na, and K in glasses).

Acknowledgments   The author gratefully acknowledges the many contributions made by Dr. G. C. Nelson of Sandia Laboratories in the preparation of this manuscript.

REFERENCES

1. Abrahamson, A. A., *Phys. Rev.*, **178**, 76 (1969).
2. *Annual Book of ASTM Standards*, American Society for Testing and Materials, Philadelphia, 1981, Part 42, p. 578.

3. Ball, D. J., T. M. Buck, D. MacNair, and G. H. Wheatley, *Surf. Sci.*, **30**, 69 (1972).
4. Baun, W. L., *Phys. Rev. A*, **17**, 17 (1978).
5. Baun, W. L., *Surf. Sci.*, **72**, 536 (1978); *Surf. Sci.*, **75**, 141 (1978).
6. Baun, W. L., private communication.
7. Baun, W. L., *Appl. Surf. Sci.*, **7**, 46 (1981).
8. Baun, W. L., *Anal. Chem.*, **48**, 931 (1976).
9. Baun, W. L., N. T. McDevitt, and J. S. Solomon; R. S. Carbonara and J. R. Cuthill, eds.; *ASTM Special Technical Publication 596*, American Society for Testing and Materials, Philadelphia, p. 86, 1976.
10. Baun, W. L., and Lieng-Huang Lee, Ed., *Characterization of Metal and Polymer Interfaces*, Vol. 1, Academic Press, New York, 1977, p. 375.
11. Bertrand, P., *Nucl. Instrum. Meth.*, **170**, 489 (1980).
12. Biloen, P., R. Bouwman, R. A. Van Santen, and H. H. Brongersma, *Appl. Surf. Sci.*, **2**, 532 (1979).
13. Bingham, F. W., *Sandia Res. Rep. No. SC-RR-66-506*, Clearing House for Fed. Sci. and Tech. Info., NBS, U.S. Dept. of Commerce, 1966.
14. Bloss, W., and D. Hone, *Surf. Sci.*, **72**, 277 (1978).
15. Brongersma, H. H., and P. M. Mul, *Chem. Phys. Lett.*, **19**, 217 (1973).
16. Brongersma, H. H., and P. M. Mul, *Surf. Sci.*, **34**, 393 (1973).
17. Brongersma, H. H., and P. M. Mul, *Chem. Phys. Lett.*, **14**, 389 (1972).
18. Brongersma, H. H., and T. M. Buck, *Surf. Sci.*, **53**, 649 (1976).
19. Brongersma, H. H., and J. B. Theeten, *Surf. Sci.*, **54**, 519 (1976).
20. Brunee, C., *Z. Phys.*, **147**, 161 (1957).
21. Buck, T. M., Y. S. Chen, G. H. Wheatley, and W. F. van der Weg, *Surf. Sci.*, **47**, 244 (1975).
22. Buck, T. M., G. H. Wheatley, and L. K. Verheij, *Surf. Sci.*, **90**, 635 (1979).
23. Buck, T. M., I. Stensgaard, G. H. Wheatley, and L. Marchut, *Nucl. Instrum. Methods*, **170**, 519 (1980).
24. Christensen, D. L., V. G. Mossotti, T. W. Rusch, and R. L. Erickson, *Chem. Phys. Lett.*, **44**, 8 (1976).
25. Christensen, D. L., Ph.D. Thesis, University of Minnesota, (1980).
26. Coburn, J. W., and E. Kay, *Crit. Rev. Solid. State Sci.*, **4**, 561 (1974).
27. Creemers, C., H. van Hove, and A. Neyens, *Appl. Surf. Sci.*, **7**, 402 (1981).
28. Czanderna, A. W., A. C. Miller, H. H. G. Jellinek, and H. Kachi, *J. Vac. Sci. Technol.*, **14**, 227 (1977).
29. Datz, S., and C. Snoek, *Phys. Rev. A*, **134**, 347 (1964).
30. DeWit, A. G. J., R. P. N. Bronckers, and J. M. Fluit, *Surf. Sci.*, **82**, 177 (1979).
31. Eckstein, W., V. A. Molchanov, and H. Verbeek, *Nucl. Instr. Methods*, **149**, 599 (1978).
32. Erickson, R. L., and D. P. Smith, *Phys. Rev. Lett.*, **34**, 297 (1975).
33. Everhart, E., G. Stone, and R. J. Carbone, *Phys. Rev.*, **99**, 1287 (1955).
34. Fluit, J. M., J. Kistemaker, and C. Snoek, *Physica*, **30**, 870 (1964).
35. Godfrey, D. J., and D. P. Woodruff, *Surf. Sci.*, **105**, 438 (1981).
36. Goldstein, H., *Classical Mechanics*, Addison-Wesley, Reading, Mass., 1950, p. 58.
37. Grunder, M., W. Heiland, and E. Taglauer, *Appl. Phys.*, **4**, 243 (1974).
38. Hagstrum, H. D., *Phys. Rev.*, **96**, 336 (1954).
39. Harrington, W. L., and R. E. Honig, *Proceedings of the 20th ASMA Annual Conference on Mass Spectrometry*, 1972.

40. Harrington, W. L., R. E. Honig, A. M. Goodman, and R. Williams, *Appl. Phys. Lett.*, **27**, 644 (1975).

41. Hart, R. G., and C. R. Cooper, *Surf. Sci.*, **82**, L283 (1979).

42. Heiland, W., and E. Taglauer, *J. Vac. Sci. Technol.*, **9**, 620 (1972).

43. Heiland, W., F. Iberl, E. Taglauer, and D. Menzel, *Surf. Sci.*, **53**, 383 (1975).

44. Helbig, H. F., P. J. Adelmann, A. C. Miller, and A. W. Czanderna, *Nucl. Instrum. Methods*, **149**, 581 (1978).

45. Hoffman, D. W., *Surf. Sci.*, **50**, 29 (1975).

46. Hofmann, S., *Surf. Interface Anal.*, **2**, 148 (1980).

47. Honig, R. E., and W. L. Harrington, *Thin Solid Films*, **19**, 43 (1973).

48. Kivilis, V. M., E. S. Porilis, and N. Yu Turaev, *Sov. Phys. Dokl.*, **12**, 328 (1967).

49. Leys, J. A., 1973 Pittsburgh Conference on Analytical Chemistry and Applied Spectroscopy, Cleveland, Ohio, March, 1973.

50. Mashkova, E. S., and V. A. Molchanov, *Sov. Phys. Dokl.*, **7**, 828 (1963).

51. McCune, R. C., J. E. Chelgren, and M. A. Z. Wheeler, *Surf. Sci.*, **84**, 1515 (1979).

52. Mccune, R. C., *Anal. Chem.*, **51**, 1249 (1979).

53. Miller, A. C., A. W. Czanderna, H. H. G. Jellinek, and H. Kachi, *J. Coll. Interface Sci.*, **85**, 224 (1982).

54. Moyer, C. A., and K. Orvek, *Surf. Sci.*, **114**, 295 (1982).

55. Muda, Y., and T. Hanawa, *Surf. Sci.*, **97**, 283 (1980).

56. Nelson, G. C., *J. Coll. Interface Sci.*, **55**, 289 (1976).

57. Nelson, G. C., *Sandia Res. Rep. No. SLA-73-0465*, Clearing House for Fed. Sci. and Tech. Info., NBS, U.S. Dept. of Commerce, 1973.

58. Nelson, G. C., *J. Appl. Phys.*, **47**, 1253 (1976).

59. Nelson, G. C., *Anal. Chem.*, **46**, 2046 (1974).

60. Nelson, G. C., private communication.

61. Nelson, G. C., *Surf. Sci.*, **59**, 310 (1976).

62. Nelson, G. C., and P. M. Holloway; R. S. Carbonara, and J. R. Cuthill, Eds.: *ASTM Special Technical Publication*, 596, American Society for Testing Metals, Philadelphia, p. 68, 1976.

63. Nelson, G. C., *J. Vac. Sci. Technol.*, **13**, 974 (1976).

64. Niehus, H., and E. Bear, *Surf. Sci.*, **47**, 222 (1975).

65. Panin, B. V., *Sov. Phys. JETP*, **42**, 313 (1962); *Sov. Phys. JETP*, **45**, 215 (1962).

66. Poelsema, B., L. K. Verhey, and A. L. Boers: *Surf. Sci.*, **56**, 445 (1976); *Surf. Sci.*, **60**, 485 (1976); *Surf. Sci.*, **64**, 537 (1977); *Surf. Sci.*, **64**, 554 (1977).

67. Rusch, T. W., and R. L. Erickson, *J. Vac. Sci. Technol.*, **13**, 374 (1976).

68. Rusch, T. W., J. T. McKinney, and J. A. Leys, *J. Vac. Sci. Technol.* **12**, 400 (1975).

69. Schubert, R., *J. Electrochem. Soc.*, **125**, 1215 (1978).

70. Sebastian, K. L., V. C. J. Bhasu, and T. B. Grimley, *Surf. Sci.*, **110**, L571 (1981).

71. Shelef, M., M. A. Z. Wheeler, and H. C. Yao, *Surf. Sci.*, **47**, 697 (1975).

72. Smith, D. P., *J. Appl. Phys.*, **38**, 340 (1967).

73. Smith, D. P., *Surf. Sci.*, **25**, 171 (1971).

74. Smith, D. P., and R. F. Goff, in 29th Annual Phys. Electron. Conf., Yale Univ., 1969; *Bull. Am. Phys. Soc.*, **14**, 788 (1969).

75. Somorjai, G. A., *Principles of Surface Chemistry*, Prentice-Hall, New York, 1972, Chapter 2.

75a. Sonda, R., M. Aono, C. Oshima, S. Otani, and Y. Ishizawa, *Surf. Sci.*, **150**, L59 (1985).

75b. Sonda, R., and M. Aono, *Nucl. Instrum. Methods Phys. Res.*, **B15**, 114 (1986).

76. Strehlow, W. H., and D. P. Smith, *Appl. Phys. Lett.*, **13**, 34 (1968).

77. Suurmeijer, E. P. Th. M., and A. L. Boers, *Surf. Sci.*, **43**, 309 (1973).

78. Swartzfager, D. G., unpublished data.

79. Taglauer, E., and W. Heiland, *Appl. Phys.*, **9**, 261 (1976).

80. Taglauer, E., and W. Heiland, *Appl. Phys. Lett.*, **24**, 437 (1974).

81. Taglauer, E., and W. Heiland, *Surf. Sci.*, **47**, 234 (1975).

82. Taglauer, E., W. Melchior, F. Schuster, and W. Heiland, *J. Phys. E*, **8**, 768 (1975).

83. Thomas, G. E., G. C. J. Van der Ligt, G. J. M. Lippits, and G. M. M. van der Hei, *Appl. Surf. Sci.*, **6**, 204 (1980).

83a. Thomas, T. M., H. Neumann, A. W. Czanderna, and J. R. Pitts, *Surf. Sci.*, **175**, L737 (1986).

83b. Tsukada, M., S. Tsuneyaki, and N. Shima, *Surf. Sci.*, **164**, L811 (1985).

84. Tongson, L. L., and C. B. Cooper, *Surf. Sci.*, **52**, 263 (1975).

85. Tully, J. C., *Phys. Rev. B*, **16**, 4324 (1977).

86. Van Den Berg, J. A., and D. G. Armour, *Vacuum*, **31**, 259 (1981).

87. Veksler, V. I., *Sov. Phys. Sol. Stat.*, **4**, 276 (1962).

88. Wehner, G. K., *General Mills Report 2309*, (1962).

89. Wehner, G. K., and D. J. Hajicek, *J. Appl. Phys.*, **92**, 1145 (1973).

90. Wheeler, M. A. Z., presented at 3rd Annual 3M ISS Users Conference, Cable, WI (1974).

91. Williams, F. L., and D. Nason, *Surf. Sci.*, **45**, 377 (1974).

92. Wu, M., and D. M. Hercules, *J. Phys. Chem.*, **83**, 2003 (1979).

93. Ziemba, F. P., G. J. Lockwood, G. H. Mogen, and E. Everhart, *Phys. Rev.*, **118**, 1552 (1960).

Subject Index

308

SUBJECT INDEX

Powders, spark source mass spectrometry
(SSMS), 223–224
Preferential sputtering, ion-scattering
spectroscopy, 292–294
Pressure, FT–ICR technique, 123–132
external ion sources, 128–132
resolution and, 123–124
two-section cells, 124–128
Pressure-dependent two-photon
photodissociation, 178
Priority pollutant analysis, 75–81
Probability-based matching system (PBM), 72–74
Product ion detection, FT–ICR, 152–154
Product (sample) ion molecules, 238, 243
Profiling techniques, ion-scattering spectroscopy,
292–294
Proton affinities:
carbon cluster chemistry, 163
chemical ionization, 18
Pseudo-first-order decay, silicon cluster
chemistry, 157
Pulsed decelerator, FT–ICR, 129–130, 133
Pulsed valve device, FT–ICR, 124, 150–151
Pulse sequences, photodissociation, 178–179

Quadrupole Fourier transform mass spectrometer
(Q–FTMS), 128–129
Quadrupole ion storage trap (QUISTOR), 135–
136
Quadrupole mass spectrometer, 32–36
Qualitative analysis:
ion-scattering spectroscopy (ISS), 274–280
spark source mass spectrometry, 213–215
Qualitative fingerprinting, plasma
chromatography, 248
Quantitative analysis:
ion-scattering spectroscopy, 280–283
mass spectrometry, 74–87
polychlorinated dibenzoioxns (PCDDs) and
polychlorinated dibenzo furane
(PCDFs), 82–87
priority pollutant analysis, 75–81
spark source mass spectrometry, 215–216
Quasiequilibrium theory (QET):
fragmentation, 66
mass spectrometry, 49, 51
Quasi-molecular ions, 17–18
"Quench-off" technique, 151–152
Quench pulse, 151–152

Radial distribution, ion cyclotron resonance
(ICR), 123
Radiation source, plasma chromatography, 241–
242

Radio frequency (rf) spark:
circuitry, 203–205
ion cyclotron resonance (ICR), 122–123
ionization mechanisms, 195–196
solids ionization, 193–196
spark source mass spectrometry (SSMS), 191,
196
Rapid scanning, time-of-flight mass
spectrometer, 38
Rate constant vs. excitation energy, 51–53
Rate process mechanism, photodissociation,
180–182
Reactant gas, plasma chromatography, 238–240
Reactant ion isolation:
FT–ICR, 145–152
plasma chromatography, 238
silicon cluster chemistry, 155
Reactant ion NICIMS, 22
Reactant neutral introduction, 150–151
Reaction region, plasma chromatography, 241–
242
Reactivity, silicon cluster chemistry, 160–161
Readout systems, plasma chromatography, 244
Reagent gas:
ionization potentials, 20
reactant ion NICIMS, 22–23
Rearrangement reactions:
mass spectrometry theory, 53–54
odd-electron ions, 70–71
Recovery standards, mass spectrometry
quantitation, 83
Reduced mobilities, plasma chromatography,
244, 248
Reference compound, MS quantitation, 74–81
Reference spectra, mass spectrometry
applications, 71–72
Reflectron, time-of-flight mass spectrometer, 38
Reionization, ion-scattering spectroscopy, 274
Relative rate constants, silicon cluster chemistry,
162
Relative response factors, 85–86
Relative sensitivity coefficient, 215–216
Resolution:
depth, ion-scattering spectroscopy, 292–294
FT–ICR technique, pressure and, 123–124
ion-scattering spectroscopy, 278–280
mass analysis, 24–28
quadrupole mass spectrometer, 36
quadrupole ion storage trap (QUISTOR),
135–136
spark source mass spectrometry, 200–201
Resolving power (RP):
mass analysis, 24–27
single-focusing magnetic analyzer, 29–30